CRITICAL PATH
METHODS IN
CONSTRUCTION PRACTICE

JAMES M. ANTILL is the author of *Civil Engineering Management,* McGraw-Hill, Sydney, 1970, revised 1973; *Construction Contract Administration,* Australian Federation of Construction Contractors, Sydney, and University of British Columbia, Vancouver, 1975, revised 1981; and co-author of *Civil Engineering Construction* by Antill & Ryan, McGraw-Hill, Sydney, 1957, revised 1965, 1967, 1974, 1981, 1988; and *Australian Concrete Inspection Manual* by Nagarajan & Antill, Longman Cheshire Australia, 1978.

RONALD W. WOODHEAD is co-author of *Frame Analysis* by Hall and Woodhead, John Wiley & Sons, New York, 1961, second edition, 1967; co-author with D. W. Halpin of *CONSTRUCTO—A Heuristic Game for Construction Management,* University of Illinois Press, Urbana, 1973, and *Design of Construction and Process Operations,* John Wiley & Sons, New York, 1976, and *Construction Management,* John Wiley & Sons, New York, 1980; co-author of *Design and Planning of Engineering Systems,* by Meredith, Wong, Woodhead and Wortman, Prentice Hall Inc, Englewood Cliffs, N.J., 1973, 1985; and co-author of *Project Manpower Management—Management Processes in Construction Practice,* by Anderson and Woodhead, John Wiley & Sons, New York, 1981, and co-author of *Project Manpower Management—Decision Processes in Construction Practice,* by Anderson and Woodhead, John Wiley & Sons, New York, 1987.

CRITICAL PATH METHODS IN CONSTRUCTION PRACTICE

FOURTH EDITION

JAMES M. ANTILL

Chartered Engineer (Australia),
Consulting Construction Engineer and
Practicing Arbitrator, Sydney,
Honorary Visiting Professor of Civil Engineering,
formerly The University of New South Wales,
Sydney, Australia

RONALD W. WOODHEAD

Chartered Engineer (Australia),
Emeritus Professor
The University of New South Wales,
Sydney, Australia

WILEY

A WILEY-INTERSCIENCE PUBLICATION

JOHN WILEY & SONS

New York / Chichester / Brisbane / Toronto / Singapore

Library of Congress Cataloging-in-Publication Data:

Antill, James M.
    Critical path methods in construction practice/James M. Antill,
Ronald W. Woodhead—4th ed.
        p.   cm.
    "A Wiley-Interscience publication."
    Bibliography: p.
    Includes index.
    1. Critical path analysis.   2. Engineering—Management.
I. Woodhead, Ronald W.   II. Title.
TA194.A57   1982
658.4′032—dc20   89-35471
ISBN 0-471-62057-2   CIP

Printed in the United States of America

10 9 8 7 6 5 4 3 2 1

PREFACE

It is now 24 years since this book first appeared. The authors are particularly gratified by its reception during this period and by its acceptance by professional engineers, contractors, and others as a standard text on the subject. Critical path methods were first introduced to the construction industry about 1960 and are now familiar to many construction engineers, managers, planners, accountants, and cost estimators. They have already proved their value in both small and large projects by highly successful results and are acceptable as technical evidence in courts of law. The technical press has written much in the past two and a half decades about this technique and its extensive ramifications, but there is still undoubtedly a need for presenting in a single volume the mechanics and procedures of this mathematical method of construction planning and works control. We think that we have fulfilled this need.

We did not write this book for the casual reader; it was prepared expressly for those who are concerned with the application of critical path methods in their work, whether in the office or in the field. It is concerned not with abstruse mathematics but with the solution of the practical problems commonly encountered by all those engaged in the management, administration, and practice of construction. It presupposes some knowledge of the conventional methods of planning, estimating, costing, and control of building and civil engineering projects, in addition to construction methods and procedures. In actual examples from construction practice, the costs have been quoted in Australian dollars, and the reader should bear in mind that Australian wage rates are about one-half of those in North America. Currently $1 Australian is equal to about $0.75 U.S.

In the second edition (1970) the subject matter was rearranged and enlarged in several places: four new chapters were introduced devoted to scheduling and resource leveling (previously part of Chapter 4), the evaluation of work changes and delays by the factual network concept introduced in 1966 by one of the authors (previously referred to in Chapter 9), the integration of project development and management, and the consideration of critical path methods (CPM) as a system. A specification for the use of CPM in practice was included as an appendix, but is now incorporated into Chapter 12 as currently revised. Problems (with answers) were provided for the benefit of students, as well as an appendix on linear graph theory and its relevance to CPM networks.

In this fourth edition, the entire text has been reviewed in the light of developments during the past decade. Since monetary examples are from actual projects, no attempt has been made to revise these costs as a result of inflation. Not a great deal of the first eleven chapters required revision; Chapter 12 has been totally rewritten to exclude PERT (now outdated, but included as an appendix), and to include new thoughts on use of CPM, and Chapter 13 has been updated to incorporate recent advances in computer technology. Additional evaluations and problems have been included throughout the text to support and encourage the reader to evaluate the concepts of the new and revised chapters. Consequential changes have been made elsewhere as required. Chapters 14, 15, and 16 needed little amendment, being concerned largely with proven procedures. In view of the world conversion to the International System of Units (metric), all measurements in the text are now quoted in SI units.

Grateful acknowledgment is made to all colleagues who have permitted us to cite examples from their projects and to those readers who have given us the benefit of their comments and constructive criticism. Finally, we greatly appreciate the reception accorded the original and subsequent editions and trust that this revision will prove of benefit to all. Like its predecessors, we dedicate this fourth edition to the personnel of the construction industry.

JAMES M. ANTILL
RONALD W. WOODHEAD

March 1989

CONTENTS

CRITICAL PATH
METHODS IN
CONSTRUCTION PRACTICE

1

INTRODUCTION TO CRITICAL PATH METHOD

1.1 A NEW TOOL FOR THE CONSTRUCTION INDUSTRY

The critical path method is a powerful tool for the planning and management of all types of projects. Essentially it is the representation of a project plan by a schematic diagram or network that depicts the sequence and interrelation of all the component parts of the project, and the logical analysis and manipulation of this network in determining the best overall program of operation. It is a method admirably suited to the construction industry, and it provides a far more useful and precise approach than the conventional bar charts and progress graphs that previously formed the basis of construction planning and control. Furthermore, it permits the ready evaluation and comparison of alternative works programs, construction methods, and types of equipment. When the best plan has been prepared in this way, the critical path diagram clearly indicates the site operations that control the smooth execution of the works. Finally, as construction proceeds, the diagram provides the project manager with precise information on the effects of each variation or delay in the adopted plan, thus permitting the identification of the operations that require remedial action.

This technique unfortunately acquired a number of names, including network analysis, critical path analysis, critical path scheduling, and least-cost estimating and scheduling; but the designation *critical path method* (abbreviated to CPM) is the most satisfactory because there is no limitation implied in its use. CPM may be employed not only for the planning and control of construction works, but for research programs, maintenance problems, sales promotion, and related operations in other industries.

1.2 HISTORICAL BACKGROUND

The critical path technique had its origin from 1956 to 1958 in two parallel but different problems of planning and control in projects in the United States.

In one case, the U. S. Navy was concerned with the control of contracts for its Polaris Missile program. These contracts comprised research and development work as well as the manufacture of component parts not previously made. Hence neither cost nor time could be accurately estimated, and completion times therefore had to be based on probability. Contractors were asked to estimate their operational time requirements on three bases: optimistic, pessimistic, and most likely dates. These estimates were then mathematically assessed to determine the probable completion date for each contract, and this procedure was referred to as *program evaluation and review technique,* abbreviated to PERT. It did not consider cost as a variable. Subsequently the inclusion of cost data (on the same sort of probability basis) has been introduced, and this system is known as PERTCO (PERT with costs). It is therefore important to understand that the PERT systems involve a "probability approach" to the problems of planning and control of projects and are best suited to reporting on works in which major uncertainties exist.

In the other case, the E. I. du Pont de Nemours Company was constructing major chemical plants in America. These projects required that both time and cost be accurately estimated. The method of planning and control that was developed was originally called *project planning and scheduling* (PPS), and covered the design, construction, and maintenance work required for several large and complex jobs. PPS requires realistic estimates of cost and time and is thus a more definitive approach than PERT. It is this approach that has since been developed into the critical path method (CPM), which is finding increasing use in the construction industry. Although there are some uncertainties in any construction project, the cost and time required for each operation involved can be reasonably estimated and all operations may then be reviewed by CPM in accordance with the anticipated conditions and hazards that may be encountered on the site.

1.3 THE TIME-COST PROBLEM

To analyze or review a construction project, whether PERT or CPM is used, it is necessary first to prepare a diagram (or "model") in the form of a schematic network that will show all the component operations in relation to one another. A simple diagram of this type is seen in Figure 1.1, in which each operation is represented by an arrow between two circles. Any construction project is easily divided into a number of processes or operations, each of which may be performed by different combinations of construction methods, equipment, crew sizes, and working hours. The major factors dominating the selection of the best combination may be cost, time, or both. At first thought,

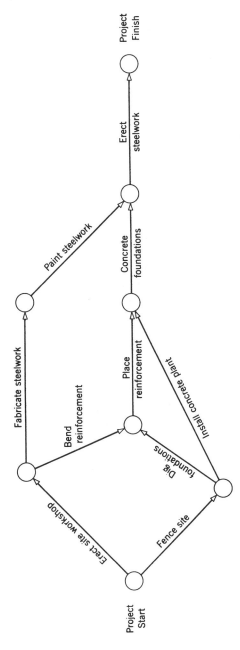

Figure 1.1 Network diagram for a simple project, showing component operations.

the direct cost of each operation may be expected to predominate, especially with the contract system, in order that the works may be completed for the lowest total cost; but the total cost of the project includes all the indirect and overhead charges associated with the complete execution of the works, and these are proportional to time. Furthermore, from the contractor's standpoint, the early release of personnel and plant for other work may be desirable, and planning for lowest direct cost alone may not be the best solution. Time is therefore an equally essential factor.

This time-cost problem has an infinite number of solutions. If time were of no consequence, each operation could be performed in the manner that gave the lowest direct cost. If cost were of no importance, each process could be speeded up to complete it in the least time. Between these two limits lies the best answer; but to find it requires consideration of a complex collection of concurrent, interrelated, and overlapping operations. Speeding up one process will increase its cost and shorten its time but may not decrease the overall time of the project unless the accelerated operation is a critical one in the long chain of activities comprising the works. It is therefore necessary to find the proper combination of operations that should be accelerated in order to produce the most economical project shortening, having in mind both direct and indirect costs.

The solution to the time-cost problem is not simple. All costs vary with time; direct costs tend to decrease if more time is available for an operation, but indirect and overhead costs will increase with time. It is the proper balance between time and total cost that gives the optimum solution. In the past, this solution was attempted by experienced appraisal of the problem, often with some trial-and-error calculations; but the construction plan adopted depended solely on judgment. Today, CPM methods can and do provide a systematic procedure for correlating the effects of cost and time to produce the optimum solution for every construction problem. In addition, the solution is presented as a visible network diagram or model showing all the essential data for each operation.

1.4 ADVANTAGES OF CPM

A fully developed critical path network is a logical mathematical model of the project, based on the optimum time required for each work process and making the most economical use of available resources (labor, equipment, finance, etc.). It has therefore been tuned to the individual problems of the particular project, and may be as detailed as desired to suit the anticipated conditions and hazards. During execution of the project it permits systematic reviewing of current situations as they arise, so that allowance can be made for the effects of uncertainties in the original planning, as well as enabling a reevaluation of future uncertainties to be made, and remedial measures initiated for those operations—and only for those operations—that require correction or acceleration.

It is indeed significant that, where the critical path method has been introduced, considerable reduction of project times and costs has resulted. In the United States its use in the construction industry has led to decreases up to 20% in project times over similar projects not employing CPM as a management tool. This has been made possible because the network diagram clearly shows the processes whose completion times are responsible for establishing the overall duration of the complete project; these *critical operations* must be kept "on time all the time." Together they form a connected pathway of operations through the network; this is the *critical path* through the project. All other operations have some leeway in starting and finishing dates, and may be arranged (within limits) to smooth out manpower and equipment requirements.

The use of CPM enables the most economical planning of all operations to meet desirable completion dates. It replaces the judgment based on experience (or trial-and-error) formerly used to select operation times, crew sizes, equipment, etc.; indeed, with CPM one readily determines with certainty the best completion date for the project. And finally, it provides a means of assessing the effect of all variations—change orders, extra work, or deductions—on the time of completion and on the cost of the works. In the past, a rational basis for calculating such effects was seriously lacking and led to many unpleasant arguments. Today the time and cost of every change from the original optimum plan can be rapidly assessed at any stage of the construction period.

An important point to realize is that CPM is an open-ended process that permits different degrees of involvement by management to suit their various needs and objectives.

1.5 BASIC DATA FOR CPM

In applying CPM to construction planning and related problems, it is necessary first to have accurate estimates of time and cost for each operation comprising the project. In practice, this simply means that the conventional direct-cost estimate, based on the itemized bill of quantities, is prepared in the usual way. The component operations of the project may be the actual items of work listed in the bill, or these items may be divided (or combined) into operations and processes more suitable to the particular works under consideration. Thus the breakdown of the project into its individual operations may be as simple or as detailed as desired; the essential requirement is that the direct cost for each operation be estimated separately.

After the direct-cost estimate has been made, the normal time to complete each operation will be calculated from the total worker-hours and/or plant-hours involved in the conventional way. There is nothing new in preparing the normal cost and time data for use with CPM except to ensure that the division into operations is suitable for the magnitude and nature of the works.

After completing and listing the normal cost and time data for each operation, it is neither a difficult nor lengthy process to make other similar lists, based on other-than-normal conditions. In this way, cost and time data are prepared for such variations as longer working hours, shift work, different crew sizes, use of alternative equipment, changes in construction methods, or any other variation or resource that may appear feasible for the project. At first this procedure may appear to be tedious, particularly when numerous variations are feasible; but with a little practice, and the acceptance of a reasonable order of approximation in the calculations, such alternative lists are rapidly prepared.[1]

The number of feasible variations to be investigated will differ with each project, and some may be rejected out-of-hand after inspection of the times and costs involved. Finally, each of the variations that are relevant will be reviewed on separate critical path network diagrams in order to find which of them will provide the optimum solution to the time-cost problems of the project. In this way the most economical overall method of carrying out the job is determined with confidence.

Details of how this is done form the substance of this book; it is sufficient to say here that the calculations involved are reasonably simple, but many steps and combinations may be required before the best solution is found. This entire procedure is known as *network analysis,* and does not require a knowledge much beyond simple arithmetic.

For projects involving a vast number of operations and alternatives, all of which must be considered in deriving a proper solution, the calculations can be done with electronic computers. In most construction projects, however, personal calculation (called ''manual methods'') will suffice, and they have the advantage that the planner gains a more detailed knowledge of the project and its problems by this personal attention to the preparation of the network diagrams and arithmetical calculations. Whether the work is to be done by computer or by manual methods, the same input data is required, and the same output information is obtained. The use and practice of manual methods will show that, for most cases, tasks that appear tedious and complex at the initial encounter become much simpler after a few attempts.

1.6   CURRENT DEVELOPMENTS

The use of critical path methods is already well established in a variety of industries, and is particularly suited to building and civil engineering construction. It is therefore of major importance to all contractors, estimators, construction engineers, works managers, and clerks of works. It is also vital

---

[1] It is a good failing for the tyro, however, to tend to be overmeticulous, both in choice of variations and in the cost and time estimates. Although meticulousness will unduly consume time at first, the tyro will more quickly grasp the fundamentals of the procedures and acquire skill and speed in the preparation of essential project information; and will also learn the art of rapid reasonable approximation of such data and the site operations for which it is required.

to others whose work is related to construction work, such as financiers, cost accountants, office managers, supervisors, and superintendents. It is becoming of significant importance to the legal profession, for it provides a sound mathematical basis on which to settle claims for delays, works variations, wage changes, and the like. Expert testimony based on CPM programs and factual networks is acceptable evidence in litigations and arbitrations concerned with contractual disputes.

Some contract documents already are specifying that the contractor submit critical path diagrams at the commencement of the works, and several contractors have been using CPM as part of their bidding procedure for some time past. Increasing use of CPM will demonstrate the importance of time performance records and cost performance information to the construction industry. Very little systematic documentation of relative speeds of operation has been yet achieved, either by construction authorities or by contractors, although it is obvious that such records are needed in order to relate actual performance rates to works costs. It is essential that cost statements should be revised to show whether the work was performed at normal speed or at a faster than normal rate (e.g., by working overtime) for any operation. Such data are of major importance in estimating the costs of developmental projects by the authorities, as well as in the operation of a construction contracting business.

Assessment of bonus and penalty provisions in contract work is readily attained with CPM both by the principal and the contractor. Similarly, the assessment of the cost of hazards may be investigated, and a rational approach made to the problem of expenditure in lessening site risks; floods, inclement weather, and similar hazards lend themselves to this approach so that the cost of the risk may be assessed and taken into account in the planning. On the basis of these considerations a contractor may adopt a strategy that permits the bidding of a range of completion times and prices; the owner may then select the bid that will prove the most economical from the overall viewpoint, allowing a realistic monetary value for early completion. CPM is the essential tool in this development.

Sophisticated computer applications have already appeared in resource leveling, cost accounting, and the design and monitoring of field operations in the construction field emphasizing the necessity for dynamic project control by management. Indeed this function has already spread in Australia to shipbuilding and repair operations, retail warehousing, and commercial operations. Other applications will undoubtedly occur, for the critical path method could be used in any activity in which planning, scheduling, comparison of alternatives, cost recording, finance, and management are essential.

Evaluation

a. Suggest a number of reasons why new ideas and techniques meet with mixed reception by engineering practitioners.

b. List some advantages and disadvantages of the use of CPM.

2

CRITICAL PATH METHOD PROCEDURES AND TERMINOLOGY

2.1 GENERAL

If critical path methods are applied to the development of a construction plan, to the preparation of an estimate, or to the control of actual operations in progress, a number of logical procedures must be followed. These procedures may be conveniently grouped into planning and scheduling, the principal tasks to which CPM is directed. They may be defined as follows.

Planning is the process of choosing the one method and order of work to be adopted for a project from all the various ways and sequences in which it could be done. The sequence of steps required to achieve the optimum result is the proper plan for the works, and this can be schematically shown on the CPM network diagram.

Scheduling is the determination of the timing of the operations comprising the project and their assembling to give the overall completion time. Scheduling can be done only after a particular project plan has been defined and modeled in such a way that it can be committed to paper in the form of a network diagram.

2.2 PROJECT BREAKDOWN

The first step in planning a job is to break it down into the separate operations or processes necessary for its completion. The degree of breakdown will vary for each project and will be influenced by the nature of work and class of labor involved, the location of the work on the site, the costing data required by management, the itemized bill of quantities or rates, and the broad general sequence of the job. Each of these separate operations or processes is called an *activity,* and the completion of an activity is an *event* that signals the successful achievement of work. Activities therefore consume time, whereas events do not, and events are separated from one another by activities.

When a list of all the activities in a project has been prepared, the next step is to determine the essential relationship between these activities. Although many of them may proceed concurrently, certain activities must be constrained to a given sequence or *chain:* for example, casting of concrete presuppposes formwork erection and reinforcement installation, and pipelaying presupposes pipe delivery. These are examples of *technology* or *physical constraints* applicable to activities, and are apparent as soon as each activity in the project is subjected to the following questions.

1. What activities *must* precede this activity?
2. What activities *may* be done concurrently with this activity?
3. What activities *must* follow this activity?

In this way each activity is examined and the necessary sequences of activities are determined. Every activity therefore has a definite event to mark its possible beginning; this event may be the start of the project itself or the completion of a preceding activity.

It should now be clear that the end of one activity signals the start of a related dependent activity. Consequently, in a diagram where each activity or operation is an entity overlapping operations are prohibited. If they occur they must be broken down into two or more activities representing those portions of the operation to be completed before later portions are commenced. The overlapping of work as shown on conventional bar chart construction programs is impossible with CPM,[1] and hence this new technique offers a much greater degree of control over all the operations on the site.

Beside physical constraints there may be other types. *Safety constraints* may necessitate the sequential separation of activities that could otherwise be concurrent: for example, ground floor concreting operations may be prohibited while steel framework is being assembled immediately overhead. *Resource constraints* can occur when it is essential to delay because resources for certain operations cannot be made available: for example, release of equipment required from another project may not be possible until a

---

[1] Nevertheless many computerized CPM packages allow the use of conditional, and overlapping, relationships to be prescribed for, or between, certain activities. This feature often enhances the consideration of resource scheduling and the representation of work management policies. Hence their use implies computer processing or conditions that must be examined in manual calculations and may lead to faulty logical relationships.

certain date, and consequently some activities must be delayed until the equipment will be available, whereas otherwise concurrent activities may proceed; or it may be essential for certain activities to be completed earlier than physically necessary in order to earn progress payments to finance otherwise concurrent activities. Such decisions influence the sequence of operations and place constraints upon the project. *Crew constraints* may also occur: for example, specialist welding crews may be hard to obtain, and all the welding activities may have to be done in sequence with a small crew, whereas they could otherwise have been done concurrently; another example is when a single operation may have to be divided into several activities to emphasize that different labor skills are required. Finally, there are *management constraints* when, for example, the sequence of otherwise independent activities is controlled by a management decision, or when normally concurrent activities are ordered to be done in a certain sequence, simply because management arbitrarily wants them done that way.

All these aspects must therefore be carefully studied by the planner when the project is broken down into its essential activities, and when considering the various chains of activities which must be maintained. To the extent that the project can be represented by a diagram, the diagram is representative of the project, and considerable skill is sometimes required in designing a diagram to satisfy all the requirements imposed by physical, safety, crew, equipment, finance, and management constraints. Obviously, all relevant constraints must be shown on the diagram.

2.3 THE NETWORK DIAGRAM OR MODEL

A *network* is a diagrammatic representation of a program or plan for a particular project (or part of a project) that shows the correct sequence and relationship of activities and events required to achieve the end objectives.

In an *activity-oriented network* or *arrow diagram* each line or arrow represents one activity, and the relation between activities is represented by the relation of one arrow to the others; each circle (or *node*) represents an event. Diagrams of this type are shown in Figure 2.1a. The length of the arrow has no significance;[1] it merely represents the passage of time in the direction of the arrowhead. Each individual activity is represented by a separate line (or arrow) and the start of *all* activities leaving a node depends on the completion of *all* the activities entering that node. Hence the event represented by any node is not achieved until all the activities entering that node have been completed. This time of achievement is called the *event time;* it is a most important concept of CPM and is discussed in detail in Section 2.5.

---

[1] Nevertheless, arrow networks are often drawn to a time scale, thereby taking on a bar chart format. In these cases, provided careful attention is paid to the consideration and portrayal of network logic, the necessity for float calculations as discussed in later chapters is obviated. The use of time-scaled networks is discussed in detail in Chapters 3, 8, and 11.

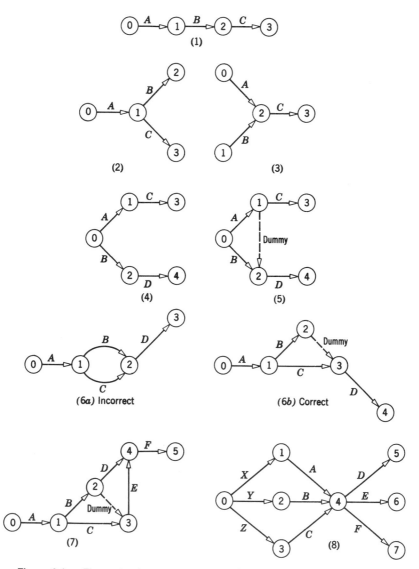

Figure 2.1a   Elements of an arrow network (ADM = arrow diagram method).

The network diagrams in Figure 2.1a illustrate some of the logical procedures adopted by CPM. In (1) it is obvious that A must precede B, and B must
precede C. In (2), A must precede both B and C. In (3), A and B must precede
C. In (4) A must precede C, and B must precede D. In (5), A must precede C
and D, and B must precede D; this necessitates using a connecting arrow
(called a *dummy*) to maintain the logical sequence of events between A and D.
Dummy activities have zero cost and zero time; they are shown by broken

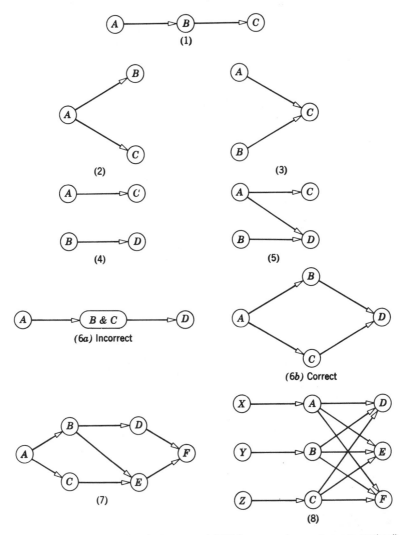

Figure 2.1b Elements of a circle network (PDM = precedence diagram method).

arrows. Dummies may also be required to maintain specific activity identification between events as shown in (6), where *A* must precede *B* and *C*, and *B*
and *C* must precede *D*. Further examples of the logical procedures used in the
compilation of network diagrams are seen in (7) and (8). In (7), *A* must
precede *B* and *C*, *B* must precede *D* and *E*, *C* must precede *E*, and *D* and *E*
must precede *F*. In (8) the situation is that *A*, *B*, and *C* must all be completed
before any of *D*, *E*, or *F* can start. Study of these diagrams will give the reader
an initial comprehension of network logic, which is described in detail in
Chapter 3.

Events and activities are numbered for identification on the network. The order of numbering is strictly immaterial, but conventionally, for various reasons, the number at the head of the arrow is always greater than the number at the tail. The project therefore begins at the first event (i.e., project start numbered as datum) and proceeds, event by event, to the completion of the works. In drafting a network layout it is axiomatic (1) that each node represents correctly the complete relation between *all* activities entering and leaving; (2) that all activities leaving a node have identical predecessors, and all those entering have identical followers; and (3) that each activity has a unique set of numbers assigned to it, with the tail number less than the head number. These conditions may be met for every job, no matter how complex, by careful drafting and the use of dummies.

There is another type of network diagram in which the nodes represent activities, and the lines or arrows represent logical time relationships between activities; the arrows therefore represent and replace the conception of events. These networks, called *event-oriented networks, precedence diagrams, and circle diagrams* are constructed in a manner similar to activity-oriented networks. Like these, the length of the arrows has no significance, for they merely point in the direction of increasing time; circle diagrams, however, eliminate the need for dummy activities, identify each activity with a single reference number, and can be readily adapted to changes in logical relationships between activities. Figure 2.1*b* shows the elements of a circle network, corresponding precisely to the situations presented in Figure 2.1*a* for an arrow network. It will be apparent that in complex relationships, of which (8) is a minor example, the circle diagram becomes difficult to follow unless resort is made to the subterfuge of dummy activities, as will be demonstrated later (see Figure 3.4*j*). Furthermore, circle networks cannot be drawn to a time scale, and events and float paths are not easily defined. However, they are useful for portraying resource leveling and scheduling problems (discussed in Chapter 7), and in the graphical portrayal of networks on computer screens; but most representations of CPM construction planning adopt activity-oriented and time-scaled networks.

Critical path methods are concerned not only with the sequence and interrelationship of activities (as disclosed by the network diagram), but also with the time and cost of completion of activities. The first draft of a network shows only the proper sequence and relationship between the various operations comprising the project; to complete the diagram, it is necessary to add the time required to carry out each activity. Hence, by definiiton, a *network diagram* is the schematic representation of a project, showing all the relevant activities and events in correct juxtaposition, and the *times required for their completion*.

For each activity there is a corresponding cost, which usually applies only to the specific time of completion shown for each activity. If the time is varied, the cost may be expected to vary also. Consequently in the final analysis of a network, it is necessary to know the effect on the cost of a change in time. Data showing this effect (called *utility data*) can also be indicated on a network for any activity. When this information is shown (that is, *when time and cost data are both indicated*), the schematic representation of the project is referred to as a *network model*.

2.4   UTILITY DATA AND TIME-COST CURVES

Utility data are the detailed time and cost information obtained from the works estimate for each project activity. Such data must be presented in a way that shows the direct cost and the time required for every possible method of carrying out each activity. This basic information is needed for the determination of optimum project cost and optimum project duration. It is the essential data for the solution of the time-cost problem referred to in Chapter 1. By the use of CPM these utility data are analyzed and correlated, and the optimum solution for the project is determined logically and systematically. This solution must be somewhere between the two extremes of *the least-cost solution* and *the least-time solution*.

The least-cost solution is commonly called the *all-normal solution,* and gives the time required for completing the project at the lowest possible direct cost.

The least-time solution is the plan necessary for completing the project in the shortest possible time, at the minimum cost for that completion time. In order to reduce the time, a large number of the activities must be speeded up (*crashed*); but it is not necessary to crash every activity in the project to attain the least-time solution. If *all* activities are fully crashed, the result is called the *all-crash solution*. This will always cost more than the least-time program, and therefore is never economical. Hence the objective of the least-time program approach is the selection of only those activities that should be crashed for the optimum solution. (See Chapters 5 and 6.)

An essential corollary to the preparation of the activity utility data from the project estimate is the production of the *direct-cost/time curves* (also called *utility data curves*) as illustrated in Figure 2.2. Here the direct cost for each method of accomplishing an activity is plotted against the duration (time) required to do it in that way. A careful examination of the figure will make it clear that, if there are a great many possible ways to crash an activity, the time-cost curve will approach the continuous ideal theoretical curve shown in 2.2*a*. In practice, however, there are normally only a limited number of ways investigated and thus only a finite number of points are defined. The practical curve shown in 2.2*b* is taken as linear between points. The features of the curve of major interest are the all-normal point, the point showing lowest crash cost for least time, and the defined points in between.

The all-normal point is by definition the time required to do the work at the lowest feasible direct cost; all faster times must cost more, because of the added expense of overtime, shift work, use of more equipment, and so on.

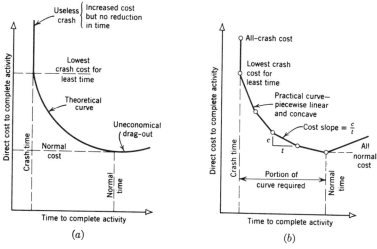

Figure 2.2   Time-cost curves for an activity. (a) Theoretical curve. (b) Practical curve.

The point labeled "crash cost for least time" shows the minimum direct cost for doing the work in the shortest practicable time. The points in between show the costs for various feasible times in which the work can be speeded up by the different practicable ways available. Curves of this type may be derived for individual activities, or groups of activities, and for a project as a whole.

The first approach to the logical planning and scheduling of a project is to find the critical path for the all-normal solution, that is, to construct the network from the utility data, using the times for the lowest direct cost of each activity. The method is described in Section 2.5.

2.5   CRITICAL PATH DETERMINATION

The following steps have now been completed:

1. The project has been broken down into unique feasible activities.
2. All these project activities have been listed.
3. All the constraints have been specified.
4. The network diagram has been sketched out, and all events numbered.
5. Utility data have been prepared for each activity.

The sixth step is to assign a time to each activity on the network, using the work times from the all-normal utility data. For example, Figure 2.3 illustrates the production of an arrow network diagram for a simple hypothetical project involving 13 activities. The first draft of the network would appear as in *a*. Assignment of all-normal times to the activity arrows is seen in *b*. Alongside each arrow is written the length of time (hours, shifts, or days, as desired) necessary to complete the work involved in this activity; this is called the *duration* of an activity. A dummy has zero duration.

Next, by proceeding through the events in numerical order from the start, simple addition will give the earliest possible time at which *all the* activities entering each event can be completed; this is then the *earliest finish time* (EFT) for the event. The EFT for each event is recorded in the left side of the oval "time-box" adjacent to that event. After proceeding right through the network, the EFT of the last event is derived; this is the earliest possible time in which the project can be completed, and it is *the sum of the durations of the longest time path through the network* from start to finish. In Figure 2.3*b* this is 63 days. If this period is accepted as the project duration which must not be lengthened, the next step is to work backward from the final event, subtracting the durations of each activity, to find the *latest finish time* (LFT) permissible for each event *if the project is to be completed by the EFT of the final event*. The latest finish time is controlled by *all* the activities starting from the event concerned, and is the minimum figure so obtained. If the event is not achieved by its LFT, the project will be delayed. The value of the LFT is entered in the right side of the oval "time box" adjacent to each event, as indicated in Figure 2.3*c*.

There are now two figures in each box, giving the EFT and LFT for each event; the difference between them is the leeway available for delays, etc., and is called *float*. For some events the same figure appears in both sides of the box, indicating the same time for latest and earliest finish; in these cases there is no float at all. These are the critical events that must be achieved on schedule if the project is to be finished in the minimum total time. *The path joining these critical events is therefore the critical path for this network, under the conditions for which it was drawn.* It is shown in heavy lines in Figure 2.3*c* and passes through events 1, 2, 4, 5, and 9. Activities along this critical path are called *critical activities*.[1]

When the critical path through a network has been established, it may be desirable to check the effect of an alternative method of construction for one activity on the network as a whole. If so, another diagram is drawn, showing the new duration for this activity. For instance, in the case of the project shown in Figure 2.3, if activity *F* (shown with a duration of 30 days) could be carried out in 20 days by using a different method or equipment (irrespective of cost), then the EFT of event 9 would reduce to 53 days via activity *F*. It

---

[1] It is axiomatic that critical activities have no float; hence the direct path between any two critical events is not necessarily a critical path. Observe activities 4–7 and 5–9 in Figure 2.4; these have float and are not critical, the critical path in this case being 1–2–4–5–7–9. The criteria for determining critical activities are fully explained in Chapter 4.

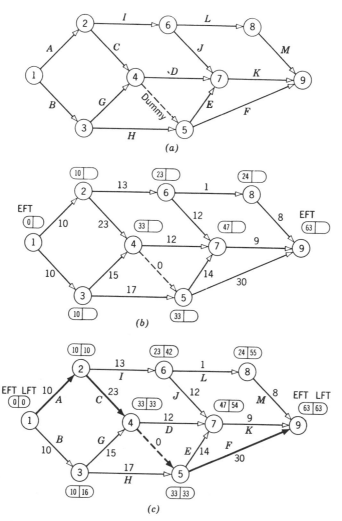

Figure 2.3 Steps in determination of critical path for a network. (a) Network diagram—first draft. (b) Timing the network. (c) Critical path for all-normal duration.

would, however, be 56 days via activity K, as seen in Figure 2.4. The total project duration thus reduces to 56 days, activity F ceases to be a critical activity, E and K become critical, and the critical path of the project changes to 1–2–4–5–7–9. Whether this is a more economical method of doing the work is not relevant at this stage; the purpose of this illustration is merely to show the effect of the network of a change in the plan for doing one activity of this project. Notice also how the float of noncritical events has been affected by this alternative proposal for activity F.

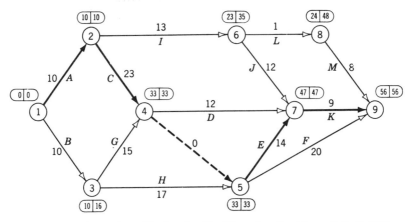

Figure 2.4 Alternative critical path with different method for activity 5–9 (*F*).

2.6 ACTIVITY TIMES AND FLOATS

Having derived the earliest and latest finish times for all the events in the project, the next step is to find all the activity times and floats.

In Figure 2.3, activity 6–8 has a duration of only one day. It is also apparent that event 6 could occur as early as the 23rd day, whereas event 8 could be as late as the 55th day, so there might be 32 days available in which this one-day operation could be done. This activity therefore has a total leeway, or float, of 31 days; it is not a critical activity. On the other hand, activity 2–4 is critical, having no float at all.

Float may be divided in several ways. The full amount of time by which the start of an activity may be delayed without causing the project to last longer, is called *total float* (TF). Such a delay may cause delays in some of the activities that follow it, but will not retard the project. Obviously, a critical activity has a total float of zero. The *free float* (FF) of an activity is the amount of time by which the start of the activity may be delayed without interfering with the start of any succeeding activity. It follows that the free float cannot be greater than the total float. *Interfering float* (IF) is the difference between the total float and the free float of an activity. Any delay in starting that involves consumption of any of the IF time of an activity will necessitate the retarding of some following activities but will not delay the overall project time. The project duration cannot be prolonged unless delays in excess of the total float occur. Other types of float are discussed in Chapter 4.

To determine the floats available to any activity it is first necessary to compute its *earliest and latest start times* (EST and LST), as well as its *earliest and latest finish times* (EFT and LFT), from the data on the network. The EST is the time at which it *can* begin, and the LST is that at which it *must* commence if the minimum project duration is to be achieved. The EFT and LFT are simply the appropriate start times plus the duration of the activity.

The relation between all these activity times and the floats defined above may be expressed as simple equations. Thus, for any activity,

$$\text{its EST} = \text{the EFT of its tail event} \qquad (2.1)$$

$$\text{its EFT} = \text{its EST} + \text{its duration} \qquad (2.2)$$

$$\text{its LFT} = \text{the LFT of its head event} \qquad (2.3)$$

$$\text{its LST} = \text{its LFT} - \text{its duration} \qquad (2.4)$$

$$\text{its TF} \;\; = \text{its LFT} - \text{its EFT} \qquad (2.5)$$

$$= \text{its LST} - \text{its EST}$$

$$\text{its FF} \;\; = \text{the EST of its following} $$
$$\text{activity} - \text{its own EFT} \qquad (2.6)$$

$$\text{its IF} \;\; = \text{its TF} - \text{its FF} \qquad (2.7)$$

It is therefore obvious that, once the network diagram has been completed and each event's EFT and LFT recorded in its "time-box," all the activity times and floats may be calculated. This is the first step in the scheduling of the project. It is emphasized that these calculations are simple and purely mechanical; no previous experience is required for the process.

2.7 SCHEDULING

Table 2.1 shows the results of all the time and float calculations for the network considered in Figure 2.3. The essential time relationships between the activities may now be analyzed and decisions made concerning the com-

Table 2.1   Scheduling of activities (see figure 2.3)

Activity									
Item	Arrow	Duration	EST	LST	EFT	LFT	TF	FF	IF	Remarks
A	1–2	10	0	0	10	10	0	0	0	Critical
B	1–3	10	0	6	10	16	6	0	6	—
C	2–4	23	10	10	33	33	0	0	0	Critical
I	2–6	13	10	29	23	42	19	0	19	—
G	3–4	15	10	18	25	33	8	8	0	—
H	3–5	17	10	16	27	33	6	6	0	—
Dummy	4–5	0	33	33	33	33	0	0	0	Critical
D	4–7	12	33	42	45	54	9	2	7	—
E	5–7	14	33	40	47	54	7	0	7	—
F	5–9	30	33	33	63	63	0	0	0	Critical
J	6–7	12	23	42	35	54	19	12	7	—
L	6–8	1	23	54	24	55	31	0	31	—
K	7–9	9	47	54	56	63	7	7	0	—
M	8–9	8	24	55	32	63	31	31	0	—

plete timing of the construction works. As a visual aid, the CPM bar chart shown in Figure 2.5 may be useful. It will be immediately apparent that this bar chart is similar to the conventional construction program, but that it shows, in addition, the critical activities and the essential information on float times. One can see at a glance which are the critical activities (no float) that must not be delayed if the project is to finish on time; and also how much delay can be tolerated in other activities. If any of these run late by an amount greater than its free float (if any), their lateness will interfere with the start of subsequent activities in the chain; but they (and subsequent activities) may be delayed by an amount of time not exceeding the total available float without prolonging the completion of the project. Float time is therefore a safety margin that may be used to offset unforeseen or deliberate delays in activities along noncritical paths.

The information obtained at the scheduling stage is of great importance to the construction manager in controlling the project in the field, and it is also equally useful to the planner. Knowledge of available float enables activities to be shifted about the program, within their float limits, to help smooth out labor and plant requirements. This is another of the major advantages that the construction industry has found in CPM: by judicious manipulation of free and interfering floats, a smoother construction plan can be achieved logically and mathematically, without extending the total project time. This aspect is fully discussed in Chapter 7, and is known as *resource leveling*. After this manipulation of noncritical activities, the revised arrangement is rescheduled

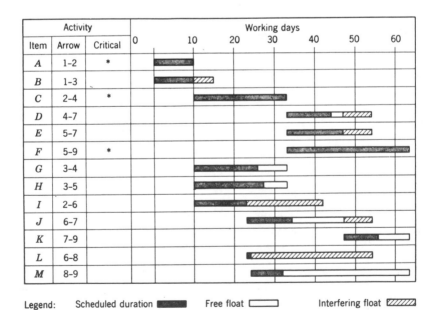

Figure 2.5   Critical path bar chart construction program for network shown in Figure 2.3.

with fixed dates to become the final project schedule for the specific duration required.

In general, with every construction project to be planned by the critical path method, all of the foregoing procedure is followed to this point, using all-normal utility data (lowest direct cost). Certainly alternative construction methods and equipment may be investigated for any desired activity, but up to this point at the all-normal level only. Thus the construction planner has arrived at the cheapest direct-cost all-normal solution for his project, and has computed its overall duration. No paper plan is perfect, but CPM provides a much more accurate picture than the former methods of planning based largely on judgment.

The next stage is determining the most favorable balance between total time and total cost, taking into account all the overhead and indirect costs, bonus and penalty clauses, and other monetary influences to obtain the most economical solution for the project.

2.8 OPTIMAL SOLUTIONS—COMPRESSION AND DECOMPRESSION

Figure 2.4 showed how project time and floats were affected by an alternative construction method for a single activity. Obviously this alternative method would be used if it resulted in a lower project direct cost. Similarly, any alternatives for other activities can be examined until the all-normal solution is obtained, irrespective of project duration. Somewhere between the all-normal and crashed least-time solutions, however, lies the overall optimum economic solution for the project. This solution must now be derived if the planning is to take the fullest advantage of critical path methods. It can only be found after a series of *optimal solutions* have been evolved, followed by a cost analysis of the project. An optimal solution requires the derivation of the lowest direct cost for the project for a specific duration; in other words, an optimal solution provides the coordinates for one point somewhere on the direct-cost/time curve for the project.

The computation of a series of optimal solutions thus provides the actual direct-cost/time curve for the construction works being planned. By supplying additional incremental resources of manpower and equipment to the all-normal solution in the most efficient way, a series of points on the curve is obtained, showing how each increment (although it adds to the direct cost) reduces the project completion time by a certain amount. This is to say, an investigation is made in the effect of successively crashing various activities. This may be continued with consecutive optimal solutions until the project can be speeded up no more. This procedure is known as *network compression*.

From the utility data, which show the feasible completion times and corresponding direct costs of each activity, it is easy to compute the *cost slope* of every activity in the project. Cost slope is the incremental direct cost per unit time, and it indicates the rate of change in direct cost of an activity with reduction of its duration. Referring to Figure 2.2, it is obvious that the cost slope itself may change as the duration decreases. It may be necessary, therefore, to compute a number of cost slopes for an activity, corresponding to ranges in the time reductions, for it is essential to know the real cost slopes of each activity that may have to be crashed during the determination of the optimal solutions. Although this may seem a formidable task, it is easily done on a calculator, being merely the direct-cost difference divided by the time saved. Tabulation of relevant cost slopes thus forms part of the utility data required for compression calculations.

Once the matter of costs is introduced and the cost data for each activity have been inserted on the network, the diagram becomes (by definition) a network model. In other words, the visible schematic representation of the project now presents all the relevant time and cost information necessary for the computation of optimal solutions; it has thereupon become a real mathematical model of the construction works under consideration.

If the all-normal network model is carefully considered, it will be apparent that *a reduction of project duration is only possible by crashing activities in turn along the critical path;* and that, furthermore, the cheapest compression is obtained by speeding up the activity with the cheapest (that is, flattest) cost slope first. Project compression, therefore, involves the selection of the critical activities that can be compressed individually, or in combination, at minimum cost per unit time, until they reach their crash limits, or until additional critical paths develop in the network; the total project direct costs and the new project durations are recorded after each network compression. With the advent of new critical paths, additional activities become available for economic compression, and these are crashed in turn, until further project compression becomes impracticable and the crashed least-time solution is finally reached.

In this way, each stage of the compression calculations provides the coordinates for one optimal solution, and it is thus possible to plot the actual curve of optimal project direct costs against project duration. On this direct-cost/time curve may now be superimposed the curve for indirect costs against project duration. Adding these two will produce the curve for total project costs against project duration. The result will look like Figure 2.6, which immediately shows the optimum overall economic solution for the project.

The details of compression procedures are fully discussed in Chapters 5 and 6. For the moment, it is only necessary for the reader to understand that, by starting with two schedules (the all-normal, and the all-crash) and the appropriate activity utility data pertaining to each, the construction planner can derive the project cost-time curve and the optimum economical construction solution. This compression technique may not be warranted in full for smaller construction works, but it will certainly be worthwhile in larger and

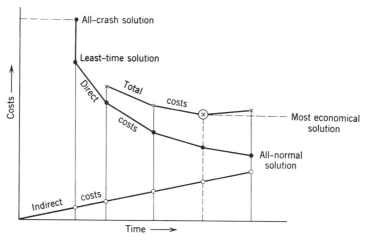

Figure 2.6 Project cost-time curves showing most economical solution.

more complex projects. Very often the all-normal solution, or a solution with one or two compressions, will prove to be the most economical overall plan; if the complete results are plotted on the cost-time graph after each compression stage, the correct answer soon is apparent, and further compressions are abandoned.

There is another approach to the problem of optimal solutions. If a planner has the utility data for the all-normal solution and the all-crash solution, it is a fairly simple matter to generate any point on the direct-cost/time curve by starting the calculations from the crashed solution and carrying out the compression procedure in reverse. This is known as *decompression,* and is fully explained in Chapter 6. It is most useful when the specified project duration obviously demands a near-crash effort; for otherwise it would be essential to carry the compression calculations right through from the all-normal to the least-time solution, thus wasting a considerable amount of mathematical effort.

Finally, a combination of decompression and compression, carried out alternately, may be adopted to find an optimal solution when one starts with a feasible nonoptimal estimate. This technique is called *optimization of nonoptimal solutions,* and is described in Chapter 6. Its major advantage is that one can begin the calculations from any nonoptimal feasible estimate, based on a project duration approximating only that specified, and use the procedure to improve the original construction plan; at every stage of the calculations the new plan is better than the previous one, so that no effort is wasted, as it may be with pure compression from the all-normal solution or decompression from the crashed solution.

Compression and decompression, although purely mechanical and repetitive processes, lead to greater insight into the problems of the project. For projects with between 200 and 300 activities, it is usually practicable and advantageous to carry out optimal solution procedures by manual methods. For larger and more complex construction works, however, the use of manual compression is often avoided because it is involved and may require some time to undestand fully; recourse is therefore made to computer processing. Decompression, on the other hand, must be carried out manually because it has not yet been provided for in computer programs.

2.9 USE OF COMPUTERS

When there is considerable complexity in project sequences, all the network calculations have to be done by computers. This necessitates the development of suitable computer programming and data processing. So much preliminary work is entailed, however, that experience has proved that it is seldom necessary to process the more usual civil engineering works (with, say, 200 to 400 activities) through computers. On the other hand, large building and engineering projects may require computer processing because of the vast number of activities involved; since is has been found necessary to adopt machine computations, many computer companies have already established suitable programs that enable projects with as many as 3000 to 4000 activities to be analyzed and solutions to be obtained in a few hours from the time the data have been set up for the machine.

Although the authors advocate manual methods for as much work as practicable, the necessity for various computer calculations cannot be ignored. Recent CPM computer packages greatly assist the planner in systematically portraying project networks and in setting up project data files; usually these networks are in circle notation (precedence network form). Major project control, accounting, and finance are essentially computer work. Ignorance of the capabilities of computers is probably the greatest limitation of construction planners; they should fully understand the types of programs available and the classes of work that computers can do to aid them in their field.

The use of computers in construction practice and project management is discussed in Chapter 13.

Evaluation

a. List, in sequential order of use, the basic processes involved in developing a CPM-based project plan.

b. Why bother with the portrayal of simple or complex relationships between project activities?

3

THE NETWORK DIAGRAM
AND UTILITY DATA

3.1 INTRODUCTION

Critical path methods require the determination of network diagrams and models that specifically and uniquely portray the features of the construction project under consideration. It is most important to develop the network in sufficient detail to show validly the true characteristics of the construction methods to be adopted. The demands of management and the depth of planning required will decide the magnitude and detail necessary for the network. The summary discussions of Sections 2.2, 2.3, and 2.4 are now expanded in detail in this chapter, and the remainder of Chapter 2 in subsequent chapters, in order to ensure that the reader will fully understand the ramifications of the CPM concept.

A network diagram or model is deterministic: the component activities require definition and identification, and they are related to each other in a specific way by logic. The logic may have its basis in technology or result from physical considerations or be derived from managerial decisions. Consequently, in this sense a network diagram or model implies or indicates a rigid representation of a specific way of doing the work. Hence it is essential that the construction methods be decided before the network can be drawn. This does not mean that flexibility in planning is denied; on the contrary, flexibility is ensured by considering as many approaches as desired for the completion of the project and reviewing each of these on separate networks so that the best may be chosen.

The outstanding advantage of the netowrk diagram is that it forces a precise and complete presentation of all activities in a project at its inception by all those concerned with its completion. Managers and planners must think their way through the works from start to finish before the job begins. As they do so, and as their networks show up relationships previously unsuspected, more efficient construction sequences and methods will be developed.

3.2   HOW TO BEGIN A NETWORK

A construction project is a collection of individual operations or activities. The order in which these activities are started and their relationships to each other constitutes the construction plan. The network diagram can only be finalized after the collection of activities and their ordering has been decided.

The first step in the preparation of a network is the division of the project into its activities. These are best arranged by listing them in columns. No specific order of precedence is necessary, but systematic listing by trades, skills, location, or plant requirements is often helpful.

The next step is to formulate the construction logic, or the specific ordering of the activities. This involves a precise statement of the relationships between them. A *general* ordering of activities within the project it not difficiult, since their description often implies a relative location within the job; the *specific* ordering, however, is more difficult, and requires very careful consideration.

A good approach to specific ordering is first to determine the obvious physical and safety constraints, then the crew and other resource constraints, and finally the management constraints. The physical constraints lead initially to chains of activities, simply determined and coupled. The consideration of other constraints, and the detailed determination of physical requirements, usually leads to the branching and intermingling of the chains into networks. It is often helpful to tabulate the activities systematically; to note those that must precede and those that must follow each activity, and those that may be carried out simultaneously. The network layout is then determined by trial and error, first satisfying some of the conditions and then refining the portions of the network that violate the remainder.

To design a network that will satisfy all the constraints requires a great deal of skill. In some cases, managerial decisions are extremely difficult to formulate in a diagram. However, it is reasonably simple to test a given network. Consequently, to determine improvements to the diagram, it is easier to start with a rough network (incorporating finer details successively) than to attempt a detailed diagram at the outset.

For example, consider the simple construction of concrete footings, which involves earth excavation, reinforcement, formwork, and concreting. A preliminary listing of activities might be

A–Lay out foundations.
B–Dig foundations.

C–Place formwork.

D–Place concrete.

E–Obtain steel reinforcement.

F–Cut and bend steel reinforcement.

G–Place steel reinforcement.

H–Obtain concrete.

The activities are listed in the order in which they are thought of and without any definite ordering in a specific construction plan. Activity identification is, in this case, given by alphabetical letters; but it may be given by any coding system for skills, areas, etc., using alphanumerical notation.

Examination of the list of activities shows that some grouping is obvious. Thus, considering physical constraints only, the following physical chains are developed:

1. From a consideration of the actual footings:

    A–Lay out foundations.

    B–Dig foundations.

    C–Place formwork.

    G–Place steel reinforcement.

    D–Place concrete.

2. From a consideration of the steel reinforcement:

    E–Obtain steel reinforcement.

    F–Cut and bend steel reinforcement.

    G–Place steel reinforcement.

    D–Place concrete.

3. From a consideration of the concrete only:

    H–Obtain concrete.

    D–Place concrete.

When the project is seen from these different viewpoints, individual chains of activities emerge; but, on viewing the job as a whole, it is obvious that interrelationships exist. For example, it is useless to pour concrete before the steel reinforcement is placed and the formwork installed. Therefore all the chains must merge before pouring the concrete. And if steps are to be taken to obtain the steel and the concrete immediately work begins (this would be a management decision or constraint), then the chains all start at the same point or event with the laying out of the foundations.

The development of a preliminary network for the project is possible at this stage because first, a list of activities has been defined and second, a rough construction logic has emerged.

The actual representation and appearance of the network depend both on the modeling form adopted and on the spatial locations of the symbols as drawn. As mentioned previously in Chapter 2, there are two basic ways in which activities can be modeled: (1) when the activities are represented by "arrows" in an activity-oriented network and (2) when the activities are represented by "nodes" in a circle network. These two basic modeling forms emerge naturally from linear graph theory. (See Appendix A.)

In Figure 3.1 a preliminary network is developed, in both arrow and circle forms, from the above information. Subsequent finer developments will be shown later after a closer examination of the actual situation, such as the relationship between activities C (formwork) and G (place steel reinforcement). Notice at this stage that the emphasis is at present on the determination of sequences or chains in the network; later their interconnection or tying together will appear.

An alternate approach is to use a tabulation into which is built the construction logic. In this case, the network is developed by concentrating first on the interconnection of the activities and later on the determination of the chains between locations at which ties occur. As discussed in Chapter 2, the construction logic is formulated by considering the following questions, which (although specifically aimed at individual activities) will enable the proper relationships between activities and events to be established:

1. What activities *must* be completed immediately before commencement of this activity?
2. What activities are independent of this activity, and *can* be done concurrently?
3. What activities *must* be started immediately after completion of this activity?

The answers to these three questions specify the physical constraints in the project. If the operative words "must" and "can" are replaced by "will," the answers to the questions specify the management constraints involved. Again, if the word "can" in the second question is replaced by "cannot," the answers may involve safety, resource, and crew constraints.

The construction logic pertaining to the above activities A to H may be formulated as indicated in Table 3.1. From this it can be seen that the constraints imposed upon the activities G (place steel reinforcement) and D (place concrete) now specify the interaction between the chains of activities. Finally, the starting of the project may be considered as an event in arrow notation from which the initial activities must branch or it may be considered as an activity in circle notation. Initial activities are determined by applying some of the above questions to the initial event. All these investigations lead to the nucleus of a network diagram, as illustrated in Figure 3.2. Once this nucleus is established, the finer definition of the various chains is relatively simple, and it results in the same network diagrams shown in Figures 3.1b and 3.1d, respectively.

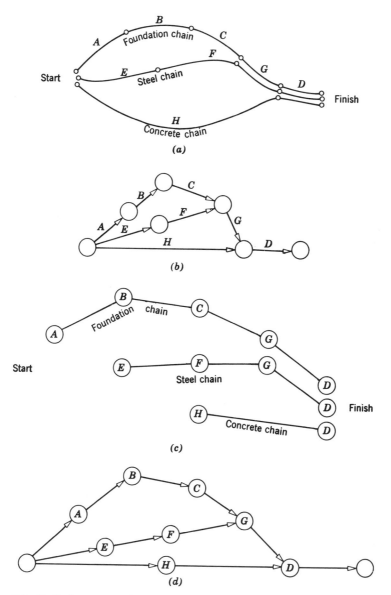

Figure 3.1 Preliminary network diagram. Arrow network: (a) initial sketch; (b) first draft. Circle network: (c) initial sketch; (d) first draft.

Important features of a network diagram follow from the information in the table. Considering a particular activity, the following deductions can be made: first, if two or more activities occur in the column headed "Preceding Activities," then two or more chains converge to the beginning of this activity; second, if two or more activities appear in the column headed "Following

Table 3.1 Tabulation of construction logic

Activity Description	Symbol	Preceding Activities	Concurrent Activities	Following Activities
Lay out foundations	A	None	E, F, H	B
Dig foundations	B	A	E, F, H	C
Formwork	C	B	E, F, H	G
Place concrete	D	G, H	None	None
Obtain steel reinforcement	E	None	A, B, C, H	F
Cut and bend steel	F	E	A, B, C, H	G
Place steel reinforcement	G	C, F	H	D
Obtain concrete	H	None	A, B, C, E, F, G	D

Activities,'' the branching of chains is implied at the completion of this activity; and finally, parallel chains exist if any entries appear in the column headed "Concurrent Activities."

It should be obvious that, if the construction logic is fully determined and the activities clearly stated, only one unique network diagram portrays correctly the construction project, provided, of course, that the list of activities is not altered. The visual appearance of this network may, of course, differ from one representation of the diagram to another, depending on the scale and the location on paper of the various lines delineating the chains of activities. For instance, Figure 3.3 shows different preliminary representations of the same arrow network diagram for this foundation construction.

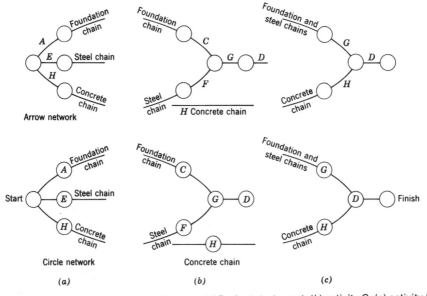

Figure 3.2 Nucleus of network diagram. (a) Project start event; (b) activity G; (c) activity D.

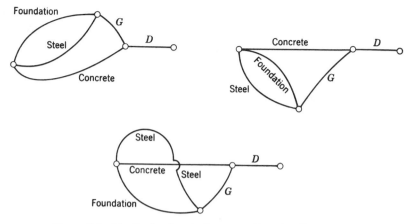

Figure 3.3 Identical network diagrams. (See also Figure 3.1.)

Very often a combination of the two approaches to the commencement of a network will be useful. The first approach—determining physical chains—is very helpful when the network is essentially composed of more or, less independent chains with few interactions. The second approach—determining the grouping of activities around certain specific activities and events—is better when a highly interconnected project is involved. The combination of visual and logical information usually leads (after a few trials) to a diagram sufficiently accurate to enable a detailed representation of the proposed construction plan to be effected. The further development of this network diagram, encompassing greater refinements, is deferred until after a detailed discussion of basic network logic.

3.3 NETWORK LOGIC

As already described in Chapter 2, a network consists of a series of arrows and nodes, arranged so that they provide a detailed visual representation of a project. Each individually specified construction activity is placed on the diagram so that its location and logical orientation in the network indicate its position and dependence in the construction project.

In the arrow diagram each construction activity is shown as an arrow. The arrow tail signifies the start of the activity, and the arrowhead the completion of the activity. Arrow length and direction (in the sense of a compass direction) have no intrinsic significance as far as network logic and graph theory are concerned. Note, however, time-scaled networks in Figure 3.9b. Finally, all arrows must start and finish at network nodes (events).

An event corresponds to the concept of a "milestone of progress" toward completing the project and is modeled in the diagram by a node. Each event is numbered to provide a simple arrow or activity identification. The arrows and events form an oriented network in such a way that progress past an event is only possible when all arrows (activities) terminating at that event are completed. Once an event has been reached, all arrows (activities) commencing at that event may be started. In some instances, however, artifical activities are invented to provide control events for specific managerial purposes. Two special events are those denoting the start and the finish of the project.

In the circle diagram each construction activity is shown as a node. Entry into the node signifies the start of the activity, and exit from the node the completion of the activity. The nodes are connected together by arrows (indicating logical precedence) to form an oriented network portraying the construction plan. An activity may be started when all the logical prerequisites for the commencement of the activity have been satisfied, that is, when all the arrows terminating at the particular activity are fulfilled. Once an activity has been completed, all arrows commencing at that activity node are validated and contribute toward the satisfying of conditions for succeeding activities. Often artifical activities denoting the start and finish of the project are included in the diagram, and frequently they are also incorporated to avoid complexities, as in Figure 3.4*j*.

In the compilation of network diagrams, a detailed understanding of network logic is essential. This is best illustrated by simple examples. Referring to Figure 3.4, the following points are of the greatest importance.

In 3.4*a*, the specific location of an activity *B* within a chain of activities requires the identification of the activities preceding and following it. Logically, *A* must precede *B*; and *B* must precede *C*.

In 3.4*b*, when a single chain branches into two chains, a complete specification of the activities at the branching node must be given; that is, *A* must precede *B* and *C*; and *B* and *C* are independent and logically are concurrent. Notice that this specification is made from the viewpoint of the branching node; nothing is implied about the branched chains, except that initially they are independent. The logical concurrency of activities *B* and *C* does not imply that they must be scheduled and worked in the same time slots.

With 3.4*c*, then, the joining of two chains into one—the reverse of *b*—may be readily formulated as: *A* and *B* must precede *C*; and *A* and *B* are independent and concurrent.

In 3.4*d*, the development of two independent chains, branched from a single chain, may be logically specified as: *E* must precede *A* and *B*; *A* and *B* are independent and concurrent; *A* must precede *C*; and *B* must precede *D*.

With 3.4*e*, however, when the development of the branched chains is not entirely independent, the simplest statement is: *A* and *B* are independent and concurrent; *A* must precede *C* and *D*, which are independent and nonconcurrent; and *B* must precede *D*. Notice in the arrow notation diagram the use of the dummy activity *F* to ensure that the beginning of activity *D* is dependent on the completion of activities *A* and *B*, but activity *C* is only dependent

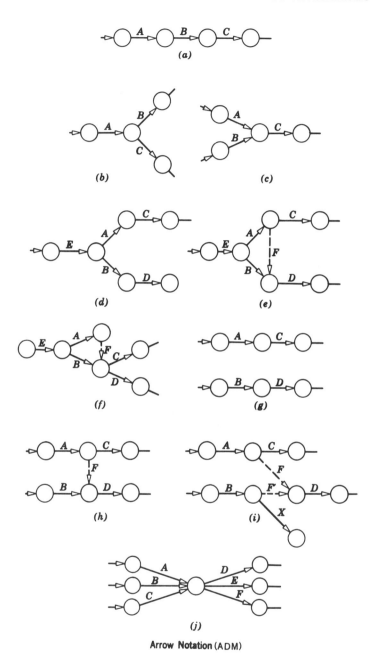

Arrow Notation (ADM)

Figure 3.4 Elementary network logic.

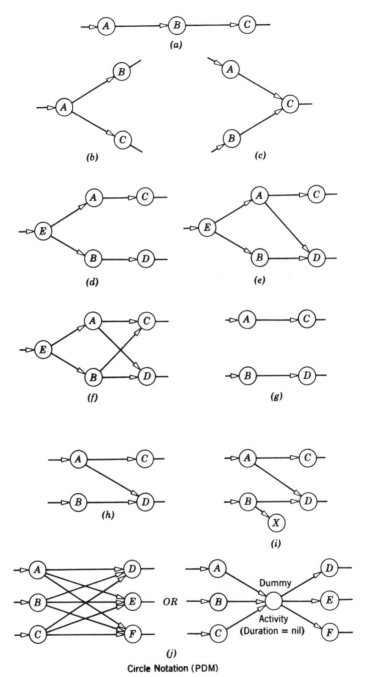

(j)

Circle Notation (PDM)

Figure 3.4 *(continued)*

on activity A. In circle notation there is no need to introduce the concept of dummy logical arrows since all branches are logical arrows.

If, however, activity C were also dependent on the completion of both A and B, the specification becomes: A and B are independent and concurrent; A must precede C and D; B must precede C and D; C and D are independent and concurrent. The network would then appear as in 3.4f.

At 3.4g, where the development of two parallel independent chains is being considered, and where specifications are required about possible simultaneous activities, it is essential to make independent statements, each formulated from the standpoint of different nodes: Chains (1) and (2) are independent and noncurrent; in (1) A must precede C; and in (2), B must precede D.

However, where the development of two parallel chains is not entirely independent, as in 3.4h, the specific ordering becomes: A and B are independent and nonconcurrent; C and D are independent and nonconcurrent; A must precede C and D; and B must precede D. But if B must also precede another activity X which is independent of A, C, and D, then the arrow notation diagram must be drawn with two dummies, as in 3.4i, to preserve the logic of the dependence of D on A and B simultaneously with the dependence of X on B only; only in this way can unintentional logical connections be avoided. In this case, the circle notation diagram requires only one additional logic connection.

The complexity of circle notation diagrams resulting from cases like Figure 3.4j can be avoided only by the use of a dummy activity, which, of course, will have zero duration. Thus the claim that precedence diagrams do not require dummies is not strictly correct.

In all networks it is essential to number the nodes so that any activity may be identified. Thus, referring to Figure 3.5a, activity 4 is uniquely identified as activity 0-1; similarly, 2-3 refers only to activity D. However, for activities B and C no unique identification is possible in arrow notation unless a dummy activity is introduced, as indicated in Figure 3.5b and c. Logically, both of these representations are correct and enable unique identification of activities B and C; however, the representation in Figure 3.5c is better for float calculations by computer. As shown in Figure 3.5d, the identification of activities in circle diagrams presents no problems.

Dummy activities therefore have two uses in arrow diagrams: first, to maintain the network logic and second, to ensure unique identification of activities. It is axiomatic that dummy activities require neither time nor resources for their completion. When a time delay is required in the network, an *artificial activity* is invented for the purpose; such artificial activities therefore require time but not resources for their completion.

The situation often arises that part of one activity can commence as soon as some of the preceding activity has been achieved. For example, referring to Figure 3.6a, the specification is that: all of A must precede any of B; and all of B must precede any of C. Suppose, however, that it is possible after

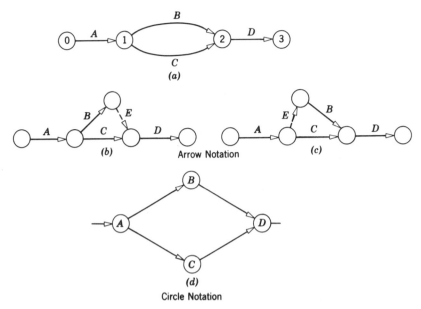

Figure 3.5   Activity identification. (*a*) Incorrect. (*b*) Correct. (*c*) Correct. (*d*) Unique node identification.

completing an amount $A1$ of activity A to begin an amount $B1$ of activity B; then the network in Figure 3.6*b* illustrates that some of A (that is, $A1$) must precede $B1$; all of A (that is, $A1$ and $A2$) must precede $B2$; and all of B must precede C.

Two further refinements of this nature are shown in Figure 3.6*c* and *d;* but in (*d*) the progress through activity A is *not* dependent on that through activity B, whereas in (*c*) it is dependent on such progress stage by stage.

Clearly, then, when overlapping activities are intended, the expedient modeling shown in Figure 3.6*e* should be shown as in Figure 3.6*f,* because activities once defined as entities must be modeled and logically related as entities.[1] Refer to Section 8.8 for a more detailed discussion of repetitive and interacting activities.

Complete comprehension of the foregoing points of network logic is essential before proceeding with the development of any diagram; otherwise, confusion results in interpreting the interrelationships between activities and constraints and the diagram will not represent accurately the problems of the

---

[1] In many practical cases, however, it may be desirable to commence an *unrelated* activity after only a portion of another has been completed. Although this situation can be readily handled by a subdivision of the initiating activity into parts, as mentioned above, some computer programs enable users to relate the two activities directly, thereby circumventing the strictness of the network logic.

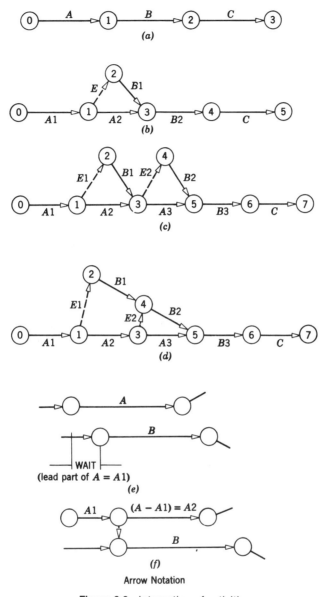

Figure 3.6 Interaction of activities.

project. The reader is advised to study these points and to grasp them thoroughly before proceeding further.

The adoption of arrow or circle notation is a matter of personal choice. In general, circle notation is favored by some construction planners because of its ease in handling logic. However, those familiar with bar charts find arrow

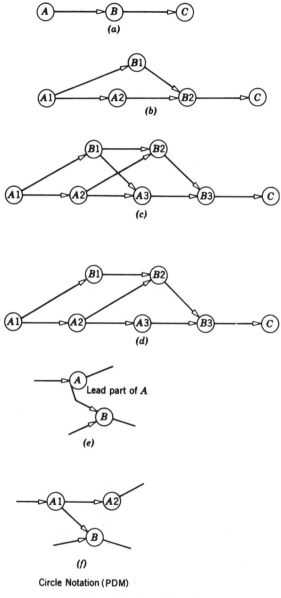

Figure 3.6   *(continued)*

notation more meaningful. It is possible in both arrow and circle diagrams to locate the activities on a drawing with a time scale. If this is done, the need for directed arrows is removed since the arrowheads and tails are implied by the time scale of the drawing. In arrow notation networks this leads to logically connected bar charts and factual networks (see Sections 11.5 and 11.6).

3.4   HOW TO DEVELOP A NETWORK

Once the nucleus of a network has been obtained, the development of the final diagram requires the following four considerations:

1. The checking of the diagram against the given construction plan to ensure that a valid network exists.
2. The inclusion of more detailed activities until the network obtains sufficient sensitivity to portray accurately the given construction plan.
3. The finer determination of basic logical decisions made by the various constraints; this implies a more detailed specification of the construction plan, and hence alterations to the network. It often happens that a careful consideration of the diagram will indicate an advantageous change in the construction method or plan.
4. The final testing of the network for internal logical consistency. This involves the correct numbering of events, unique activity identification, deletion of loose ends within the network, and so on.

The development of the finished diagram from the initial nucleus requires a great deal of thought and skill. The interaction of the above four considerations demands constant attention from a planning team preferably composed of those involved in the project. It is obvious that the refinement of a network is essentially a trial-and-error process, which is therefore difficult to specify. In the following examples arrow notation will be used unless specifically stated otherwise.

Figure 3.7 shows two network diagrams for the footing construction, (*a*) being identical with the network of Figure 3.1. This diagram implies that all the formwork (activity C 3-4) must be completed before any steel reinforcement (activity G 4-5) can be placed. The second diagram, (*b*), implies that all the steel reinforcement (G 3-4) must be placed before any formwork (C 4-5) can be erected. Which of these diagrams is correct depends on specific construction details. It may be that neither is right and that an entirely different diagram is required, based on the relationship between these two activities.

One possible development of this relationship is illustrated in Figure 3.8, where two networks are shown, both implying that a certain quantity ($G1$) of the activity "place steel reinforcement" can be carried out independently of the formwork, whereas the quantity $G2$ of reinforcement depends on completing the formwork activity C. The basic difference between these two diagrams is the relationship between the two subactivities $G1$ and $G2$ into which the previously single activity G has been split. The network in Figure 3.8*a* implies that all the reinforcement may be done with the one crew; on the other hand, Figure 3.8*b* implies that two separate crews may be used (which would be the case if different skills were involved). Whichever crew con-

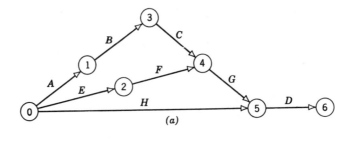

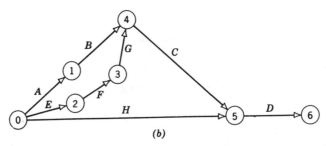

Figure 3.7 Construction logic: (a) All formwork (C) before any steel placed (G). (b) All steel placed (G) before any formwork (C).

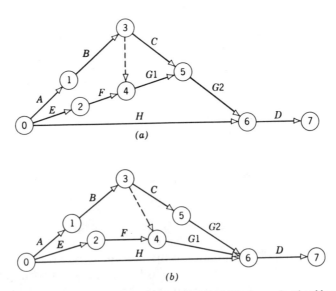

Figure 3.8 Construction logic. (a) Some steel placed (G1) independently of formwork (C). (G2 follows G1: crew constraint?) (b) Some steel placed (G1) independently of formwork (C). (G1 independent of G2: two crews?)

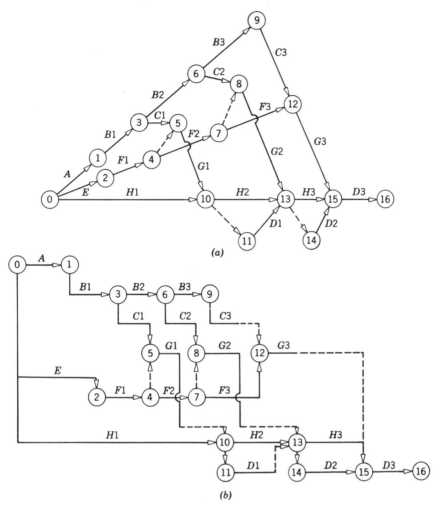

Figure 3.9 Detailed network diagrams for footings (constructed in three stages). (a) Conventional network diagram. (b) Time-scaled form of network diagram.

straint applies, the first network prohibits simultaneously carrying out G1 and G2, whereas no such prohibition exists in the second diagram.

The consideration given to the relationship between these two activities can, of course, be extended to other activities as well. Thus it may be considered advantageous to construct the footings in three stages, with consequent refinement of a whole chain of activities. For example, in Figure 3.9 some of the activities have been subdivided into three stages (not necessarily equal or comparable in magnitude or difficulty), as follows:

*B*1, *B*2, *B*3–Digging of foundations
*C*1, *C*2, *C*3–Placing of formwork

$F1$, $F2$, $F3$–Cutting and bending of steel reinforcement
$G1$, $G2$, $G3$–Placing of steel reinforcement
$H1$, $H2$, $H3$–Obtaining of concrete
$D1$, $D2$, $D3$–Placing of concrete

The constraints imposed by this subdivision may have originated from physical, safety, crew, resource, or management sources. The result of their imposition is clearly seen in the refined network diagrams of Figure 3.9 as compared with the original nucleus of Figure 3.1a.

In Figure 3.9 the refined network diagram is shown in two ways. In a, the various activities are conventionally represented by arrows whose lengths are entirely arbitrary and on which the activity durations will be indicated by numerical information alongside each arrow; this is the usual method of presenting the diagram.

It may be a freehand sketch or a neatly ruled drawing. But it has been shown earlier that, once a network is finalized, its physical appearance can be altered without affecting the network logic (see Figure 3.3). Therefore it is simple to rearrange the diagram and to present it as in (b), where each activity is drawn to scale according to its duration, yet with the same number of events as in (a), so that the same unique activity identification is preserved. Each activity has been plotted to suit its earliest start time in (b) so that any float appears at its end. It is simply an arrow diagram drawn to a horizontal time scale. The arrowheads are not really necessary except on dummies because the obvious left-to-right direction of time flow may be inferred. A time-scaled network is often helpful in visualizing the relative times of execution and the amounts of float of the various activities, and is indeed the initial step toward the production of a critical path bar chart construction program of the type shown in Figure 2.5.

In time-scaled networks the horizontal lines represent the passage of time (activity duration shown by solid lines and float by broken lines), whereas vertical lines have no time significance, being required only to maintain the logic. The number of nodes should be kept to the minimum required for the corresponding network diagram, thus avoiding unnecessary dummies. The nodes should be numbered in chronological sequence while still maintaining the conventional requirements of event numbering (see Figure 3.9b).

The general development of Figure 3.9 from Figure 3.1 should be noticed closely, for this indicates the relative ease with which networks may be refined and presented. Careful attention should be given to the use of dummies so that doubtful logic does not occur.

There is indeed no limit to the amount of refinement that may be applied to a network. As the definition of activities becomes finer, the diagram will contain more and more arrows because the construction is broken down into more and more individual operations. It soon becomes apparent that the numerical identification of activities (based on event numbering) is essential.

For this reason, a network diagram has associated with it a tabular layout of activities, cross referenced to the network by the numerical identification, as shown in Table 3.2 and Figure 3.10. These show another possible detailed development of this footing construction project, based on the original nucleus of Figure 3.1.

In developing a project network it is usually simpler to assume unlimited resources initially when devising the nucleus layout; fixed and obvious limitations in resources are then incorporated later as the diagrams become more refined. These limitations may be crew or equipment constraints; thus, in Figure 3.10, the equipment chains for the excavator (1–3, 3–4, 4–17), the concrete mixer (9–11, 11–13, 13–17), and the calfdozer (12–13, 13–14, 14–17) may imply difficulty in obtaining more than one of each piece of equipment

Table 3.2 Detailed project breakdown

Identification	Description of Activity
0–1	Preparation time
1–2	Move onto site
1–3	Get excavator
1–5	Design concrete mix
1–6	Get steel reinforcement
1–9	Get concrete materials
1–10	Arrange inspection of steel reinforcement
2–3	Lay out foundations
2–4	Get formwork
3–4	Dig foundations
4–7	Erect formwork
4–17	Move excavator off site
5–8	Test concrete mix
6–7	Bend steel reinforcement
7–9	Dummy
7–10	Place steel reinforcement
8–11	Get concrete mix approved
9–11	Get concrete mixer
10–11	Inspection of steel reinforcement
11–12	Dummy
11–13	Place concrete
11–15	Arrange final inspection
12–13	Get calf-dozer
13–14	Backfill
13–17	Move concrete mixer off site
14–15	Dummy
14–16	Clean up
14–17	Move calf-dozer off site
15–16	Final inspection
16–17	Leave site

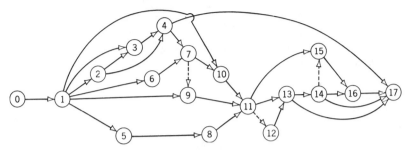

Figure 3.10 Detailed development of network diagram (more activities) in accordance with Table 3.2.

for this job rather than just a physical constraint of utilizing them in this way in the construction plan.

It is, of course, essential that the complete network contain all relevant activities that may affect the duration of the project. In this case, the chain 1-5-8-11, involving the acceptance of the concrete mix, and the chain 1-10-11-15-16, involving inspections, may drastically affect the project completion. Similarly, it may also be necessary to introduce chains for financial or other control purposes when these are critical to project duration or activity sequence. Thus the network diagram must be thoroughly checked from all relevant viewpoints.

The final network plan is then ready for scheduling. The advantages obtained from the careful definition of the construction and its proposed method of completion will alone repay the effort, time, and cost involved in its preparation.

3.5 THE UTILITY DATA CURVE

It is an interesting feature of critical path methods that, if initially a range of feasible utility data is provided, not only are optimal solutions in terms of time and cost obtained for the entire project, but also complete specifications for the time and cost of each activity. It is therefore important that utility data, in the form of completion times and direct costs, be provided for each construction activity. With these data, it is possible to draw a curve showing the relationship between direct cost and time of completion (duration) for each of the construction operations, as shown in Chapter 2 and Figure 2.2. The points on the curve are, and must always be, the minimum direct costs for completing the activity in the given durations.

This direct-cost curve divides the time/cost area into two regions: that above the curve represents the area of physically possible or feasible time-cost solutions for the completion of an activity, and the area below represents physically impossible solutions. It should be observed that both the theoretical and the practical curves of Figure 2.2 are concave, when viewed from the

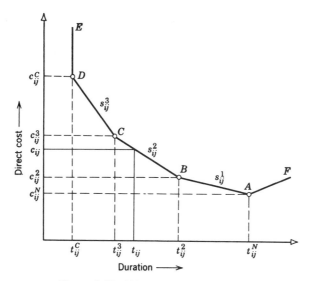

Figure 3.11   Utility data for activity A_{ij}.

region of feasible solutions, and that the slope of the curve increases in magnitude negatively with reduction in activity completion time. (The mathematician would say that the cost slope increases monotonically with decrease in activity duration.) If the utility data are, in fact, such that the curve would be partly convex, then an approximation must be made to produce a concave curve, or at least a straight line between "normal" and "crash" values;[1] this problem of nonconcave curves is considered in detail in Chapter 6.

A typical activity utility data curve is shown in Figure 3.11. By convention, any activity is designated by the notation A_{ij}; the subscripts "ij" identify the activity as that represented by the arrow in the network diagram whose tail is at event i and whose head is at event j. In construction projects the utility data usually take the form of time-cost curves—or, more specifically, as the direct cost c_{ij} for completing the activity A_{ij} in the time t_{ij}. For this particular case, four ways have been calculated for completing the activity, using different resources, labeled A, B, C, and D. It may be that many alternatives exist between A and D, in which case the straight lines AB, BC, and CD represent an approximation to the complete utility curve; if this is so, any point lying on these straight lines anywhere between A and D is assumed to be a valid (and optimal) solution for the particular completion time t_{ij} under consideration.

If, however, the points A, B, C, and D represent the only possible alternative ways of completing the activity, then points lying on the lines AB, BC, and CD—other than these four points—have no physical meaning and are not

---

[1] This approximation is necessary for the automatic generation of successive optimal solutions (see Section 5.2). Manual calculations are possible with any form of utility data, but a compression logic yielding optimal solutions would be difficult to formulate (see Section 6.4).

valid. It is nevertheless important to assume initially that the curve *ABCD* is continuous, since it is necessary for carrying out compression calculations later; naturally, steps are taken finally to ensure obtaining only valid solutions.

The points *A* and *D* are of special interest since they refer to two particularly important ways of completing activity A_{ij}. The "normal" point *A* has coordinates t_{ij}^N and c_{i2}^N, and it represents the least-cost method of completing the activity. This point *A* depicts the normal approach, where there is no slacking in the crews and where the work is carried out efficiently; if slacking and inefficiency occur, points are obtained lying on or above the line *AF*. At the other end of the curve the "crash" point *D* has coordinates t_{ij}^C and c_{ij}^C and represents the least-time method of completing the activity. This point therefore depicts the fastest possible approach when cost is no hindrance and all available resources are utilized; naturally, unnecessary costs are eliminated from the determination of c_{ij}^C, otherwise points on the line *DE* are obtained.

These two points *A* and *D*—the normal-time/normal-cost and the crash-time/crash-cost—are so important that they are identified by the superscripts *N* and *C*, respectively, on any utility curve. The remaining identifiable points (in this case, *B* and *C*) are simply specified by an appropriate numerical superscript (see Figure 3.11).

It is often convenient to indicate how close to the crash-time an activity is being completed; by convention this is expressed as a "percentage crash." If t_{ij}^N has zero percentage crash, and t_{ij}^C has 100 percentage crash, the percentage crash of a completion time t_{ij}^x is given by the time $t_{ij}^N - t_{ij}^x$ expressed as a percentage of the time $t_{ij}^N - t_{ij}^C$; therefore

$$Percentage\ Crash\ of\ t_{ij}^x = \frac{t_{ij}^N - t_{ij}^x}{t_{ij}^N - t_{ij}^C} 100 \qquad (3.1)$$

Another important concept in CPM compression and decompression calculations is the slope of the time-cost curve. This is defined as the "cost slope" and designated by the notation s_{ij}; it simply expresses the magnitude of the slope of the curve at the time t_{ij}. Physically, it represents the increment of cost required for a unit reduction in activity completion time. It will, of course, vary for different parts of the utility curve. Mathematically, for the lines *AB, BC,* and *CD* in Figure 3.11, the cost slopes are:

$$s_{ij}^1 = \frac{c_{ij}^2 - c_{ij}^N}{t_{ij}^N - t_{ij}^2} = -\frac{c_{ij}^2 - c_{ij}^N}{t_{ij}^2 - t_{ij}^N}$$

$$s_{ij}^2 = \qquad -\frac{c_{ij}^3 - c_{ij}^2}{t_{ij}^3 - t_{ij}^2} \qquad (3.2)$$

$$s_{ij}^3 = \qquad -\frac{c_{ij}^C - c_{ij}^3}{t_{ij}^C - t_{ij}^3}$$

Naturally, it follows that, at any point along the line AB, the cost slope is constant and equal to s_{ij}^1. At the point B a unit reduction in the activity duration costs an additional amount s_{ij}^2, whereas a unit increase in the activity duration gives a saving in costs of an amount s_{ij}^1; similarly, at point C, where the cost slope changes from s_{ij}^2 to s_{ij}^3. The significance and important of cost slopes are more fully discussed in Chapters 5 and 6.[1]

Special cases arise when only one or two ways of completing an activity are available. If, for example, only two excavators are accessible to dig a particular trench (say a $\frac{1}{2}$-cu yd and a 4-cu yd drag shovel), the normal cost and duration then refer to the use of one excavator and the crash cost and duration refer to the other. Manual labor might be employed, working various shifts, to produce a number of points on the utility curve for this activity, but this could be valid only if a concave curve is obtained similar to that in Figure 3.11; however, it is far more likely that both the costs of shovel excavation are cheaper than manual excavation, and so a convex curve would result. Hence it is necessary to produce separate utility curves when different methods of construction are being considered; the cheaper method is adopted for obvious reasons. The valid utility data in this case are therefore the two points applicable to the two excavators only.

When there is only one way of completing an activity, the entire utility curve collapses to a single point: the crash duration and the normal duration are identical.

Utility data for dummy activities are automatically zero, since neither resources nor time is required to complete a fictitious activity. Similarly, artificial activities having time but no cost cannot have any valid utility data.

3.6 PREPARATION OF TIME-COST CURVES FOR ACTIVITIES

The total cost of completing an activity includes (1) the material, labor, and equipment costs and (2) administration and supervision charges, site expenses, interest, and penalty payments. The first group of costs are those directly related to the individual activity and are generally variable with its duration; they are thus classed as direct costs. The second group comprises the indirect costs; they are assessed for the entire project, are not related to individual activity durations, and generally vary approximately linearly with project time.

It has been emphasized previously that the utility time-cost curves are for direct costs only, and therefore they include labor, material, and plant charges, plotted against the time for completing an activity by the various

---

[1] Some sophistication in the treatment of cost slopes is possible where it is undesirable to choose certain alternatives and artificial cost slopes are introduced. Also, some activities with multiple-point utility curves are best handled on computers by artificially splitting the activity and costs slopes into pseudo-activities with pseudo-cost slopes.

feasible available methods. Usually full utility curves will have to be prepared for only a small proportion of the project activities, as will be seen later when dealing with compression calculations (Chapters 5 and 6). Experience will soon show the planner how much utility data will be required for the analysis of any particular project. When needed, it is easily prepared in the following manner.

Consider a simple activity, such as earth excavation for a pipeline. Various construction methods are available, including manual labor only, as well as several mechanical means of excavation. Assume that, in this case, the pipeline is short, with many bends, and that only manual labor is considered practicable.

The first step is to assess correctly the magnitude of the job: earth quantities, timbering requirements, efficient crew sizes, and so on. These estimated quantities and suitable working capacities enable a reasonably accurate firm estimate of the "normal duration" and "normal direct cost" to be made. Normal working time is 8 hours per day, 5 days per week.

Suppose that in this case the situation requires 300 worker-days of labor and a crew size of ten workers. If the normal wage is 4 units of money for an 8-hour worker-day, then

$$\text{The normal duration } t^N = \tfrac{300}{10} = 30 \text{ working days}$$

$$\text{The normal cost } c^N = 10 \times 30 \times 4 = 1200 \text{ units}$$

If two shifts of ten men are used and the second shift has a penalty wage rate of one extra cost unit, then the same quantity of work requires more resources. Thus

$$\text{The two-shift duration} = \tfrac{300}{20} = 15 \text{ working days}$$

$$
\begin{aligned}
\text{The two-shift cost} = \quad &\text{1st shift,} \quad 10 \times 15 \times 4 = \ \ 600 \text{ units} \\
&\text{2nd shift,} \quad 10 \times 15 \times 5 = \underline{\ \ 750 \text{ units}} \\
&\hspace{6.2em}\text{Total} \quad 1350 \text{ units}
\end{aligned}
$$

Similarly, if three shifts, each of ten workers, are used, with a penalty wage rate of two extra cost units for the third shift, then

$$\text{The three-shift duration is 10 working days}$$

$$\text{The three-shift cost is 1500 units}$$

Of course, there are many other possibilities, such as working overtime each day at specific penalty wage rates, working weekends with other penalty wage rates, more efficient work output by offering incentive payments, and so on.

Suppose now that a duration of much less than 10 working days is essential. The first inference is that more workers are needed for each of the three shifts and that (since ten workers was selected as the best crew size) some inefficiency will result from larger crews. This inefficiency will increase the 300 original worker-days of work; the amount of this increase can be estimated only from previous experience, and there will be some lower limit of time in which the work can be physically done. For example, 300 workers working for 1 day may be quite impossible. When a minimum duration is essential, the availability of resources affects the solution and costs become a secondary consideration. In this case it may be that 90 workers are available in three crews of 30 workers each; the activity, through inefficiency, may now be estimated to require 360 worker-days of labor. Hence the utility data for crashed least-time conditions are

$$\text{The crash duration } t^C = \frac{360}{3 \times 30} = 4 \text{ working days}$$

$$\begin{aligned}
\text{The crash cost } c^C = \quad & \text{1st shift, } 30 \times 4 \times 4 = 480 \text{ units} \\
& \text{2nd shift, } 30 \times 4 \times 5 = 600 \text{ units} \\
& \text{3rd shift, } 30 \times 4 \times 6 = 720 \text{ units} \\
& \phantom{\text{3rd shift, }}\text{Total} \quad \overline{1800 \text{ units}}
\end{aligned}$$

This time-cost information is plotted in Figure 3.12 to give the utility data curve for this activity. The cost slopes may now be computed; since almost

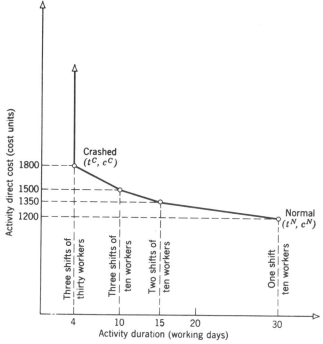

Figure 3.12   Utility data for pipeline earth excavation (construction method = manual labor).

any duration between 4 and 30 days is possible, this curve may be used for any intermediate time of completion.

The activity duration was expressed in working days. If work is not done on weekends, the normal duration of 30 working days, starting on a Monday, would cover 10 weekend days, so that a total elapsed time of 40 calendar days are required; starting in midweek, however, 12 weekend days are involved, so the elapsed time becomes 42 calendar days. Thus the number of calendar days is doubtful, unless float is available so that the activity can begin on a specific week day. For this reason, durations shown on a network diagram or model are *always* expressed in working days (or shifts, hours, etc.). Also, a period of 30 working days will generally cover some unworkable days because of rain, and thus the elapsed time required for completing the activity will be still greater. The methods of dealing with these normal lost times in practical planning are discussed in Chapter 8. The important point is that *the cost slopes are not affected, provided that no additional direct costs are incurred on the days not worked.* Such items as watchmen, dewatering, and wet weather payments may be included in the indirect costs of the project.

If the pipeline excavation activity discussed above could alternatively be carried out with a trenching machine and three workers in 12 normal working days, and if the direct cost of the equipment is 120 money units per 8-hour working day, then the utility data for this alternative construction method are as follows:

The normal duration t^N = 12 working days.

The normal cost c^N = machine, 12×120 = 1440 units
                + labor, 3 $\times 12 \times$ 4 = 144 units
                                Total   1584 units

The two-shift duration = 6 working days.

The two-shift cost = machine,   12 shifts $\times$ 120 = 1440 units
                    labor, 3 $\times$ 6 shifts $\times$ 4 =   72 units
                        +3 $\times$ 6 shifts $\times$ 5 =   90 units
                                Total   1602 units

The three-shift duration = 4 working days = t^c

The three-shift cost = machine, 1440 units
                    labor,   180 units
                        Total   1620 units = c^C

In this instance the three-shift operation represents full crash, because the machine must work nonstop to complete the task in 4 days. Unless odd working hours are adopted, this utility curve is not continuous, since only either one-, two-, or three-shift operation is practical; this is an example of discrete point utility data (see Section 6.2).

3.7 COMBINATION OF ACTIVITIES

It often happens in a network that two or more activities in a chain always appear in sequence or in conjunction, unaffected by any other portion of the diagram. Here the activities concerned may be grouped into one activity, thus simplifying the network. The combined activities are replaced by a single activity whose utility data are compounded from those of the individual original activities.

An illustration of this is shown in Figure 3.13, where the original network for the footing project is seen at *a*). If activities E and F are combined into a single activity, the diagram looks like (*b*). The utility curves for both these activities appear in Figure 3.14, together with the combined curve. Because the two original activities are continuous and strictly sequential, their separate normal times and costs and their individual crash times and costs are merely added together to obtain their combined normal and crash times and costs. The remaining points on the combined curve are the *cheapest* for intermediate activity durations. Thus the point marked $(E^2 + F^N)$ on the combined curve represents the lowest cost for completing the combined activity in 31 days; it is obtained by finishing the E portion of the combined activity in 5 days less than E's normal time and the F portion in its normal duration.

In Figure 3.14 there are two points not on the optimal combined curve as plotted: $(E^C + F^N)$ and $(E^N + F^C)$, with durations of 26 and 28 days, respectively. Although both are feasible and valid solutions, they are nonoptimal, since their incorporation would lead to a convexity in the utility curve. (The treatment of utility curves of this type is discussed in Section 6.4). These two

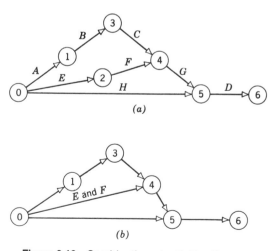

Figure 3.13 Combination of activities E and F.

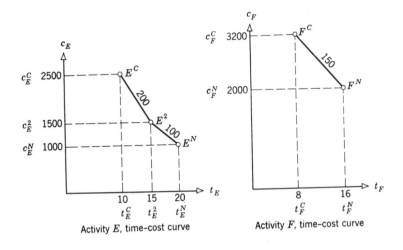

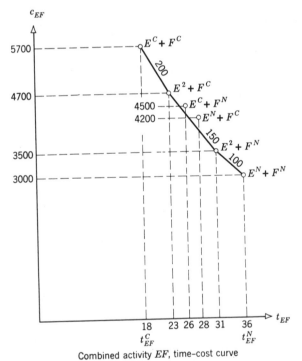

Activity E, time–cost curve

Activity F, time–cost curve

Combined activity EF, time–cost curve

Figure 3.14   Utility data for combined activities.

points would, however, be optimal solutions if both individual cost curves were of the discrete point type (see Section 6.2).

The optimal combined curve may be determined simply by considering the cost slopes of the individual activities. Activity E has two cost slopes (100 and 200), but activity F has only one (150). If these cost slopes are arranged in increasing order of magnitude (that is, 100, 150, 200), the combined activity utility curve is generated automatically from the "all normal" $(E^N + F^N)$ coordinate by laying off the cost slopes in that order with their relevant effective durations.

It can therefore be seen that the combination of sequential activities produces simplification in network models but generates more involved utility curves. The combination of parallel activities and groups of activities forming small networks in themselves is not as simple, and this operation is considered in Section 5.9.

Evaluation

a. How would you check the validity and accuracy of utility data and cost slopes of major construction activities?
b. How would you determine the level of breakdown and detail required for the portrayal of a particular project?
c. How carefully should project logic be determined?

PROBLEMS

3.1 Draw a circle notation diagram for the arrow network diagram Figure 3.9 in the text.

3.2 Figure P3.2 shows a schematic diagram for two activities A and B where the start of activity B is delayed for a certain duration X after the start of activity A. Draw logically correct arrow and circle diagrams when:
  (i) X is a specific duration (for example, 6 days).
  (ii) X is a specific duration made equal to an estimated fraction of the duration of activity A.
  (iii) X is an unknown duration and the intent is to commence activity B only when a certain portion of activity A has been completed.

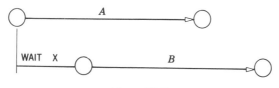

Figure P3.2

3.3 Draw logically correct arrow and circle diagrams for the schematic diagrams in Figure P3.3.

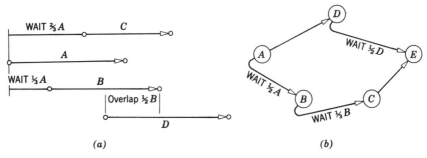

(a)   (b)

Figure P3.3

3.4 Figure P3.4 shows two activities *A* and *B*, each of which contains identifiable portions of the activity located within the activities. Draw logically correct arrow diagrams when the identified portions of *A* and *B*:

(i) are completely independent.

(ii) are logically concurrent.

(iii) both require the previous portions of both activities to be finished before they can be started.

(iv) must be worked together jointly at the same time.

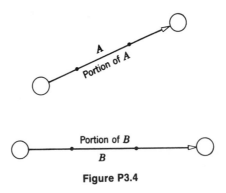

Figure P3.4

3.5 Draw arrow and circle diagrams to illustrate the following logic:
B precedes *C* and *P*.
C and *P* are independent and *concurrent* logically.
L cannot start until both *C* and *P* are finished.
J follows *L*.
E follows *L*.

F follows *J* and *E*.
and where

 (i) *J* and *E* are independent and concurrent logically.

 (ii) *J* and *E* have certain portions which can be worked jointly.

 (iii) *J* and *E* can be totally worked jointly.

3.6 Find and correct the errors in the arrow notation of Figure P3.6.

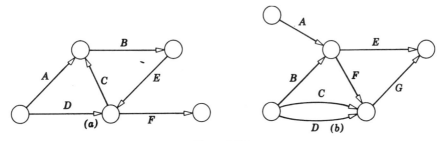

Figure P3.6

4

NETWORK CALCULATIONS
I: CRITICAL PATHS
AND FLOATS

4.1 EXAMPLE OF A NETWORK DIAGRAM FOR
PIPELINE CONSTRUCTION

Once a network diagram for a construction project has been formulated and the utility data prepared for the constituent activities, all the important information relating to the project may be determined by simple calculations. Activities can be classified into those that do (or do not) affect the duration of the project, and each activity can be given the status of "critical" or "noncritical." Recognition of chains of critical activities as critical paths within the network is the first step in the effective control of the construction, for these are the chains that determine the project duration. The assessment of how nearly critical are the other noncritical activities and chains of activities enables effective scheduling of the rest of the project. It is therefore obvious that float calculations yield immediate benefits, and that a knowledge of critical paths and floats is essential in determining optimal project durations. The graphical significance of paths, chains, and floats and their relevance to CPM is discussed in Appendix A.

Since a network diagram is essential in determining project durations and critical paths, calculations are best done on the network itself, for only selective individual activity durations need be added. Since "arrow" notation is more meaningful to the construction worker because of the analogy that one is moving along the arrow as the work proceeds, and because a time-scaled arrow diagram bears a marked resemblance to the familiar bar chart, it is considered in more detail than "circle" notation. However, Section 4.6 considers "circle" notation calculations in detail for comparison purposes. The necessary calculations may be easily illustrated by a specific problem, such as the construction of a pipeline.

In this example of pipeline construction, a list of all activities, together with their associated costs and durations, is given in Table 4.1; these utility data are arbitrary and were prepared for illustration only. The network diagram, which is based on this classification of activities, is shown in Figure 4.1; it is also arbitrary, and the reader is invited to list and classify the various types of constraint implied by this diagram. Whether this is done or not, it is important to realize that all the calculations which follow depend on the precise arrangement of this network and on the utility data supplied in Table 4.1.

The total cost of the project (C_T) is the sum of the direct (C_D) and indirect (C_I) costs of the completed works. Since all activities must be completed before the project is finished, the project direct cost is the sum of all the direct

Table 4.1   Utility data for piepline project

No.	i–j	Activity A_{ij}	Normal Duration, t_{ij}^N	Normal Cost, c_{ij}^N	Crash Duration, t_{ij}^C	Crash Cost, c_{ij}^C	Cost Slope, s_{ij}
1	0–1	Lead-in time	10	200	10	200	0
2	1–2	Move onto site	20	200	20	200	0
3	1–3	Obtain pipes	40	1800	40	1800	0
4	1–6	Obtain valves	28	500	20	580	10
5	2–4	Lay out pipeline	8	150	8	150	0
—	3–5	Dummy (logical)	0	0	0	0	0
6	3–6	Cut specials	10	100	6	260	40
7	4–5	Dig trench	30	3000	10	6600	180
8	5–6	Prepare valve chambers	20	2800	8	3400	50
9	5–7	' Lay and joint pipes	24	1000	14	1650	65
10	6–8	Fit valves	10	200	6	520	80
11	7–8	Concrete anchors	12	400	8	520	30
—	8–9	Dummy (identification)	0	0	0	0	0
12	8–10	Backfill	10	200	5	500	60
13	8–11	Finish valve chambers	6	200	3	320	40
14	9–10	Test pipeline	6	150	4	290	70
15	10–11	Clean up site	4	300	4	300	0
16	11–12	Leave site	4	100	2	180	40

Project totals $C_D^N = 11,300$ $C_D^C = 17,470$

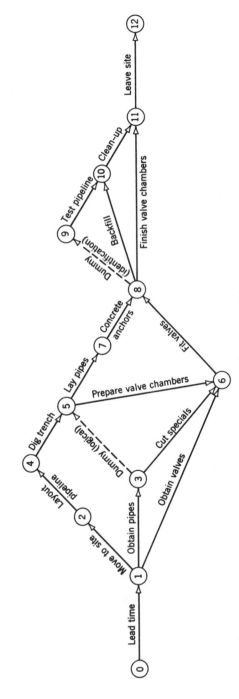

Figure 4.1 Network diagram for pipeline construction.

costs of the individual activities. By referring to Table 4.1, it can be seen at once that, if all activities A_{ij} are completed in their normal durations t_{ij}^N, each for its direct cost c_{ij}^N, the project direct cost becomes

$$C_D{}^N = \sum c_{ij}^N = 11{,}300 \text{ units} \qquad (4.1)$$

If, on the other hand, all activities are completed in their crash durations t_{ij}^C, the project direct cost becomes

$$C_D{}^C = \sum C_{ij}^C = 17{,}470 \text{ units} \qquad (4.2)$$

The optimum construction plan will have a direct cost somewhere between these two extremes.

Before any solution can be determined, it is necessary to find the critical paths through the network and to calculate the floats in the noncritical chains. The first step toward this determination is to derive the "earliest possible occurrence times" and the "latest permissible occurrence times" for every event in the network diagram. In linear graph theory this is equivalent to determining the earliest start and latest start trees (see Appendix A).

4.2   EVENTS: EARLIEST POSSIBLE OCCURRENCE TIMES (T^E)

When all activity durations t_{ij} have been specified, the project duration T_P may be found by determining the earliest possible occurrence time T_I^E for each event (I) in the network in turn, thus ultimately concluding at the last event in the diagram (project completion).

An event is satisfied only when all activities terminating at that event have been completed. Network logic demands that progress past an event is impossible if any of the activities included in the specification of that event (that is, terminating in this event) is unfinished. Hence the earliest possible occurrence time for an event is the time taken to finish all the activities on the most time-consuming path from the start of the project to the event under consideration. These calculations correspond to determining an earliest start tree rooted at the initial node (see Appendix A). This time is referred to in Chapter 2 as the *earliest finish time* (EFT).

Figure 4.2*a* shows the EFT for each event in the pipeline network, based on the assumption that each activity is completed in its normal duration, and as soon as possible. The activity duration is shown numerically along the shaft of the arrow representing the activity: for example, activity 0–1 (lead-in time) has a normal duration of 10 days, and activity 1–6 (obtain valves) has a normal duration of 28 days. Each event has beside it an associated "time box" of oval shape, where the left side is reserved for recording the earliest finish time T^E of the event.

The time T_J^E for an event (J), which has a single activity A_{ij} terminating at

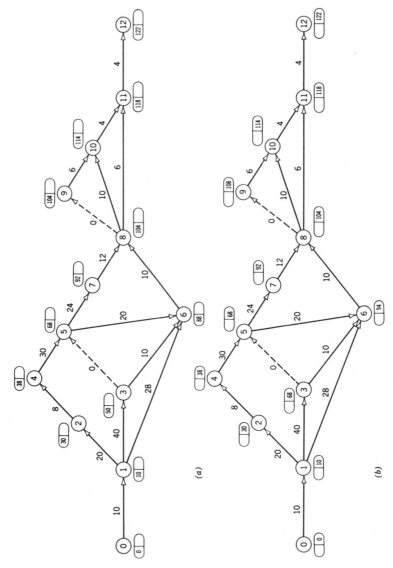

Figure 4.2 (a) Earliest possible event occurrence times (all-normal). (b) Latest permissible event occurrence times (all-normal).

that event, is determined by adding its duration t_{ij} to the time T_I^E for the preceding event (I). Thus, referring to Figure 4.2a and starting with event 0,

$$T_1^E = T_0^E + t_{0-1} = EFT_{0-1} = 0 + 10 = 10$$
$$T_2^E = T_1^E + t_{1-2} = EFT_{1-2} = 10 + 20 = 30 \qquad (4.3)$$
$$T_4^E = T_2^E + t_{2-4} = EFT_{2-4} = 30 + 8 = 38$$
$$T_3^E = T_1^E + t_{1-3} = EFT_{1-3} = 10 + 40 = 50$$

As progress is continued through the network diagram, T^E is required for events that have two or more activities terminating at the event. For event 5, with two paths terminating at the event,

$$T_5^E \text{ is the maximum of } T_4^E + t_{4-5} = EFT_{4-5} = 38 + 30$$

and
$$T_3^E + t_{3-5} = EFT_{3-5} = 50 + 0 \qquad (4.4)$$

that is,
$$T_5^E = \max \{EFT_{*-5}\} = 68$$

where EFT_{*-5} is the earliest finish time of an activity terminating at event 5. For event 6, with three paths terminating at the event,

$$T_6^E \text{ is the maximum of } T_1^E + t_{1-6} = EFT_{1-6} = 10 + 28$$

and
$$T_3^E + t_{3-6} = EFT_{3-6} = 50 + 10 \qquad (4.5)$$

and
$$T_5^E + t_{5-6} = EFT_{5-6} = 68 + 20$$

that is,
$$T_6^E = \max \{EFT_{*-6}\} = 88$$

and so on, for all events in turn.

When all the activities are completed in their normal durations, these calculations show that the project duration T_P^N is 122; that is the "all-normal" solution.

If all the activities were completed in their crash durations, the "all-crash" solution is derived; the project duration T_P^C under these circumstances is 83; the calculations proceed in exactly the same way as before, with numerical values indicating crash durations for each activity and crash times for each event.

4.3   EVENTS: LATEST PERMISSIBLE OCCURRENCE TIMES (T^L)

Once the project duration has been determined, it is possible to calculate the latest permissible occurrence time T_I^L for each event (I) in the network. This is done by working backward from the last event in the project, using a time origin equal to the project duration; obviously, *this calculation assumes that the project is to be completed in the earliest possible time.* In other words, referring to Figure 4.2b, $T_{12}^L = T_{12}^E = 122$.

The calculations are similar to those in the preceding section, except that they commence at the project completion event and go back to the project start event (that is, a latest start tree rooted at the final node is determined. See Appendix A.) The right side of each "time-box" is used for recording its event's latest permissible occurrence time, referred to in Chapter 2 as the *latest finish time* (LFT).

The time T_I^L for an event (I), which has a single activity A_{ij} starting at that event, is determined by simply subtracting its duration t_{ij} from the time T_J^L for the succeeding event (J). Thus, referring to Figure 4.2b and commencing with event 11:

$$
\begin{aligned}
T_{11}^L &= T_{12}^L - t_{11-12} = \text{LST}_{11-12} = 122 - 4 = 118 \\
T_{10}^L &= T_{11}^L - t_{10-11} = \text{LST}_{10-11} = 118 - 4 = 114 \quad\quad (4.6) \\
T_9^L &= T_{10}^L - t_{9-10} = \text{LST}_{9-10} = 114 - 6 = 108
\end{aligned}
$$

Next, T^L is derived for events which have two or more activities starting at the event. For event 8, with three paths starting at the event,

$$
\begin{aligned}
T_8^L \text{ is the minimum of } \quad T_9^L - t_{8-9} &= \text{LST}_{8-9} = 108 - 0 \\
\text{and} \quad\quad\quad\quad\quad\quad\quad\quad T_{10}^L - t_{8-10} &= \text{LST}_{8-10} = 114 - 10 \\
\text{and} \quad\quad\quad\quad\quad\quad\quad\quad T_{11}^L - t_{8-11} &= \text{LST}_{8-11} = 118 - 6 \quad (4.7) \\
\text{that is,} \quad\quad\quad\quad\quad\quad\quad\quad\quad T_8^L &= \min\{\text{LST}_{8-*}\} = 104
\end{aligned}
$$

where LST_{8-*} is the latest start time of an activity commencing at event 8.

The calculations proceed backward through the network, event by event, until the LFT of every event has been derived; the result is shown in Figure 4.2b.

In practice, the two computations shown separately in Figures 4.2a and b are recorded on the one network diagram, using both sides of the event "time-box" to show the EFT and LFT for each event. Figure 4.3 illustrates this presentation for both the "all-normal" and "all-crash" networks.[1]

4.4 CRITICAL PATHS

When the EFT and the LFT for each event have been calculated, the critical path or paths through the network diagram can be determined. An event lies

---

[1]The symbol X is used to mark the arrow of an activity which is being completed at its crash duration.

on the critical path if its EFT (T^E) and its LFT (T^L) are identical, for any delay in satisfying this event automatically violates its latest permissible occurrence time (which was based on the EFT for project completion), and the project is consequently delayed.

If, however, the EFT for any event is less than its LFT, a certain delay is tolerable in completing the event without affecting the project duration. For example, referring to Figure 4.2a and b, event 3 has an EFT of 50 days and an LFT of 68 days; event 3 is therefore not a critical event, since its completion is possible at any time between 50 and 68 days. On the other hand, event 5 is a critical event lying on the critical path, because T_5^E and T_5^L are equal, and no tolerance is permissible in satisfying this event if the project completion is to be achieved in 122 days.

An analysis of Figure 4.2a and b indicates that events 0, 1, 2, 4, 5, 7, 8, 10, 11, and 12 are critical events lying on the critical path. Events 3, 6, and 9 are noncritical events and therefore do not lie on the critical path.

Activities lying on the critical path are not so easily determined. First, to be eligible for consideration, they must start and finish at critical events; second, the duration of the activity (t_{ij}) must equal the difference in event times between the head event (J) and the tail event (I), that is, $T_J - T_I = t_{ij}$.

Activities 0–1, 1–2, 2–4, 4–5, 5–7, 7–8, 8–10, 10–11, and 11–12 satisfy both conditions and are therefore critical activities lying on the critical path. However, activity 8–11, although satisfying the first condition (events 8 and 11 *are* critical events), is not a critical activity because it violates the second condition ($t_{8–11}$ of 6 days *is not* equal to the 14 days of $T_{11} - T_8$). All other activities are noncritical, since they satisfy neither of the stipulated conditions.

Figure 4.3a shows the "all-normal" critical path, 0-1-2-4-5-7-8-10-11-12, as a heavy line with solid arrowheads. The time required to complete all the activities on this critical path equals (and, in fact, determines) the project duration of 122 days; any delay in one of these critical activities results in a delay in the whole project and its duration is increased to more than 122 days.

Noncritical activities are shown with finer lines and hollow arrowheads; they lie on noncritical chains or paths, where some delay is tolerable without prolonging the overall project duration.

The critical path within a network diagram is a function of the specific logic of the network and of the activity durations pertaining to it. Figure 4.3b shows the critical path for the "all-crash" solution to the same network; although the layout is identical, the activity durations have all been altered to crash times. As a result, the critical path has changed so that it follows a different chain of activities, now giving a project duration of 83 days. Any delay to an activity on this critical path will extend this minimum crash duration beyond 83 days.

It should now be clear that, within any given network, the critical path is dependent solely on the durations of its activities, so that, by suitable planning, it can be made to follow specific chains of activities. The constructional advantages of this manipulation of critical paths are obvious.

64

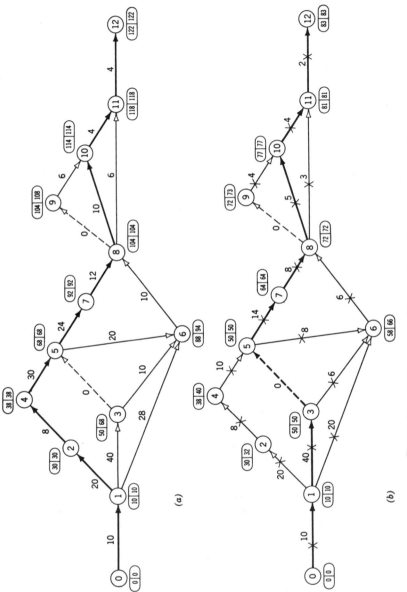

Figure 4.3 (*a*) All-normal critical path. (*b*) All-crash critical path.

4.5 FLOATS

Although critical activities must be completed as fast as possible to prevent prolonging the project duration, this is not true with the noncritical activities, for they have more time available for their completion than is strictly necessary. Thus their starting time and finishing times, limited, of course, by the timing of their starting and finishing events, can be altered without affecting the project duration; these noncritical activities and noncritical chains are therefore able to "float" about within the total time available for their completion. In linear graph theory, "float" is related to the concepts of tree and link branches and circuits (see Appendix A). As was pointed out in Chapter 2, the excess time available, beyond that actually necessary for their completion, is called *total float*.

In Figure 4.3a, activity 8–11, which normally takes 6 days to complete, has, in fact, 14 days available. This derives from the timings shown for events 8 and 11: $T_8 = 104$ and $T_{11} = 118$. Activity 8–11 therefore has 8 days float. Again, the noncritical chain of activities 1–3, 3–6, and 6–8 has 94 days available in which to do 60 days work; this chain has 34 days float. Similarly, it will be seen that the chain 1–6–8 has 56 days float, whereas the chain 1–3–5 has 18, and the chain 5–6–8 has 6.

In noncritical chains the float may either be used entirely before beginning any activity in the chain, kept in hand until all activities in the chain have been finished, or interspersed between the various activities in the chain, as suited to the planning. Obviously, some interdependence exists in the float times available for the noncritical chains all passing through event 6, as will be shown later.

Since floats can be used as safety margins, to delay the application of resources, to average out manpower requirements, etc., it is not surprising that a variety of float measurements exist including total float (TF), free float (FF), interfering float (IF), independent float (Ind.F.), and scheduled float (SF).

Total Float

$$
\begin{aligned}
\text{TF} &= T_J^L - (T_I^E + t_{ij}) \\
&= \text{LFT}_{ij} - \text{EFT}_{ij} \\
&= \text{LST}_{ij} - \text{EST}_{ij}
\end{aligned}
\tag{4.8}
$$

Note that, in the context of arrow notation, the single subscript entities (e.g., T_J) refer to node quantities, whereas the double subscript entities (e.g., EST_{ij}) refer to arrow quantities, that is, activity times.

Total float is the maximum additional time that can be made available to complete a particular activity and cannot be exceeded without delaying the project. It follows that critical activities have zero total float; in fact, critical paths may be defined as those chains of activities having zero total float. Noncritical chains will always contain total float; the larger the number of activities in the chain, and the smaller the total float, the closer the chain comes to being critical. Noncritical chains with small total float must be carefully watched during construction and are spoken of as *near-critical paths*.

Figure 4.4 indicates the total floats for the activities of the pipeline network. For example, activity 5–6 has a total float of 6 days, calculated as follows:

$$TF_{5-6} = T_6^L - (T_5^E + t_{5-6})$$
$$= 94 - (68 + 20) = 6 \qquad (4.9)$$

Similarly, activity 6–8 also has a total float of 6 days.

Consider now the chain 5–6–8. It has a total float of 6 days, because the 6 days TF of activity 5–6 is the same 6 days TF of activity 6–8. If, therefore, the activity 5–6 consumes the entire 6 days TF available to the chain 5–6–8, the activity 6–8 will have no total float left and will consequently become critical. In other words, the total float available to the chain can be used only once. It may be consumed partially by each activity, if desired, but it cannot be separately used by both, since only one total period of 6 days is available.

Consider next the noncritical chain 1–6–8, with 56 days TF available; activity 1–6 has 56 days TF, whereas activity 6–8 has a TF of 6 days. If activity 1–6 consumes the full 56 days, none is left for activity 6–8, which then becomes critical; but activity 1–6 may use up to 50 days and still leave 6 days TF for activity 6–8.

Free Float

$$FF = T_J^E - (T_I^E + t_{ij})$$
$$= EST_{j*} - EFT_{ij} \qquad (4.10)$$

where EST_{j*} is the earliest start time of the following activities.

Free float is the additional time available to complete an activity, assuming that all other activities commence and finish as early as possible. Full use may be made of available free float without disturbing the following activities, which may still begin at their earliest starting times. Thus free floats are usually shown concentrated at the end of noncritical activities or chains, where they become a safety margin to offset any unavoidable delays. Free float can be shared only by previous activities in the chain and is that quantity of the total float that may be consumed without affecting subsequent activities. Since critical activities have no total float, they automatically have zero free float.

In addition to total floats, Figure 4.4 indicates the free floats available in

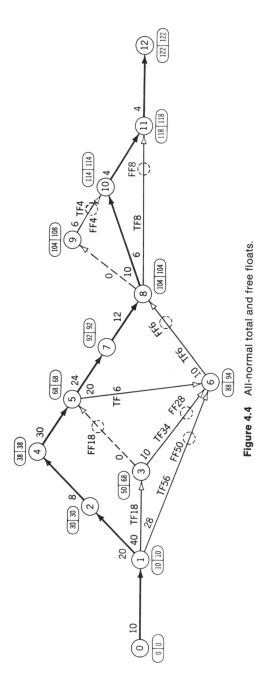

Figure 4.4 All-normal total and free floats.

the pipeline network. For example, the free float of activity 3–6 is computed as follows:

$$FF_{3-6} = T_6^E - (T_3^E + t_{3-6})$$
$$= 88 - (50 + 10) = 28 \qquad (4.11)$$

The free floats are shown in the figure at the ends of the noncritical chains or activities, with a dummy event symbol (a dashed circle) to indicate the possibility of earlier satisfaction of the event (as far as this particular chain is concerned). Free floats are therefore as useful as total floats in showing how close to critical a noncritical chain is becoming. Free floats are a most important feature of network compression calculations (see Chapters 5 and 6).

The free float of a noncritical chain ending in a dummy activity may be indicated on the dummy. For example, the chain 1–3–5 has 18 days FF, which is shown on the final activity 3–5 (a dummy) in the conventional way in Figure 4.4.

In cases where a noncritical chain is logically terminated by two dummies to two different and more critical chains, then two different free floats will be assigned to these dummies, and the smaller free float is the only one available to the preceding chain. Equation 4.10 is still valid.

The determination of suitably connected chains is made by studying free floats. In the chain 1–3–6–8, the activity 3–6 has a free float of 28 days and at event 6 meets with activity 6–8, which has only 6 days FF. This free float of activity 6–8 limited by the normal duration of activity 5–6 in the more nearly critical chain 5–6–8; hence it is more dependent on activity 5–6 than on the chain 1–3–6. It is therefore more correct to class activity 6–8 with 5–6 and delineate 5–6–8 as one chain, terminating the chain 1–3–6–8 at event 6. See Appendix A for earliest start trees and free floats.

Interfering Float

$$IF = TF - FF \qquad (4.12)$$

Interfering float is the difference between total and free floats for any activity. If an activity is delayed by an amount greater than its free float, but less than or equal to its total float, the lateness of this activity will not, of course, delay the project; it will, however, interfere with the start of some subsequent activity. Thus, if any part of the interfering float is consumed, it will be necessary to reschedule all the activities following in that chain. If the IF is fully used, subsequent activities in the chain will become critical; if it is exceeded, the project duration will be increased.

In Figure 4.4, since activity 1–3 has a TF of 18 days and no FF, the IF is 18 days. If any float is consumed by this activity, it will alter the time T_3^E of 50 and hence interfere with the freedom of scheduling activity 3–6. For this latter activity $TF_{3-6} = 34$ and $FF_{3-6} = 28$, so that its IF is 6 days; if this activity consumes float in excess of 28 days (its free float), the scheduling of activity 6–8 is interfered with.

Independent Float

$$\text{IND.F} = T_J^E - (T_I^L + t_{ij}) \qquad (4.13)$$

Independent float is the amount of time that an activity may be delayed or displaced, regardless of the state of the preceding or following activities within the project, without affecting the project duration. The independent float of an activity cannot be shared with any other activity.

In Figure 4.4, activity 1–3 has no independent float; but activity 3–6 has an independent float of 10 days, 1–6 has 50 days, and 8–11 has 8 days. Independent float exists in single isolated noncritical activities which start and finish on critical paths (e.g., activity 8–11), or in the final activities in noncritical chains which terminate on other more critical noncritical chains (for example, activities 1–6 and 3–6).

The relationships between the various types of float are indicated in Figure 4.5 within a bar chart format.

Scheduled Float

Scheduled float is that amount of float the planner specifically assigns to an activity for suitable scheduling of the project after arranging the resources

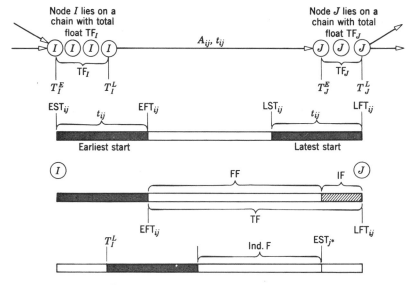

Figure 4.5 Relationships between floats.

satisfactorily by "activity shifting," as discussed in Section 7.2. It may be arbitrary, but it cannot exceed the total float or the project duration will be prolonged. Scheduled float is used mainly in computer processing, and only appears when events are allotted specific times (so that the EFT and the LFT of an event are replaced by a single scheduled time).

The use of floats in scheduling and the technique of activity shifting in order to smooth out the resource requirements of a project are discussed in Chapter 7.

4.6   CIRCLE NETWORK CALCULATIONS

In circle networks, nodes model activities, whereas in arrow networks they model events. Since activities require a time period for completion, and events are instantaneous in time, it is obvious that nodes in circle networks have different time attributes to nodes in arrow networks and may require different labeling.

In arrow network calculations, the objective of the forward pass is to label each event node with its earliest possible event time (T^E), and of the backward pass to label each event node with its latest permissible event time (T^L). Thus in arrow networks each node is labeled with two time values.

In circle network calculations, the objective of the forward pass is to determine the earliest period of time during which the activity may be carried out. Thus the forward pass identifies the earliest start time (EST) and the earliest finish time (EFT) for each activity node. In the backward pass, the calculations identify the latest finish time (LFT) and the latest start time (LST) for each activity node. Thus in circle networks each activity node may be labeled with four time values. Whether each node is so labeled or whether known relationships (see equations 4.14 and 4.15) are used to reduce the labeling to two time values is immaterial since all four values must be computed in any case.

Typical circle network calculations are shown for the pipeline construction project of Table 4.1 in Figure 4.6. As shown in the legend each node is identified and labeled with the activity duration (t_I) within the node circle. Associated with each activity node are two oval time-boxes; the one associated with earliest times in the forward pass is located above each node, and the one associated with the latest times in the backward pass is located below each node.

Specifically in the forward pass calculations each activity node I is labeled with its earliest start time (EST_I) and its earliest finish time (EFT_I) according to the following equations:

$$EST_I = \max EFT_*$$

$$EFT_I = EST_I + t_I \tag{4.14}$$

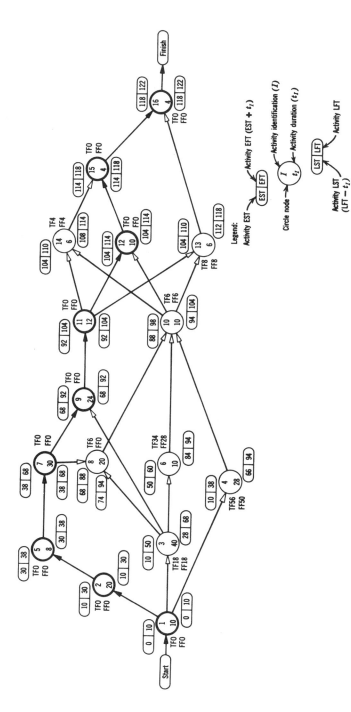

Figure 4.6 Circle network calculations for pipeline construction.

where EFT* is the earliest finish time of a logically preceding activity to activity I (that is, of any activity node with a branch or arrow leading to node I). The first of equations (4.14) ensures that activity I cannot be commenced before all its logically preceding activities have been completed; this it does by taking the maximum value of all the possible EFT* values relative to the node being considered.

Similarly, in the backward pass calculations each activity node I is labeled with its latest finish time (LFT$_I$) and its latest start time (LST$_I$) according to the following equations:

$$LFT_I = \min LST_*$$

$$LST_I = LFT_I - t_I \qquad (4.15)$$

where LST* is the latest start time of a logically following activity for activity I. Calculations proceed similarly to arrow network calculations. Thus, for example, in the forward pass calculations for activity node 8:

$$EST_8 = \max \left\{ \begin{array}{l} EFT_3 = 50 \\ EFT_7 = 68 \end{array} \right\} = 68 \qquad (4.16)$$

$$EFT_s = 68 + 20 = 88 \qquad (4.16)$$

And in the backward pass calculations for activity node 10:

$$LFT_{10} = \min \left\{ \begin{array}{l} LST_{12} = 104 \\ LST_{13} = 112 \\ LST_{14} = 108 \end{array} \right\} = 104 \qquad (4.17)$$

$$LST_{10} = 104 - 10 = 94$$

If an activity lies on the critical path, its EST is equal to its LST; similarly, its EFT is equal to its LFT. Thus in circle networks the identification of critical activities requires simply the location of nodes with identically labeled earliest and latest time boxes.

In referring to Figure 4.6, activities 1, 2, 5, 7, 9, 11, 12, 15, and 16 are critical, and any path that passes through critical nodes only is a critical path. In this case, one critical path (or chain of critical activities) exists.

From equations 4.8 total float in circle notation becomes for activity I:

$$TF_I = LFT_I - EFT_I$$

$$= LST_I - EST_I \qquad (4.18)$$

Thus circle notation calculations give activity total floats directly from time labels in the earliest and latest time boxes at each node. Examples are shown in Figure 4.6.

Free float calculations in circle networks, however, are not so simply determined as in arrow networks. From equations 4.10 free float in circle notation becomes for activity I

$$FF_I = \min \text{EST}_* - \text{EFT}_I \qquad (4.19)$$

where EST_* refers to the earliest start time of a logically succeeding activity node.
Referring to Figure 4.6 at activity node 6,

$$FF_6 = \text{EST}_{10} - \text{EFT}_6 = 88 - 60 = 28$$

and at activity node 10

$$FF_{10} = \min \left\{ \begin{array}{l} \text{EST}_{12} = 104 \\ \text{EST}_{13} = 104 \\ \text{EST}_{14} = 104 \end{array} \right\} - 98 = 6$$

It will become apparent that the only essential difference between circle and arrow network calculations is the form of notation and arrangement employed. In succeeding chapters of this book, treatment and emphasis will be placed on arrow diagrams because they are more commonly used in practice.

Evaluation

a. Do you prefer arrow or circle notation? Why?

b. What float concepts do you consider the more important? Why?

PROBLEMS

4.1 Determine the arrow and circle network diagrams for the following project, and hence calculate the total float, free float and interfering float for each activity.

Activity	Duration	Immediately Following Activities
A	22	D J
B	10	C F
C	13	D J
D	8	—
E	15	C F G
F	17	H I K
G	15	H I K
H	6	D J
I	11	J
J	12	—
K	20	—

4.2   Draw an arrow network diagram for the following project, calculate the earliest and latest starting and finishing times, and the total and free floats. Indicate critical activities by an asterisk.

Activity	Duration	
1–2	5	
1–3	10	
1–4	12	
2–4	0	Dummy
2–5	14	
3–4	6	
3–6	13	
4–5	7	
4–7	11	
5–7	17	
5–9	9	
6–7	0	Dummy
6–8	8	
7–8	5	
7–9	13	
7–10	8	
8–10	14	
9–10	6	

4.3   From the following network data, determine the critical path, starting and finishing times, total and free floats

Activity	Description	Duration
1–2	Excavate stage 1	4
1–8	Order and deliver steelwork	17
2–3	Formwork stage 1	4
2–4	Excavate stage 2	5
3–4	Dummy	0
3–5	Concrete stage 1	8
4–6	Formwork stage 2	2
5–6	Dummy	0
5–9	Backfill stage 1	3
6–7	Concrete stage 2	8
7–8	Dummy	0
7–9	Dummy	0
8–10	Erect steel work	10
9–10	Backfill stage 2	5

4.4·   Draw a bar chart for the project of Problem 4.1 assuming each activity commences on its EST time. Show floats as indicated in Figure 2.5 in the text.

5

NETWORK CALCULATIONS II: SIMPLE COMPRESSION

5.1 WHAT IS NETWORK COMPRESSION?

The completion of each activity within a project requires the utilization of a certain quantity of resources and a specific amount of time. With a minimum of resources and a maximum of time, the activity is completed at normal cost and normal duration. If faster and more expensive ways are available, additional resources enable the activity to be finished in less than the normal duration but at greater cost. This expediting of an activity, which may be described as "activity duration compression," is dependent only on the availability of resources, the form of the utility curve, and the desire to speed up completion of the activity. The compression of individual activity durations is independent of their position within the project, and thus is independent of the network model arrangement; this is not to imply that it is economical to expedite any activity, but merely that it is possible to do so independently of other activities.

To begin the compression of any activity, it is essential to know the full utility data pertaining to that activity. For convenience, these data may be shown by some convention along the arrow of the network corresponding to the activity under consideration.

Figure 5.1a shows a simple activity network model (complete with utility data) for a project comprising three activities, A, B, and C, arranged in two chains. The utility data are here shown in the form of direct-cost/time curves (with cost slopes) for each activity; they are assumed to be physically continuous and feasible for all activity durations between normal and crash. Figure 5.1b shows the "all-normal" solution for a 30-day project duration and a project cost of 6300 units. The all-normal solution is, of course, the cheapest.

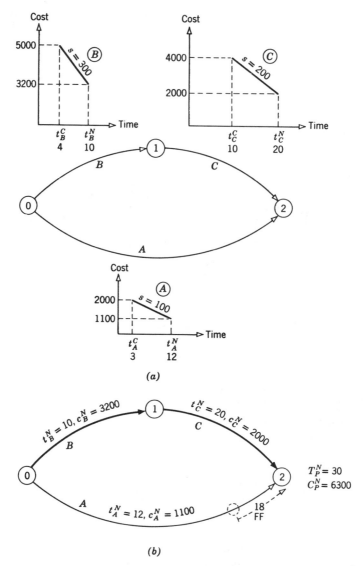

Figure 5.1   Utility data and all-normal solution. (a) Network model. (b) All-normal solution.

Suppose that the 30-day duration is unacceptable for some reason and that an additional 600 units of financial resource is available for expediting the project. Obviously, it is quite possible to sink the 600 units into any one of the three activities, since the crash cost clearly requires more total financial resource than is available. The problem is to find the most economical way of expediting this project within the available extra expenditure of 600 units.

If only activity A is considered, the cost slope of 100 permits its duration to be compressed by 6 days for an expenditure of 600 units; this gives a duration of 6 days at a total activity direct-cost of 1700 units. The situation is shown in Figure 5.2a, where it is obvious that there has been no reduction in the project duration but only an increase in the float available for this activity A.

If, however, activity B alone is considered (cost slope = 300), an activity duration compression of 2 days is possible, giving a completion time of 8 days at an activity cost of 3800 units. Figure 5.2b demonstrates that a reduction of 2

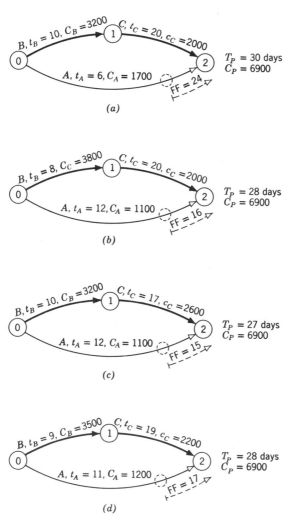

(a)

(b)

(c)

(d)

Figure 5.2   Activity compression and network compression. (a) Compression of activity A. (b) Compression of activity B. (c) Compression of activity C. (d) Compression of all activities.

days is thus obtained in the project duration for a slight decrease in the float available for activity A (compared with its all-normal float).

Finally, if only activity C is considered (cost slope = 200), a compression of 3 days can be obtained, giving a new activity duration of 17 days at a cost of 2600 units, as illustrated in Figure 5.2c; the project duration is reduced by 3 days, again for only a slight reduction in the float available for activity A. This then is the greatest project reduction time if the total extra funds are expended on one activity only.

There is, of course, another possibility: to compress each activity, in this case by an amount of 1 day each. Thus activity A takes 11 days at a cost of 1200 units, B takes 9 days at a cost of 3500, and C takes 19 days at a cost of 2200. The project cost now totals 6900, and the duration is 28 days, as seen in Figure 5.2d.

A study of these four situations shows that the best solution is the one in Figure 5.2c, where the compression of activity C gives the cheapest cost per day reduction in the project duration. A consideration of the cost slopes of the activities and of their status (critical or noncritical) in the network will convince the reader that this is the optimal solution for the added investment of 600 cost units. Noncritical activity A does not affect the project duration, so its cost slope (although the cheapest) is ineffective; of the two other activities whose compression will reduce the overall project time, the compression of activity C is preferable because of its cheaper cost slope. Hence one may conclude that *compression is started with the critical activity having the cheapest cost slope.* In this and in more complicated situations, a selective and logical approach is better than an exhaustive set of calculations.

Network compression calculations involve systematic and progressive reductions in project duration through the application of increments of additional resources to the construction plan. These calculations take into account the overall network model, the current status of the individual activities, and the specific data of the activity utility curves concerned. The consequent alteration to the durations of individual activities leads to a revision of the project schedule, a process commonly called *schedule compression.*

The basic procedure for compressing the network model is to crash activities along the critical path, starting with the activity having the flattest cost slope and then considering successively those having steeper and steeper cost slopes. Care must be taken that the amount of compression proposed does not interfere with the remainder of the network; if it does, limitations on the amount of crash must be imposed in order to maintain the compression logic.

At each stage of the network compression calculations, a logical analysis must be made in accordance with the following rules:

1. List the activities on the critical path.
2. Delete those with zero potential for compression; these will include

activities whose normal and crash durations are identical, as well as those already fully crashed in previous stages.

3. Select that activity with the flattest cost slope, since this will give the cheapest compression.

4. Determine the amount by which this activity can be compressed and its relevant cost.

5. Determine if any network limitations to this compression exist and the reasons for their existence.

6. Carry out the compression within the limitations imposed.

7. Compute the new project duration and the corresponding project direct cost.

Some insight into network compression can be found from linear graph theory (see Appendix A). Rule 1 corresponds to the identification of the only subgraph of the network model that must be considered at each stage of compression. Cut sets and their associated values in this subgraph give the graph theory rationale for rules 2 and 3.

As pointed out in Chapter 2, each stage of network compression provides an optimal solution and the coordinates of one point on the project direct-cost/time curve. Therefore this curve can be plotted point by point as the compression calculations proceed stage by stage.

5.2 NETWORK COMPRESSION LOGIC

Network compression calculations belong to that branch of mathematics known as parametric linear programming, and some insight into the logical theoretical approach of the calculations may be of interest. It is emphasized, however, that no knowledge of linear programming is necessary for network compression calculations. Some readers may therefore prefer to bypass this section if they are not interested in the mathematical background; they can still employ the critical path technique.

Ordinary linear programming problems require the successive determination of feasible (that is, possible) solutions that satisfy the logical constraints imposed so that the specified objective function is continuously improved in value. The sequence of feasible solutions terminates when no further improvement in the objective function is possible; the final answer is then known as the optimal solution. This solution minimizes or maximizes the objective function.

The characteristic which distinguishes parametric linear programming problems from ordinary linear programming problems is that one variable (called the *parameter*) defines a whole family of linear programming problems as it takes on different values from time to time. The parametric problem therefore becomes a linear programming problem for each specific value of the parameter; hence a series of optimal solutions is derived for the sequence of linear programming problems, conforming with the variations of the parameter over its range of feasible values.

In construction planning, *the objective function is minimum cost for a specified project time,* and it is convenient to select the project duration (a variable, depending on the amount of resources applied to each activity) as the parameter to the problem. After one optimal solution is obtained for a given value of the parameter, it is comparatively easy to generate other optimal solutions for other values, because it is possible to supply further increments of resources for the completion of the activities, thus altering the value of the parameter (project duration). In this way, each optimal solution represents one point on the direct-cost/time curve of the project.

The all-normal solution always produces an optimal solution (i.e., minimum cost) for the specific project duration T_P^N associated with it. Because of this, it is now possible to recast the parametric linear programming problem. Given the all-normal solution to the project network model, with its associated direct cost and duration, it is required to determine the series of optimal (i.e., cheapest) solutions for shorter project durations. The complete range of the parameter T_P must necessarily lie between the limits of normal project duration T_P^N and crashed project duration T_P^C.

Network compression calculations are thereby concerned with deriving the optimal project direct-cost curve with time. This is obtained by supplying additional resources to the all-normal solution in the way that will most efficiently reduce the project duration by a certain amount. Having obtained one optimal solution, the process is continued until further compression of project duration is impossible. The results of each succeeding optimal solution are then plotted to produce the optimal project direct-cost/time curve.

The mechanics of the simple compression calculations will be clear from the examples which follow. The variety of limitations, which may inhibit further compression, will also be clear from the following sections.

5.3   COMPRESSION LIMITED BY CRASHING

The simplest case in network compression calculations occurs when an activity A_{ij}, lying on the critical path, can be fully expedited from its present duration t_{ij} to its crash duration t_{ij}^C. For this purpose, it is assumed that only one critical path exists in that portion of the network containing the activity A_{ij}, that its cost slope s_{ij} is the smallest available in the critical path, and that compression (when carried out) will not affect the status of any other activity in the model.

The reduction in the project duration is then obviously the amount by which the activity's current duration can be reduced by crashing it, namely, $t_{ij} - t_{ij}^C$. The additional resources necessary for the compression of this activity is then represented by the direct cost required to expedite the activity duration at a cost slope of s_{ij}, which is $s_{ij}(t_{ij} - t_{ij}^C)$.

Figure 5.3a illustrates a situation where network compression is possible. It is assumed that activity D has been previously fully crashed; any activity that cannot be compressed any further is conventionally marked with the symbol X on the arrow. Activities B, C, and D, lying on the critical path, are critical in status, whereas activities A and E are noncritical.

Notice the conventional manner in which the utility data are presented on the network model. The present project cost is assumed as datum, with the present activity durations indicated as usual beside the arrows of the relevant activities. The potential alterations in duration times and the applicable cost slopes are shown as superscripts to the present durations; those above and to the left indicate potential reduction of activity duration, whereas those above and to the right indicate potential extension of duration of an activity. The utility data for activities A, B, and C are the same as those shown in Figure 5.1a. Thus, for activity B, the data "$^{6/300}10$" mean that the present duration is 10 days and that this activity can be physically compressed by 6 days at a uniform cost increase of 300 units per day. Similarly, for the already crashed activity D, the utility data "$X\,6^{15/120}$" on its arrow indicate that this activity is currently fully crashed for a duration of 6 days but can be extended by 15 days if required at a uniform cost reduction of 120 units per day.

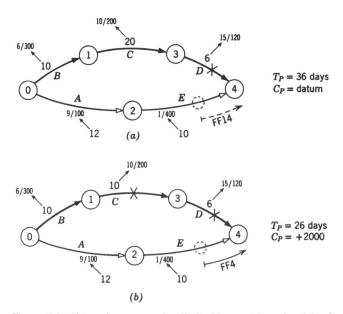

Figure 5.3   Network compression limited by crashing of activity C.

The logical analysis for network compression is as follows.

1. *List the activities on the critical path, B, C, D.*
2. *Delete those with zero potential for compression;* in this case, those already crashed, leaving *B, C.*
3. *Select that activity with the flattest cost slope,* hence *C.*
4. *Determine the amount by which this activity can be compressed, and the relevant cost; C* may be crashed by 10 days at a cost of 200 units per day.
5. *Determine if any network limitations to this compression exist. C* is independent of *B* and *D* in the critical path and the parallel chain *A–E* has a float of 14 days (which is in excess of the proposed compression of 10 days). Hence there is no limitation to this compression *other than that pertaining to activity C itself.*
6. *Carry out the compression within the limitations.* Therefore compress activity *C* to full crash by 10 days at a cost of 2000 units, as shown in Figure 5.3*b.*

Notice that the float available in the chain *A–E* is now decreased by 10 days. The new project duration is 26 days, at an extra cost of 2000 units above that for the network status of Figure 5.3*a.* Notice also that the activity with the cheapest cost slope (activity *A; s_{ij} = 100*) was excluded because it does not lie on the critical path and hence cannot (at this stage) affect the project duration.

5.4 COMPRESSION LIMITED BY FLOAT

A situation often arises when the full compression of an activity lying on the critical path is logically prevented because it would cause another chain of activities to become critical. The consequent introduction of a new critical path into the network obviously invalidates the basis for the full compression contemplated. Hence this situation permits only a partial compression of the relevant activity. Its crashing is limited to a point where the disappearance of the available float on the other chain creates a new critical path. This situation is always indicated when the potential for crashing the activity duration exceeds the float of the chain of activities about to become critical.

This situation occurs in Figure 5.4*a,* which is the same as that in Figure 5.3*b.* Following the rules for network compression, it is found that

1.
2. }Only activity *B* on the critical path is available for compression.
3.

4. *B* may be compressed by 6 days at a cost of 300 units per day.

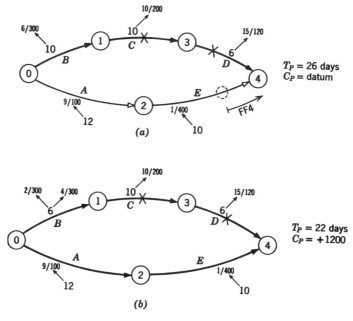

Figure 5.4 Network compression limited by float.

5. There is a chain *A–E,* starting and finishing on the critical path, which has a float of only 4 days. Hence a network limitation exists, restricting the compression of *B* to 4 days only, *because of a float limitation.*
6. Therefore compress activity *B* by 4 days at a cost of 1200 units, as seen in Figure 5.4*b.*

Notice now that there are two critical paths between events 0 and 4. Notice also the way the utility data of activity *B* are presented, showing that it is physically possible to both compress and extend the duration of this activity.[1]

5.5 COMPRESSION LIMITED BY PARALLEL CRITICAL PATHS

After two critical paths develop in a network model, all further compressions must involve equal decreases along both critical paths because otherwise the project duration will not be reduced.

This is precisely the position illustrated in the network of Figure 5.5*a,* where both chains are critical and hence both determine the project duration

[1] Network decompression calculations, involving activity duration extensions, are discussed in Chapter 6.

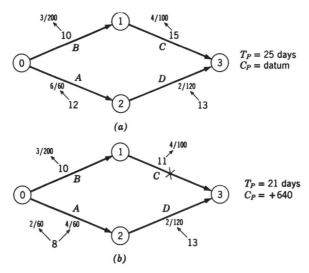

Figure 5.5 Network compression limited by parallel critical paths.

of 25 days. Following the rules for network compression, the cheapest activity available for compression in the chain 0–1–3 is activity C, with a potential of 4 days at a cost of 100 units per day; in the chain 0–2–3, activity A has the cheapest cost slope of 60 units per day and a potential compression of 6 days.

Obviously, the crashing of C limits the overall compression to 4 days, so that in the parallel chain there cannot be more than a 4-day compression *because of the limitation of the parallel critical paths*. Therefore both activities C (to full crash) and A are compressed together by 4 days, at a combined cost of: $100 + 60 = 160$ units per day. The state of the network after compression is indicated in Figure 5.5*b*.

5.6   COMPRESSION LIMITED BY CRASHED CRITICAL PATHS

Once a critical path develops, optimal compression logic demands that it remain in the network model. Eventually, with the continual application of additional resources, all the activities on the critical path must reach their crashed durations, and it is then physically impossible to compress further the critical chain. When this stage is reached, the network analysis terminates because no advantage is gained in crashing noncritical activities, for they can have no effect on the project duration if the critical path is fully crashed.

It should be clear by now that the successive optimal compressions of a network model, to the final state of a fully crashed critical path, produces a cheaper solution than the all-crash solution because in the former many noncritical activities remain still uncrashed. This final solution, limited by crashed critical paths, is, in fact, the "crashed least-time" solution–the cheapest solution for the shortest feasible project duration, based on direct costs only.

A parallel situation develops when two critical paths exist. Consider Figure 5.6a, which is repeated from Figure 5.4b. Here activity B is available for a further 2-day compression to full crash, forcing a similar 2-day compression in the parallel critical path 0–2–4 (where activity A has the cheaper cost slope). Thus a dual compression of 2 days may be made at a total cost of 400 units per day. The final state of the network is shown in Figure 5.6b. The critical path 0–1–3–4 is now completely crashed and thereby prevented from further compression. The other critical path 0–2–4 still contains activities with a compression potential of 8 days, which is, however, *inhibited by the limitation of the crashed critical path* 0–1–3–4: the entire network between events 0 and 4 has become rigid and no further compressions are possible; the crashed least-time solution has been reached.

5.7 COMPRESSION STAGES IN PIPELINE CONSTRUCTION MODEL

To illustrate the usual procedures in analyzing a compression problem in practice, network compression calculations will now be made on the pipeline construction model previously discussed in Chapter 4, whose utility data

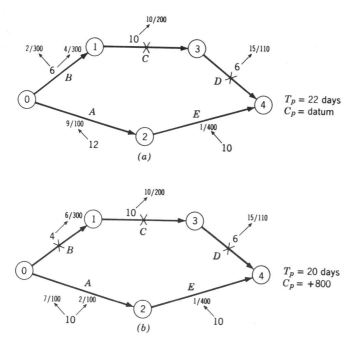

Figure 5.6 Network compression limited by crashed critical paths.

were given in Table 4.1. It will be remembered that the all-normal solution has a duration of 122 days and a direct cost of 11,300 units, whereas the all-crash duration is 83 days at a direct cost of 17,470 units.

The Network Model Ready for Compression

The initial state of this network model, shown in Figure 5.7a, is the all-normal solution of Figure 4.4, in which the critical path follows the chain 0–1–2–4–5–7–8–10–11–12 and determines the normal project duration T_P^N of 122 days; the free floats are shown, but irrelevant data are omitted. For compression calculations it is necessary only to indicate activity durations and utility data for those activities lying on the critical path. Furthermore, any such activity with zero compression potential may be considered as already fully crashed and may therefore be marked with the fully crashed symbol X; these activities have their crash durations equal to their normal durations, and, since their utility data curves thus consist of a single point, no utility information is required on the model. For simplicity, the utility data for all activities are assumed to be continuous, any duration between normal and full-crash being feasible.[1]

The free float available on the noncritical chains is needed to indicate network limitations to the compression of the critical path activities. No utility data nor individual activity durations are required for noncritical activities.

The compression calculations determine the reduction in project duration and the increase in project direct cost at each stage, thus enabling the progressive total durations and total direct costs to be shown on each successive network model of the project during the compression procedure.

First Compression of the Network Model

Of all the activities available for compression along the critical path, the one with the flattest cost slope is activity 7–8. It is possible to compress it by 4 days to full crash. If this full compression is made, event 8 can be satisfied 4 days earlier and so can all succeeding events. The noncritical chains terminating in event 8 (1–6–8, 1–3–6–8, and 5–6–8) all commence at events which will be unaffected in time and all have free floats exceeding 4 days. Hence no network limitations exist, and the full compression is possible. Therefore

Compress activity 7–8 by 4 days to full crash, at a cost of 120 units. Project duration = 118 days. Project direct cost = 11,420 units.

This provides one point on the project direct-cost/time curve, which is plotted in Figure 5.8 on p. 94. On this curve the point marked N represents

---

[1] Noncontinuous utility data are discussed in Chapter 6.

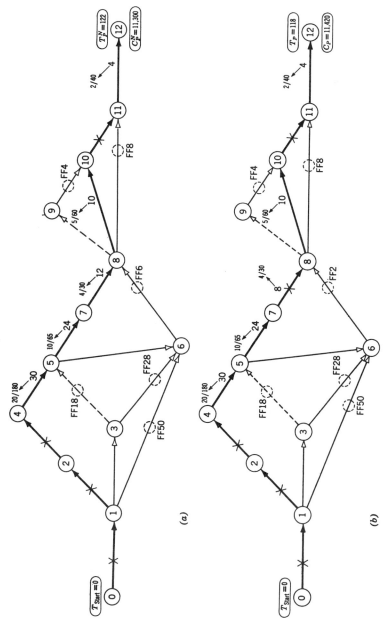

Figure 5.7 Compression of pipeline network. (a) All-normal, ready for compression. (b) Network after first compression.

the all-normal solution and point C the all-crash solution; data for these points have already been computed. The point marked 1 is plotted from the data just obtained from the first compression of the network. Succeeding compressions provide further points, enabling the complete direct-cost curve to be plotted.

Second Compression of the Network Model

The state of the network after the first compression is shown in Figure 5.7b. Since the timing of event 8 is altered, the free float in chain 5–6–8 is now reduced to 2 days, and it is therefore nearly critical; this chain will eventually impose a network limitation on future compression of the critical chain 5–7–8. Also, since the timings of events 5 and 6 are unaltered, the free floats in the chains 1–3–5, 1–3–6, and 1–6 are unaltered. Activity 7–8 (now crashed) is marked with an X; but, since it is still available for duration extension (if this should be required), its utility data are so indicated.

The next critical activity available for compression with the cheapest cost slope is 11–12, which has a potential of 2 days crash. Since no network limitations are possible for this isolated portion of the model, full compression is available. Therefore

Compress activity 11–12 by 2 days to full crash, at a cost of 80 units. Project duration = 116 days. Project direct cost = 11,500 units. This is plotted as point 2 in Figure 5.8.

Third Compression of the Network Model

Figure 5.7c shows the network after the second compression. Activity 8–10 is now the critical activity with the cheapest cost slope, and a 5-day compression is possible. If this is allowed, event 10 will occur 5 days earlier. The chain 8–9–10, however, has a free float of only 4 days, thus creating a network limitation.

It should be realized in this case that a 5-day compression of activity 8–10, with a 1-day compression simultaneously of activity 9–10 is possible and would give a compatible timing for event 10. This double compression would, however, alter the logical basis for selecting activity 8–10 on the critical path as the cheapest: first, because the compression of activity 9–10 involves an additional cost slope of 70 units, and secondly because it is currently a noncritical activity. Furthermore, this dual compression (4 days at a cost slope of 60 units, plus 1 day at 130) is not the cheapest 5-day compression available; on the contrary, the first 4 days at 60 units is now the logical and optimal compression, and a fifth day can be achieved at a cost of 65 units (instead of 130 units), as will be seen later. Thus it will always pay to follow the network compression logic. Therefore

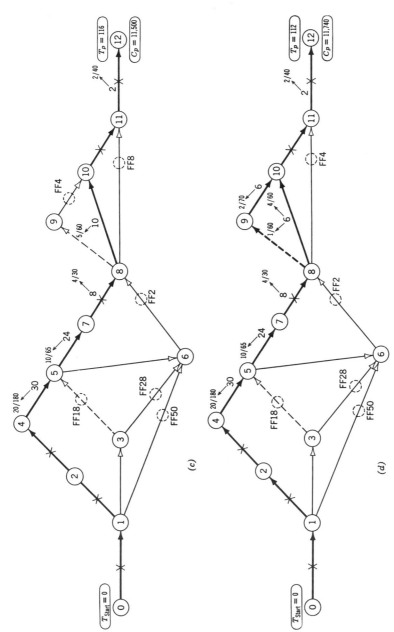

Figure 5.7 Compression of pipeline network. (c) Network after second compression. (d) Network after third compression.

Compress activity 8–10 by 4 days only, at a cost of 240 units. Project duration = 112 days. Project direct cost = 11,740 units. This enables point 3 to be plotted in Figure 5.8.

Fourth Compression of the Network Model

The third compression is shown in Figure 5.7*d,* where a new critical path now exists due to the consumption of the free float of a noncritical chain by the compression of a critical path activity. The status of activity 9–10 is now changed from noncritical to critical, and its utility data are therefore added to the model; this activity is, of course, now available for consideration in future compressions. Furthermore, since activity 8–10 is not fully crashed, it is now indicated on the network that this activity is available for both compression and decompression.

Although activity 8–10 has the cheapest cost slope of all the activities on the two critical paths, any effective compression must entail simultaneous compression of activity 9–10; hence the combined cost slope for such a compression is 130 units. Because of this, activity 5–7, with a cost slope of 65 units per day, is the cheapest effective compression available. It has a potential of 10 days. However, an earlier time for event 7 means an earlier time for event 8. But the noncritical chain 5–6–8 has only 2 days free float available, and hence a network limitation exists that restricts the compression to that amount. Therefore

Compress activity 5–7 by 2 days only, at a cost of 130 units. Project duration = 110 days. Project direct cost = 11,870 units. Point 4 may now be plotted in Figure 5.8.

Fifth Compression of the Network Model

The fourth compression adds another critical path (5–6–8) to the network model, as illustrated in Figure 5.7*e.* Consequently there are now additional activities as candidates for compression in later stages. Notice that, although event 6 now lies on a critical path, its earliest finish time has not yet been affected; consequently the free floats in the noncritical chains 1–3–6 and 1–6 remain unaltered. Examining Figure 5.7*e* reveals the following features.

1. The new critical paths form two successive loops in the main critical path.
2. Outside these loops, activity 4–5 is the only activity not crashed and still available for compression.
3. Both loops are available for compression, for in each branch of the two loops there are activities available for compression.

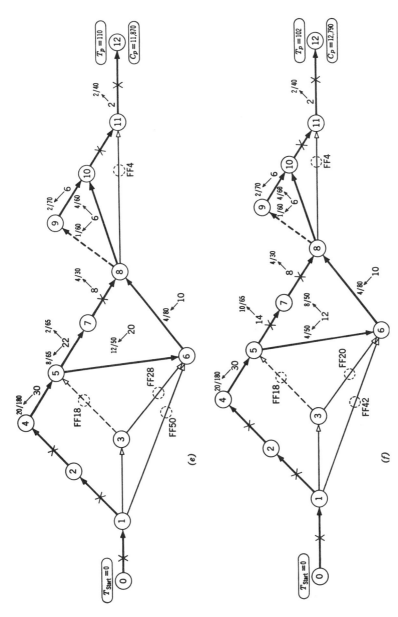

Figure 5.7 Compression of pipeline network. (e) Network after fourth compression. (f) Network after fifth compression.

4. The cost slopes of activities now available for compression are (a) the actual cost slope of 180 for activity 4–5, (b) the effective cost slope of 115 for simultaneous compression of activities 5–7 and 5–6, and (c) the effective cost slope of 130 for simultaneous compression of activities 8–10 and 9–10.

The simultaneous compression of activities 5–7 and 5–6 is now selected, for this yields the cheapest overall cost. Activity 5–6 has a compression potential of 12 days, but activity 5–7 is fully crashed after an 8-day compression; hence a limitation of 8 days exists. Therefore

Compress activities 5–6 and 5–7 simultaneously by 8 days, at a cost of 920 units. Project duration = 102 days. Project direct cost = 12,790 units. This provides point 5 in Figure 5.8.

Sixth Compression of the Network Model

The state of the network model after the fifth compression is shown in Figure 5.7*f*. Event 6 can now occur earlier, with the result that the free floats in the chains terminating at that event (1–3–6 and 1–6) are reduced by 8 days. Since event 5 is not affected, the same free float remains in chain 1–3–5.

It will be apparent that, in the looped critical paths between events 5 and 8, the critical sequence 5–7–8 is now completely crashed; consequently no further compression is possible for any activities on chains starting at event 5 and finishing at event 8, notwithstanding that activity 5–6 has the cheapest cost slope. Hence the effective cost slope of 130 units per day for simultaneously compressing activities 8–10 and 9–10 (in the looped paths between events 8 and 10) is the next selection (it is cheaper than the actual cost slope of 180 for activity 4–5). The 2 days possible compression of activity 9–10 is limited by the 1 day to full crash of activity 8–10. Therefore

Compress activities 8–10 and 9–10 simultaneously by one day, at a cost of 130 units. Project duration = 101 days. Project direct cost = 12,920 units. Point 6 is now plotted in Figure 5.8.

Seventh and Final Compression of the Network Model

Figure 5.7*g* indicates the effect of the sixth compression on the network model. The free float in the chain 8–11 has been reduced and the looped network between events 8 and 11 has one critical path completely crashed. Both the looped critical paths are now totally compressed and the only remaining critical activity available for compression is 4–5, which has a potential of 20 days at a cost slope of 180; there is, however, a limitation on its compression to 18 days provided by the free float in the chain 1–3–5.

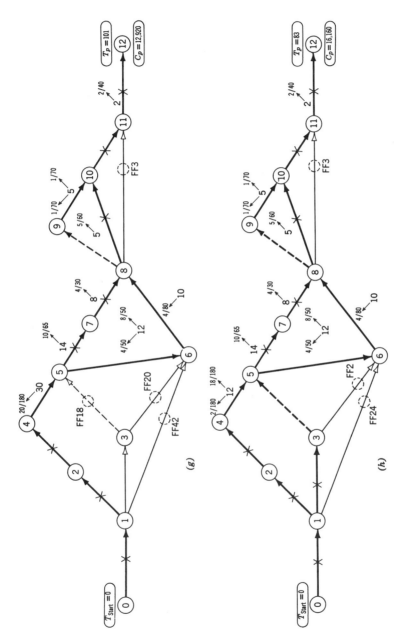

Figure 5.7 Compression of pipeline network. (g) Network after sixth compression. (h) Network after seventh compression (final position).

93

Therefore

> *Compress activity 4–5 by 18 days only, at a cost of 3240 units. Project duration = 83 days. Project direct cost = 16,160 units.*

A study of Figure 5.7*h*, showing the network model after the seventh compression, confirms that no further compression is possible. Another critical path (1–3–5) has appeared; one would expect that its activities (now critical) would become possibilities for compression, and this would be so were it not for the fact that the normal state of the chain 1–3–5 is also its crashed state, as reference to Table 4.1 shows.

Hence a completely crashed program for the critical path exists between event 0 (project start) and event 12 (project completion). The optimal least-time solution for this project has been evolved and is plotted as point 7 in Figure 5.8.

On considering the seven compression stages, it is observed that the first two stages involve simple crashing of activities (see Section 5.3); the next two stages have compressions limited by the free floats of noncritical chains (see

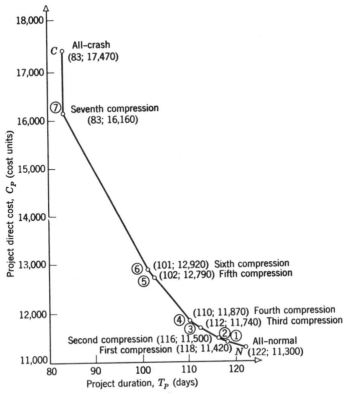

Figure 5.8 Direct-cost and duration curve for pipeline project.

Section 5.4); the fifth and sixth compressions involve parallel critical paths (see Section 5.5); and the seventh stage involves a compression limited by float which introduces the final limitation of crashed critical paths (see Section 5.6).

In network models composed predominately of chains with few cross connections, the only compressions encountered are similar to those already discussed; these are therefore easily handled. Sometimes, however, especially in highly interconnected networks, difficult situations develop that require special treatment; these are considered in Chapter 6.

Sometimes a network model may be simplified temporarily by eliminating chains of activities having unusually large free floats in the all-normal state. Reference to activity 1–6 in the pipeline model, with an initial free float of 50 days, shows that its final state still provides 24 days FF; it could therefore have been entirely ignored by deletion from the network as far as compression calculations are concerned. Such deletions must, of course, be done with caution, backed by experience; for example, chain 1–3–5, with its large initial FF of 18 days, did, in fact, control the termination of the compression calculations in this project, and therefore could not have been ignored.

In some cases, and indeed in this example, the network model may be split into two separate models and analyzed independently. For instance, that portion between events 0 and 8 and that portion between events 8 and 12 could be considered separately and later combined. This technique is discussed in Section 5.9.

5.8 COST ANALYSIS OF THE PROJECT: MOST ECONOMICAL SOLUTION

The network compression calculations just completed have enabled the salient points (or coordinates) of the project direct-cost curve to be determined and plotted in Figure 5.8. Because all individual activities were assumed to have valid continuous utility curves, the curve of Figure 5.8 implies that any project schedule is feasible between the least-time duration of 83 days and the all-normal duration of 122 days.[1] The curve is concave and the magnitude of its slope increases monotonically with reduced project duration; in other words, the cost slope is negative throughout, which is, in fact, an indication that the network compression calculations were in the logically correct order. This is a characteristic of the parametric linear programming problem.[2]

An interesting and most important feature of Figure 5.8 is the relation between the all-crash and the optimal least-time solutions for the project. It

[1] The special case of utility data with discrete points and its effect on the project direct-cost curve is discussed in Chapter 6.
[2] Any convex portion in the curve usually indicates either an arithmetical error or a logically incorrect order in the network compression calculations. There are special cases where artificial cost slopes are assigned to activities; this will result in a convex portion being generated in the direct-cost curve so that a nonoptimal curve is obtained (see Chapter 6).

will be remembered that the former requires the full crashing of every activity within the project, whereas the least-time solution crashes only sufficient activities to obtain the same minimum duration of 83 days. As pointed out in Chapter 2, the all-crash solution incorporates needless and uneconomical compression of some activities; as a result, there are two direct costs for the same project duration.

Figure 5.9 indicates the difference between the two solutions. In *a* the all-crash network is shown, with a single critical path 0–1–3–5–7–8–10–11–12 fully crashed and a minimum duration of 83 days. In *b,* is the least-time network (this is, in fact, the final seventh compression seen in Figure 5.7*h*), with its many critical paths. It has the same limiting critical path fully crashed, but many of the other activities still have compression potential; their crashing would not reduce the project duration, but would merely increase cost needlessly and introduce or increase free floats.

The least-time solution represents a saving of 1310 units of cost over the all-crash solution for the same project duration of 83 days. The details of the individual sums comprising this saving may be computed by relaxing the unnecessarily crashed activities in the all-crash solution, thus:

Relax activity 1–6	by 8	days, to save 80	cost units
3–6	4	160	
4–5	2	360	
5–6	4	200	
6–8	4	320	
8–11	3	120	
9–10	1	70	
Total saving		1310	

As previously discussed, however, the project direct-cost curve does not represent the full cost of the works: to the total direct cost must be added all the other indirect costs of the project. These indirect costs are estimated by the planner in a conventional way and vary with different project completion times. Usually they vary almost directly with project duration and so need be computed only for the all-normal solution and for one other duration (say the least-time solution) and a straight line variation assumed between these two points. It is not intended here to discuss the computation of indirect costs, but merely to indicate how they are used to make the correct decision as to the most economical project duration and total cost.

To determine the correct optimum schedule for the most economical overall project cost, the indirect cost curve is plotted on the same graph as the direct cost curve and then summed to give the total cost curve. In Figure 5.10 this has been done for the pipeline project; the project direct-cost curve has been reproduced from Figure 5.8, whereas the project indirect-cost curve has been assumed for the purpose of illustration. The project total cost curve is

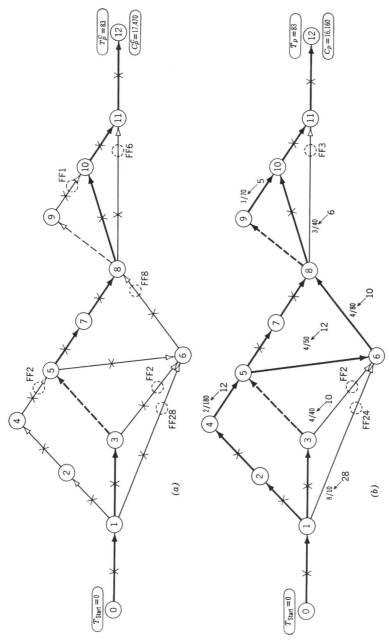

Figure 5.9 (a) All-crash solution. (b) Least-time solution: network after seventh compression.

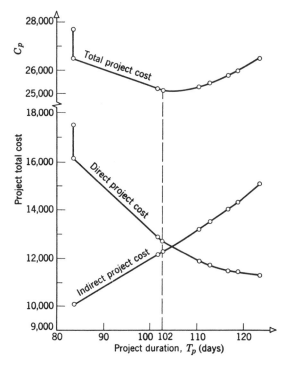

Figure 5.10   Most economical solution for pipeline project.

plotted by simple addition of the two other costs at any desired duration on the graph.

It is immediately seen that a minimum total project cost exists. This is the most economical solution for the project—the optimum solution, the duration for the least total cost. Here the total project cost reaches a minimum in the range of duration between 100 and 110 days. A practical duration in this case would be 102 days, for a plan for its execution already exists (see Figure 5.7f, showing the network after the fifth compression).

In conclusion, the minimum region of the total cost curve may be taken, for practical purposes, as being spread over a range of durations amounting to several days. Hence, by selecting a project duration at the left end of this range, one keeps a few days leeway in hand for late completion (if necessary) at little or no change in project cost.

5.9   SIMULATION OF GROUPED ACTIVITIES

In Chapter 3 it was shown that a sequence of activities can be combined into one activity, provided that these combined activities are unaffected by the remainder of the network (except for the commencement of the first activity and the termination of the last activity). Naturally, this one activity must be given a utility data curve that is equivalent to the combination of all the individual utility data curves for these activities.

In many complex projects, there may be sections of the work that are individually independent of the remainder. On the project network model these sections usually appear as separate chains or groups of activities connected to the main network at their starting and finishing events; or it may be that the project network model can be divided into separate sections by virtue of the fact that at certain events (such as milestone and interface events—see Sections 7.3 and 15.3) the relevant activities come together at a common node. Since these independent sections are in fact fragments of the main network, they are generally referred to as *subnetworks* or (colloquially) *fragnets*. Provided that connection into the main network occurs only at its starting and finishing nodes, any such subnetwork (fragnet) may be grouped into a single replacement activity with its own corresponding utility data, in a manner similar to the combined activities discussed in Section 3.7. Figure 5.11*a* shows schematically the various parts of a complex network that may be replaced by equivalent single activities, as shown in Figure 5.11*b*. It is essential that no intermediate network connections exist between the main network (shown shaded in the figure) and the individual fragnets to be replaced; if these connections are present, the substitution is not valid because the whole network characteristics will be affected.

Each of the replaced individual fragnets may then be considered as an entirely separate project, and the usual calculations can be made to determine their respective project direct-cost curves. These curves then become the utility data for the respective individual replacement activities. The pipeline construction model of Section 5.7 provides a simple illustration, for it is obvious that this network can be separated at event 8 into two subnetworks I and II, as shown in Figure 5.12. These subnetworks being entirely independent can be considered separately for compression calculations; hence, using the activity utility data previously tabulated (Table 4.1), the utility data for each subnetwork will be as seen in Figure 5.13. (The derivation of these cost curves is left to the reader as an exercise.) The network model for the pipeline construction of Figure 5.7*a* is now replaced by the simple network model of Figure 5.13 as far as its *direct-cost/duration characteristics* are concerned. From this it is simple to determine the resultant direct-cost/duration curve for the entire project; all that is involved is the combination of two sequential simulated activities (each with multistage utility data), which was previously considered in Section 3.7. The answer will appear as already determined in Figure 5.8, which the reader should compare with the utility data of Figure 5.13. It is now undoubtedly clear that, if grouped activity utility data are applied to an isolated subnetwork, full compression calculations of the entire network are avoided, and attention may be then focused only on the remaining portion of the original network.

The replacement of a fragnet group by a single activity, which simulates

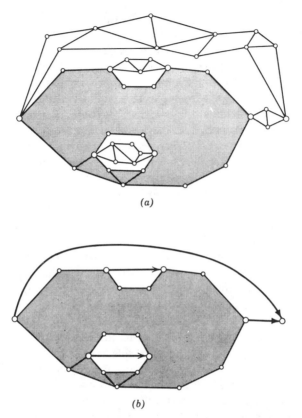

(a)

(b)

Figure 5.11 Simulation of grouped activities (fragnets). (a) Network model showing four individually independent fragnets. (b) Network model showing the four equivalent simulated activities.

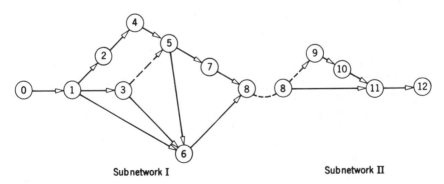

Figure 5.12 Subnetworks (fragnets) for pipeline construction model.

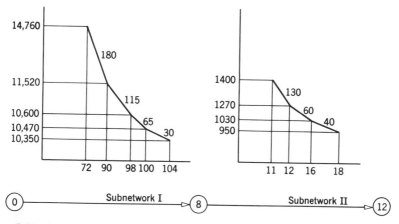

Figure 5.13 Simulated subnetworks and utility data for pipeline construction model (compare with Figures 5.7 a–h).

the properties and characteristics of the group, is desirable from several viewpoints. First, there is an obvious simplification of the overall network model; second, certain parts of a network fall automatically into special fields of interest, where, in any case, special costing may be desirable; third, the calculations can be considered separately, without affecting the validity of the final calculations for the entire network, in whatever form the model eventually appears. If individual simulated portions are not later logically altered in their attachment to the primary model, alterations can be made to this primary network without affecting the calculations already made on those simulated portions.

In this way significant savings can be made both in computations and in the time required to assess variations in the network model. In particular, the effect of using different methods for completing any single activity may be easily determined (without repeated exhaustive calculations) if the remaining section of the project model is itself considered to be an individual portion whose characteristics are simulated by a single replacement activity.

The simplification of complex network models, the flexibility in considering alternative construction methods, and the reduction of computational effort are all important features which make the simulation of independent fragnets a worthwhile consideration.

Evaluation

a. Why is there the need to determine the "most economical solution" for a particular project?

b. Why is the simulation of individual fragnets a worthwhile consideration?

PROBLEMS

5.1 Carry out optimal compression of the following network, all utility data being continuous:

Activity	t_N	c_N	t_C	c_C
0–1	20	$1000	10	$1300
0–2	30	1500	10	2860
1–2	20	2000	5	3005

5.2 Carry out complete compression of the following network, all utility data being continuous:

Activity	t_N	c_N	t_C	c_C
0–1	10	$ 200	5	$ 300
1–2	20	200	10	300
1–3	40	1800	30	2700
1–6	28	500	20	580
2–4	8	150	8	150
3–5	0	0	0	0
3–6	10	100	6	260
4–5	30	3000	10	6600
5–6	20	2800	8	3400
5–7	24	1000	14	1650
6–8	10	200	6	520
7–8	12	400	8	520

5.3 The continuous utility data for a certain project is the following.

Activity i–j	Normal		Crash	
	Duration	Cost	Duration	Cost
1–2	6	100	4	120
2–3	9	200	5	280
2–4	3	80	2	110
3–4	0	0	0	0
3–5	7	150	5	180
4–6	8	250	3	375
4–7	2	120	1	170
5–8	1	100	1	100
6–8	4	180	3	200
7–8	5	130	2	220

(a) Compute the minimum project direct cost for each possible project duration.

(b) If crash data were available for the normal critical activities only, what would be the maximum amount of compression possible?

(c) If the indirect cost rate is 20 money units per unit time, what is the optimum project duration?

6

NETWORK CALCULATIONS III: COMPLEX COMPRESSION AND DECOMPRESSION

6.1   COMPRESSION USING MULTISTAGE UTILITY CURVES

The network compression calculations of Chapter 5 were concerned with simple construction projects in which the network models were basically a series of parallel chains simply interconnected. The utility data were of elementary form, a continuous straight line between normal and crash points. The network compression logic was directed toward determining network limitations once the activity with the cheapest cost slope was selected. The only activity limitation was that imposed by the physical conditions of crashing.

In practice, however, an activity often has a multistage utility curve made up of a series of continuous straight lines, each with its associated cost slope (see Figure 3.11). This curve may be used as an approximation to the continuous theoretical curve (Figure 2.2) or it may have some other significance. Sometimes sufficient accuracy can be obtained by a straight-line approximation to this multistage curve; but if this involves intolerable errors, a piecewise linear curve must be used.

Multistage curves affect the selection of activities for compression, since once the compression proceeds past the selected cost slope range, a new cost slope must be taken into account. This does not make the calculations any more difficult; it merely requires more compression stages between the all-normal and the all-crash solutions. It also leads to a more accurate direct-cost curve since more points are determined.

Figure 6.1 shows the multistage utility curves for two activities, 2–3 and

1–3, which are involved in a project whose network model is seen in Figure 6.2a. Activity 2-3 has cost slopes of 50 and 100. If the activity duration lies between 10 and 20 days (say 12 days), the utility data become $^{2/50}12^{8/50}$; whereas if the duration lies between 5 and 10 days (say 7 days), it becomes $^{2/100}7^{3/100}$. Obviously, the only difference involved, when the duration is 10 days, is that the indication of the relevant cost slopes is $^{5/100}10^{10/50}$. Hence the conventional representation of utility data on the network model is adequate for multistage cost curves.

Figure 6.2 shows the various stages in the network compression when activites 2–3 and 1–3 have the multistage utility data of Figure 6.1. The project direct-cost curve, based on this information, is plotted in Figure 6.3, together with the results (shown in dashed line) when single straight-line approximations to the utility data of these activities are made (as indicated in Figure 6.1). Obviously, the multistage data have produced a more accurate representation of the direct-cost curve. Whether this additional accuracy is warranted need not be discussed here; it must depend on the shape of the utility curve and the amount of resources involved in the project.

The compressed schedule of 20 days shows a maximum disparity in the two curves of 246.67 cost units. This is due to the combined approximate cost slopes of $(68 + 66.67) = 134.67$ units being greater than the actual combined cost slope of $(60 + 50) = 110$ units for a 10-day compression; that is, $(134.67 - 110)10 = 246.67$ cost units.

It is interesting to see that the least-time direct cost is also in error because activity 1–3 has not been fully crashed; the difference of 60 cost units represents the divergence of the two curves for activity 1–3 for a duration of 15 days. The general conclusion can be drawn that, if the utility curves are

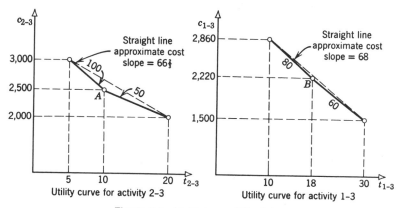

Figure 6.1 Multistage utility data.

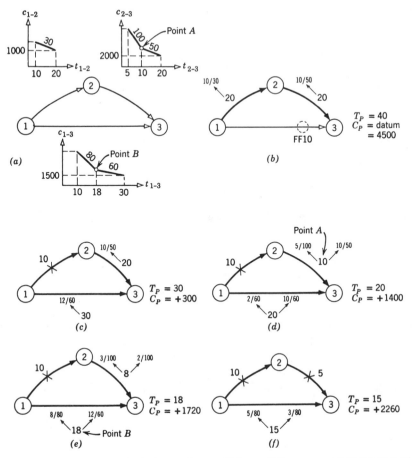

Figure 6.2 Compression calculations using multistage utility curves. (a) Network model and utility data. (b) All-normal. (c) First compression. (d) Second compression (compression limited by point A). (e) Third compression (compression limited by point B). (f) Fourth and final compression.

conservatively linearized, the project direct-cost curve will also be conservative; that is, it will indicate greater costs than actually apply.

6.2 COMPRESSION USING DISCRETE POINT UTILITY DATA

Some activities have only a limited number of ways of achieving their completion; therefore their utility curves are discontinuous and consist of isolated points. The special significance given to cost slopes does not apply because only specific reductions in duration are allowed in these discontinuous activity durations. Discrete point utility data curves force attention away from the

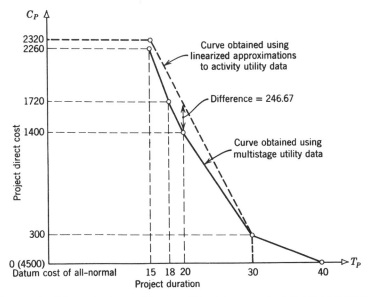

Figure 6.3 Effect of using multistage utility data on project direct-cost curve.

simple consideration of actual or effective cost slopes and demand the determination of optimal investment for a specific amount of network compression.

It may be that, considering the "apparent cost slope" of a discrete point utility curve, a certain activity is the cheapest for network compression; but if the full "jump" compression of this activity is not permitted by the network model, a more expensive activity gains precedence for optimal investment. When specific "jumps" are permitted in network compression calculations, no intermediate schedules are possible, and the project direct-cost curve has a discontinuity. In addition, the steady compression of the network model and the steady development of critical paths are interrupted. It often happens that certain critical paths vanish, and are replaced by entirely different critical paths when optimal solutions are required for specific "compression jumps." A simple example follows.

Figure 6.4 shows a small network model of three activities, in which activity 1–2 has a discrete point utility curve. This activity may be completed either by method A or method B; the dotted line between A and B indicates the "apparent cost slope" of 30 units. Normally method A is used, unless an activity duration of 10 days is possible, and then method B is used. Hence, unless a network compression of 10 days is permitted, method A stands and no activity compression is feasible. The utility data referring to this activity are given in the usual way $^{10/30}20$, but with a slash across the small pointer (see Figure 6.5) to indicate the "jump" of 10 days.

The all-normal solution is shown in Figure 6.5a. The first compression

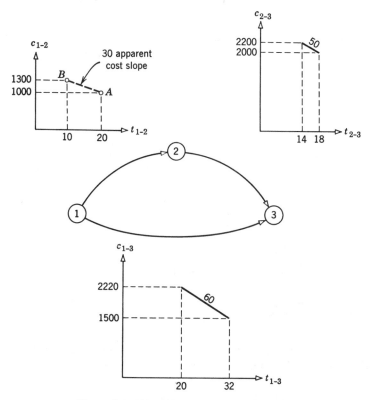

Figure 6.4 Network model and utility data.

shown in *b* involves the rejection of activity 1–2 (with the lowest apparent cost slope) and the crashing of activity 2–3 with a cost slope of 50. The second compression, in (*c*), permits the full "jump" compression of activity 1–2; but, since the critical path now switches to chain 1–3, it permits also the decompression (uncrashing) of activity 2–3, with a resultant cost saving. The project duration of 32 days can be achieved for the cost of crashing activity 1–2 alone (300 units); but the critical path 1–2–3 of the first compression is now a noncritical path with 4 days FF. Also the first compression permits any schedule between 34 and 38 days, but the "jump" compression of activity 1–2 in the second network compression gives an isolated project duration of 32 days.

The third and fourth compressions follow as usual, with all schedules possible between 24 and 32 days. The final direct-cost curve for the project then appears as in Figure 6.6, with a discontinuity of 2 days between project durations 32 and 34 days.

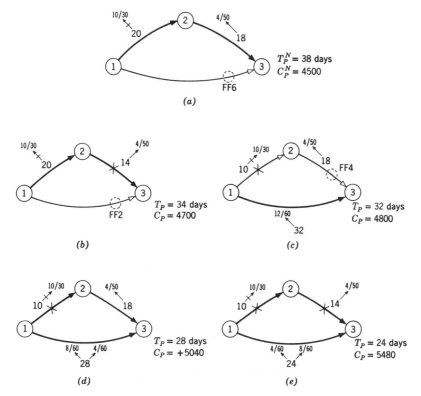

Figure 6.5 Compression stages in project with discrete-point utility data for one activity. (a) All-normal. (b) Second compression. (d) Third compression. (e) Fourth and final compression.

6.3 COMPRESSION USING MULTIPLE CRITICAL PATHS: ACTIVITY DECOMPRESSION

The compression of network models consisting of highly interconnected events will sometimes lead to the development of multiple critical paths. The compression calculations follow the same general principles outlined previously, but the determination of the best combination of activities on the critical paths, for minimum effective cost slope (S_E), becomes more involved. Activities previously compressed can often be decompressed (that is, their durations are extended) with consequent savings in cost, in conjunction with other activities now more heavily compressed, to ensure an overall network compression. This combination of decompression and overcompression can lead to the correct combination of activities yielding the cheapest effective cost slope.

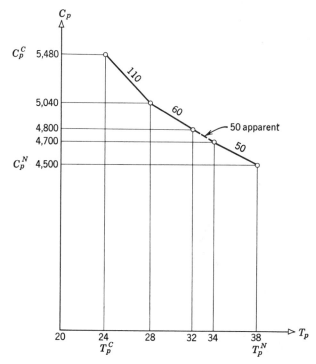

Figure 6.6   Project direct-cost curve.

The situation is made possible when an expensive activity is de-compressed and a relatively inexpensive activity is overcompressed. The network and utility data permit this situation to develop because of features such as the crosslinking of critical paths and the switching of critical paths through "jump" compressions of activities with discrete point utility data. The general approach to this problem is illustrated by the simple example shown in Figure 6.7. In highly interconnected systems, optimal solutions (if attempted) will most certainly require the use of computers; for manual calculations therefore it is best to adopt the procedure used in computer programs, namely, the systematic testing of all possibilities.

In the portion of the network model presented in Figure 6.7, all paths are critical, namely, 1–2–4, 1–2–3–4, and 1–3–4. To generalize the problem, all activities are assumed to be already partially compressed and are therefore available for further compression or for decompression, as desired; further-more, to enable simple solutions, each activity is given a simple numerical cost slope. The problem then becomes that of determining the effective cost slope (S_E) for a *one-day* compression between events 1 and 4; this unit problem is further idealized by assuming that the intermediate events 2 and 3 may have their times (T_2 and T_3) changed forward or backward at will by 1 day.

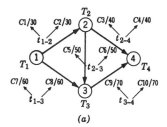

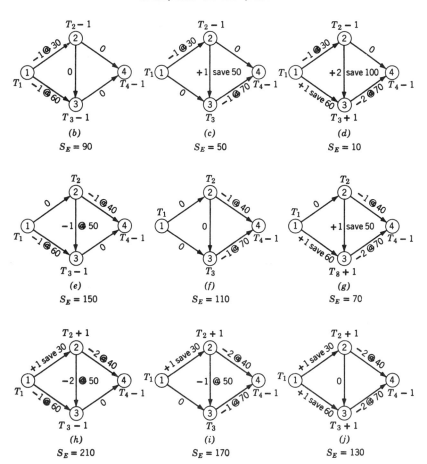

Figure 6.7 Effective cost slopes in multiple path compression.

If event 4 is to be timed a day earlier, and if T_1 is fixed, each critical path between events 1 and 4 must be effectively compressed by 1 day. This compression in each critical path may lead to a local decompression and a consequent overcompression in some other part of the same critical path.

Figure 6.7*b* to *j* shows all the possibilities based on these limitations. A systematic approach has been used, where for *each* possible timing of event 2 ($T_2 - 1$, T_2, and $T_2 + 1$) the three possible timings of event 3 are considered; thus there are nine possibilities. When all the event times are decided, the alterations necessary to the activity durations are made and the costs evaluated.

For example, Figure 6.7*c* requires that events 2 and 4 be a day earlier than in *a*, with event 3 scheduled as before. Therefore activities 1–2 and 3–4 must be compressed by 1 day; but activity 2–3 now has 1 day available and can be decompressed to provide less cost. The effective cost slope becomes $S_E = 50$, computed as 1 at 30, plus 1 at 70, minus 1 at 50.

The cheapest effective cost slope is shown in (*d*), which also includes both compression and decompression of activities. Six of the nine possible schemes for network compression involve activity decompression with cost savings, and three of these are cheaper than any of the schemes that do not consider activity decompression.

The actual scheme adopted depends on the real values of the permissible duration changes $C1$ to $C10$, for these specify the potential compression and decompression available to the activities; if, for instance, activity 1–2 is in its normal state, then $C2$ is zero, whereas if it is fully crashed, $C1$ is zero. It could be that the theoretically cheapest solution (*d*) is impossible because any one of the values ($C1$, $C6$, $C8$, or $C9$ may be zero. If this happens, then the solution with the next cheapest effective cost slope, that of (*c*), is investigated; and so on until either a physically possible scheme is selected or all are rejected as physically impossible.

Once a feasible scheme is obtained for the *unit problem,* the actual magnitude of the network compression can be determined. The maximum possible network compression is, of course, limited by the specific values of $C1$ to $C10$ actually available. Suppose that solution (*c*), with $S_E = 50$, is the cheapest scheme physically possible, with $C1 = 10$ days, $C6 = 16$ days, and $C9 = 2$ days; then only 2 days total network compression is possible because of the limitation imposed by $C9$. The position will be as shown in Figure 6.8. Similarly, the solution for each *unit problem* can be used for determining the cheapest maximum possible network compression, depending on the values of $C1$ to $C10$. Naturally, if one of these values (say, $C1$) is zero, some of the schemes—in this case, (*b*), (*c*), and (*d*)—are automatically rejected.

6.4 NONOPTIMAL COMPRESSION: EFFECT OF ACTUAL VARIATIONS IN UTILITY CURVES

The fully continuous utility curves illustrated in Figure 2.2*a* represents an idealized situation that can be only imperfectly approximated in practice. Individual activities (and those representing a group of activities) often require for their completion various labor skills and equipment, each with

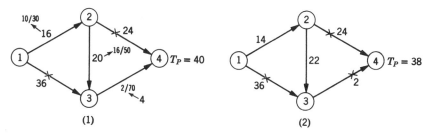

Figure 6.8 Compression of multiple critical paths. (1) Before compression. (2) After compression: one path fully crashed.

different rates of pay, optimum crew sizes, effective work rates, and so on. When an activity requires such a continuous intermingling of various skills and crew sizes, rarely can this be done so efficiently that there is no waste of time. This feature of inefficient skill and equipment utilization means that real utility curves differ considerably from the theoretical curve.

Because of the variability of this inefficiency as activity duration is compressed, the real utility curve, although generally concave, will sometimes contain undesirable convex portions and may therefore look like Figure 6.9a. This activity could well be the laying of steel pipes, involving trench trimming, pipe placement, and in-situ welding. The rate of pipe laying depends on the individual efficiency of each team, and consequently on each team's size and rate of working. To some extent this can be controlled, thus enabling the selection of desirable combinations giving those portions of the utility curve emphasized in Figure 6.9b. If the magnitude of the activity warrants investigation, optimal solutions may be based on these desirable portions of the utility curve only; in this way, inefficient combinations would be avoided in the scheduling of the work.

Alternatively, it often happens that an activity, such as trench excavation, allows different methods of construction (for example, manual or mechani-

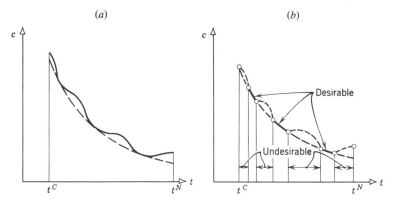

Figure 6.9 Utility curve showing interaction of various skills, and so on, producing spasmodic inefficient durations and costs. (a) Actual. (b) Desirable durations.

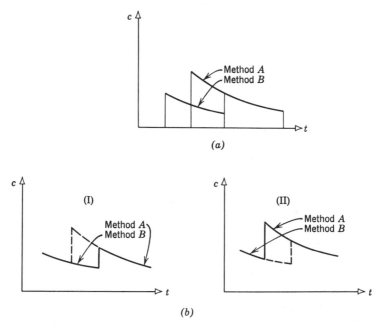

Figure 6.10   Utility curves for different construction methods. (*a*) Individual utility curves. (*b*) Combined utility curves.

cal), each with a variety of resources and durations. Figure 6.10*a* illustrates a common situation in which the different methods are more economical for different durations, whereas for some durations they are incompatible. Common approximations in this case might be the two utility curves seen in Figure 6.10*b*. If network compression calculations are based on curves of this type, nonoptimal solutions result. The correct approach for optimal solutions is to compare method *A* with method *B*, using two different sets of calculations.

Consider the simple project shown in Figure 6.11, where activity 1–2 has two possible construction methods, *A* and *B*, whose utility data are also shown. Assume that the normal cost for completing activities 1–3 and 2–3 is a total of 1000 units. Network compression calculations may proceed in three ways.

1. By method *A* alone.
2. By method *B* alone.
3. By combined methods *A* and *B*.

The resulting three project direct-cost curves are plotted in Figure 6.12. The cruve for the combination of methods *A* and *B* indicates correctly the use of method *A* alone in the 38- to 50-days range, but not for less durations because method *A* is then uneconomical. For the shorter times, 18 to 35 days, it indicates correctly the use of method *B* alone; but the curve for method *B*

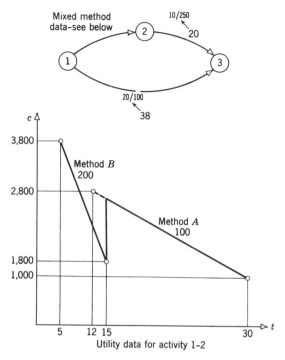

Figure 6.11 Network model with data for mixed construction methods in activity 1–2

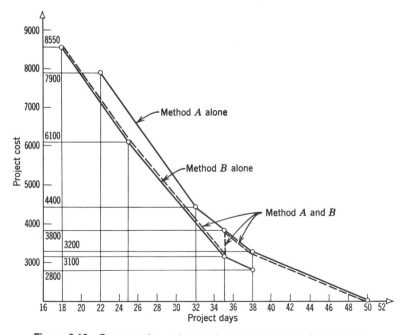

Figure 6.12 Compression solutions for mixed construction methods.

alone is the one which shows correctly the optimal solution between 35 and 38 days duration. It is apparent from this, therefore, that compression calculations for the combined methods are not reliable and that each should be computed separately. Also there is an immediate cost saving of 700 units once the sudden change is made from one method of construction to the other. If the reader makes these calculations, it is possible to find that different critical paths develop and different compression stages result in the three different analyses.

Another factor which must be borne in mind in practice is that the compression of one critical activity may well require crashing of another related activity (which may indeed be noncritical), primarily because the two activities are performed by the different members of the same crew or because of practical limitations in the regular overtime working by some crews and not by others of the same trade on the same project. These are actually industrial or political constraints that are nevertheless very real in many instances. In these cases the true cost slope of the critical activity to be compressed should be increased to include the cost slope of the related activity which must also be simultaneously crashed for practical reasons; this is necessary for the compression to occur in its proper logical sequence for optimal solutions. This problem is not unrelated to the use of artificial cost slopes, as discussed in the following section.

6.5   NONOPTIMAL COMPRESSION: USE OF ARTIFICIAL COST SLOPES

It has been demonstrated that, in network compression calculations, activities are selected for compression in their logical order of cheapness, once certain qualifying conditions have been satisfied. If the logical analysis is correct, optimal compressions are obtained because the activity with the cheapest cost slope available is *always* compressed. If, however, the cheapest available activity is not compressed in its logical sequence, nonoptimal solutions are obtained. In such cases, it must be accepted that the specific increment of resources applied to the project, at that stage, could have been used elsewhere to obtain a greater reduction in the project duration. Despite this fact, however, management may dictate the compression of an activity other than the cheapest available for reasons that are expedient, if not logical; for various reasons the choice of certain activities may seem less attractive for compression than their utility data suggest.

With activities of this type, a subterfuge is necessary in order to prevent their selection in logical order, either by the estimator using manual calculations or by the computer working automatically. Instead of falsifying the utility data, a better technique is to supply the activity with an artificial cost slope, whose magnitude is a measure of the undesirability of completing the activity in alternative durations. Eventually, as the network compression calculations proceed, this particular activity may be chosen for compression on the basis of its artificial cost slope; the obvious implication at this stage is that compression of this activity is now acceptable to management. When compression has taken place, the new schedule *must* be costed on the basis of the *real* utility data for the activity; otherwise completely arbitrary results are obtained.

Figure 6.13 indicates a simple problem, in which the activity 1–2 is assigned an artificial cost slope of 80 (real value is 30), as shown in (*a*); this will ensure that it will not be compressed before any of the other activities. The activity could have been given an artificial slope of, say, 45 if it were merely desired to ensure that activity 2–3 would be compressed before it. The effect of this on the project direct-cost curve is seen in (*b*), where the results of the artificial cost slopes are compared with the true direct-cost curve (computed from the real cost slope of activity 1–2). It is important to remember that, as

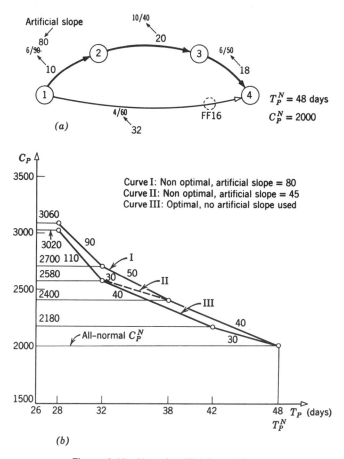

Figure 6.13 Use of artificial cost slopes.

shown in this case, all schedules using artificial cost slopes are more expensive than those based upon real slopes, at least until after compression of the activity with the artificial cost.

Project direct-cost curves computed from artificial cost slopes are not truly concave, a portion usually being convex; this, in itself, indicates a nonoptimal solution.

With manual calculations the use of artificial cost slopes is not necessary (except as a reminder) because the planner or estimator is able to omit any activity from logical compression at will; and can, of course, reconsider it at any future compression stage. Thus the use of artificial cost slopes is usually confined to computer processing.

As a common example of the selection of an artificial cost slope, consider the three activities of formwork, reinforcing, and concreting in a long retaining wall; each is performed by a separate crew, whose size is chosen so that each one is evenly matched for productivity. The wall is built in consecutive sections longitudinally, in a smooth sequence. It is obvious that, irrespective of the individual cost slopes of the different activities, one activity cannot be crashed (from a practical viewpoint) unless the others are speeded up also. Therefore it will be necessary to assign to the appropriate two activities artifical cost slopes identical to that of the most expensive one so that the computer will select all three together for simultaneous compression. In the event that one of the three could sustain a compression different from the other two, an appropriate slope (relative to the other two) could be assigned to it.

6.6   CONSTRAINED COMPRESSION CALCULATIONS

Optimal solutions so far considered have been minimum cost solutions. Any deviation from the logical ordering of activities for compression, or in the amount of their compression, automatically implied a nonoptimal solution which would cost more than the optimal (or minimum cost) solution. In preceding sections of this chapter, the use of mixed-method utility data and of artificial cost slopes led to nonoptimal solutions as far as costs were concerned.

The emphasis on costs as a measure of optimality results in a series of solutions or schedules in which, for many activities, no possibility exists for maneuver by management. This situation may be intolerable under certain circumstances and its rejection implies that optimization should have a more general meaning than minimum cost. For instance, management may decide to apply a purely policy decision or strategy to the project in selecting activities for compression; as a result, this strategy becomes an additional constraint on the compression calculations. Indeed, the technique of deferring an activity for compression by using an artificial cost slope is itself a strategy that yields optimal results on a basis other than cost. Within the framework of this or any other strategy, optimal solutions are possible since solutions previously considered as nonoptimal (from the cost viewpoint) may now be optimal solutions satisfying the constraints imposed by the strategy; the normal compression calculations are now rejected because they violate the additional constraints.

The implementation of various strategies, constraining the usual compression calculations, is a very common and important feature of the actual handling of projects, and their effects may be readily determined and costed. To do so, it is necessary only to perform two sets of network compression calculations: one with no strategy, to provide the correct optimal direct-cost curve for the project, and the other with the strategy. The plot of both direct-cost curves for the project will show at once by the difference in the cost ordinates, the price paid for implementing the strategy for any given project duration. This approach further enhances the value and importance of the usual minimum-cost optimal solution.

A generalization of this approach leads to adopting a strategy, where the arbitrary (illogical) ordering and selection of activities for compression becomes automatically an optimal solution. This method is often used in projects planned essentially on an all-normal solution, with local compressions to counter certain situations such as hazards; examples are in Chapter 8.

Nevertheless, it must be realized that this optimization represents an inefficient application of resources from a purely cost viewpoint.

6.7   NETWORK DECOMPRESSION CALCULATIONS

As explained in Chapter 5, the project duration T_P may be considered as the parameter in the parametric linear programming formulation of the project network model. This parameter ranges in value between the two limits of all-normal duration T_P^N and all-crash duration T_P^C.

Network compression is based on continuous optimal decrease of the project duration from all-normal to all-crash; in other words, *the cheapest way of reducing the project duration is determined at each stage by using the activity with the flattest cost slope.* It is usually convenient to adopt this procedure. Mathematically, however, the calculations can be just as easily based on adopting the crashed least-time solution as the starting point, and carrying out the procedure in the reverse order; this technique is conveniently known as *decompression.*

Network decompression is based on continuous optimal increase of the project duration from the least-time (or all-crash time) to the all-normal duration; in other words, *the cheapest way of increasing the project duration is determined at each stage by using the activity with the steepest cost slope.*

Given the utility data for all activities, the all-crash solution is just as easily obtained as the all-normal solution. However, whereas the all-normal solution is the optimal solution for T_P^N, and is thus the starting basis for compression calculations, the all-crash solution is *not* the optimal solution for T_P^C and is *not* the starting basis for decompression calculations. The first step in network decompression is to derive the optimal crashed least-time solution and subsequently to use this as the starting point for pure decompression calculations.

A useful byproduct of this trend of thought is to use a simple approximation to the project direct-cost curve, which requires the determination of only three sets of coordinates: the all-normal, the all-crash, and the crashed least-time, as illustrated in Figure 6.14. Here the actual project direct-cost curve is approximated by the straight line BC; this gives a much better approximation than that provided by a straight line joining the all-crash to the all-normal points (AC). This linearization of the direct-cost curve into the straight line BC is very helpful in obtaining quick (and slightly conservative) estimates of intermediate schedule costs for a project, although it is not accurate enough for determining truly optimal solutions.

6.8   DECOMPRESSION CALCULATIONS: DETERMINATION OF THE OPTIMAL LEAST-TIME SOLUTION

The problem of calculating the optimal crashed least-time solution for the project duration T_P^C from the nonoptimal all-crash solution is clear from Figure 6.14: given the network model, the utility data, and the coordinates of point A, it is required to derive the coordinates of point B. No alteration is to

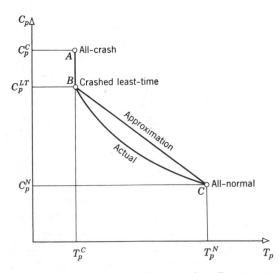

Figure 6.14   Simple approximation to project direct-cost curve.

be made to project duration, only a saving in cost. As indicated previously (in Section 5.8), the difference in costs between the all-crash and the crashed least-time solutions is entirely due to the needless crashing of noncritical activities, thus producing fully crashed noncritical paths with free float.

Hence to derive the crashed least-time solution, all that is required is *the elimination by activity decompression of as much free float as possible from noncritical activities in the all-crash network model.* Naturally, preference is given to available activities with the *largest* cost slopes for this decompression in order to obtain the greatest possible cost savings per unit extension of the duration of these noncritical activities. No decompression of critical activities is to be considered at this stage, of course, since no alteration is required in the project duration.

As an example, consider Figure 6.15*a* which shows the all-crash solution for the pipeline construction project previously considered in Chapter 5 (Figure 5.9) and Chapter 4 (Table 4.1). Utility data are required in this case for all noncritical activities capable of decompression. Since events on the critical path are fixed in time, noncritical paths consisting of single activities will (if possible) be decompressed by the full free float available.

The logical steps in the decompression are as follows.

1. *Decompress activity 8–11 by 3 days, at a cost saving of 120 units; decompression limited by normal duration of the activity and not by free float.*

2. *Decompress activity 9–10 by 1 day, at a cost saving of 70 units; decompression limited by free float.* New critical path introduced. Notice that the chain of activities 8–9–10 consists essentially of one activity 9–10 and that all the decompression is in that activity.

3. *Decompress activity 4–5 by 2 days, at a cost saving of 360 units; decompression limited by free float.* New critical path introduced. Notice that the chain 1–2–4–5 permits only one activity decompression since all the others are in a fixed state.

4. The situation involving the noncritical chains 5–6–8, 3–6, and 1–6 is not so simple. First, the chain 5–6–8 can be decompressed 8 days; second, the most expensive cost slope is that of activity 6–8, which is therefore given preference for decompression; but, third, it has only 4 days potential decompression available; so, last, only 4 days decompression is possible for activity 5–6. Therefore:

   (*a*) *Decompress activity 6–8 by 4 days, at a cost saving of 320 units; decompression limited by normal duration.* And also:

   (*b*) *Decompress activity 5–6 by 4 days, at a cost saving of 200 units; decompression limited by free float. New critical path introduced.*

5. Notice that now event 6 is timed for 4 days later, and hence activity 3–6 now has a free float of 6 days, and activity 1–6 now has a free float of 32 days. The reader should carefully consider the implications involved in

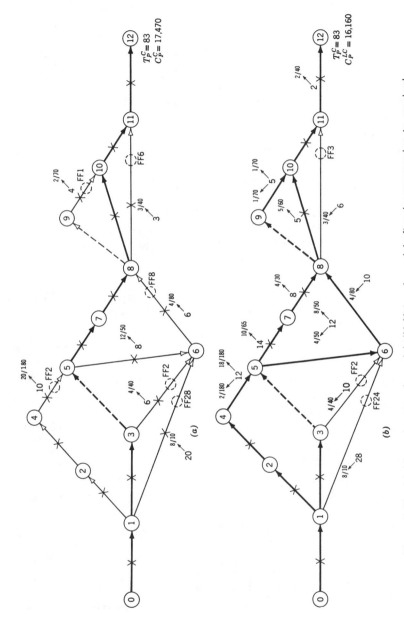

Figure 6.15 (*a*) All-crash network model. (*b*) Network model after decompression to crashed least-time solution.

defining the chains 5–6–8 and 3–6 (*not* 3–6–8), which was discussed previously.[1] Therefore:

Decompress activity 3–6 by 4 days only, at a cost saving of 160 units; decompression limited by normal duration and not by free float. Free float becomes $(2 + 4) - 4 = 2$ days.

6. *Decompress activity 1–6 by 8 days, at a cost saving of 80 units; decompression limited by normal duration.* Free float becomes $(28 + 4) - 8 = 24$ days.

Since no other noncritical activities are available, decompression now ceases. The crashed least-time solution has been reached and is cheaper than the all-crash by the summed cost savings in decompressions (1) to (6), that is, 1310 units. The condition of the network model after this initial decompression to cheapest least-time status is shown in Figure 6.15*b*. This network should be carefully compared with the seventh and final compression stage of Figure 5.9*b*, discussed in Section 5.8.

6.9 GENERAL DECOMPRESSION STAGES IN PIPELINE CONSTRUCTION MODEL

After the optimal least-time solution has been obtained from the all-crash solution, the general decompression calculations can proceed. *Since increases in project duration are required, only critical activities are considered for general decompression.*

In those portions of a network model containing a simple critical path whose activities are available for decompression) the first activity selected is that with the steepest cost slope. In those portions of a network where multiple critical paths exist, activities are selected (from those available for decompression) that have the highest effective cost slope. In single activities decompression is limited either by the normal duration of the activity or by a change in the cost slope of its utility curve (thus altering the logic of its selection). Similar limitations exist for the multiple critical path decompression, except that the limitation is introduced by the first activity violating the overall logical selection.

General decompression of the pipeline construction model is easily carried out and only a few of the calculations are given.

First Decompression of the Network Model

Activity 4–5 has the greatest single cost slope of 180 units per day increase in duration. The best combination of two activities, in the loop 5–7–8 and 5–6–8, is 5–7 and 6–8 with an effective cost slope of 115; whereas for the loop 8–9–10 and 8–10, the effective cost slope is 130. Therefore:

---

[1] See Section 4.5, Free Float.

Decompress activity 4–5 by 18 days, at a cost saving of 3240 units. Project duration = 101 days. Project direct cost = 12,920 units.

The network after this first general decompression is identical with Figure 5.7g.

Second Decompression of the Network Model

The effective cost slope of 130 units for the looped critical paths 8–9–10 and 8–10 is now the largest available. Activity 8–10 has 5 days available for decompression, but activity 9–10 has only 1 day before it reaches its normal duration. Therefore:

Decompress activities 8–10 and 9–10 each by 1 day, at a cost saving of 130 units. Project duration = 102 days. Project direct cost = 12,790 units.

The network after the second decompression is identical with Figure 5.7f.

Third Decompression of the Network Model

The effective cost slope of 115 units for the looped critical paths between events 5 and 8 is the largest available; decompression is limited by the normal duration of activity 5–6 to 8 days. Hence:

Decompress activities 5–7 and 5–6 each by 8 days, at a cost saving of 920 units. Project duration = 110 days. Project direct cost = 11,870 units.

The network after the third decompression is now identical with Figure 5.7e.

The general decompression stages, if correctly performed, reproduce exactly (but in reverse order) the compression stages fully described in Chapter 5; the reader should complete the preceding decompression calculations back to the all-normal solution to ensure his grasp of the procedure. After each decompression, the event times, floats, project duration and cost, and so on, must be recomputed; but, of course, floats do not play any part in decompression (except for the derivation of the optimal least-time solution). Once one side of a looped critical path reaches its all-normal duration further decompression between the looped events will produce float in that path; thus this path changes from critical to noncritical.

Depending on the preference of the planner, decompression may be used instead of compression to derive optimal solutions, particularly since fewer considerations enter into the calculations. Its adoption is axiomatic when a project duration close to the least-time solution is required, thus saving computational effort.

In practice, this situation arises whenever a very fast construction time is specified. With the all-crash and all-normal utility data, the planner proceeds to derive the optimal least-time solution, from which he decompresses the network until the specified duration is reached. Thus he is able to plan the job for the specified time at the minimum relevant direct cost.

6.10 OPTIMIZATION OF NONOPTIMAL SOLUTIONS: ALTERNATE DECOMPRESSION AND COMPRESSION

In some projects manipulation of the duration (contract time) by applying additional resources is not possible, especially when the time is fixed by the terms of the contract. This specified duration will lie somewhere between the all-normal and the all-crash solutions. In such cases, determination of the optimal solution, to suit the duration desired, is either by compression from the all-normal solution or by decompression from the all-crash (via the optimal least-time solution). When either of these approaches is used, calculations are continued until the specified duration is reached, the optimal solution being automatically obtained.

A more practical approach (mentioned at the end of Section 2.8) is to prepare initially a network model from a conventional estimate aimed at the specified duration. This solution will almost certainly be nonoptimal and hence more expensive than the optimal solution for its same duration. This desired optimal solution may be derived by successively improving, and hence cheapening, the original nonoptimal network model. The improvements are affected by successively decompressing and compressing the network within a small variation about the specified duration. The critical paths (and partly crashed activities) are thus gradually altered from the original conception to those of the optimal solution.

It will be found that a small amount of analysis yields immediate cost savings, even if the optimal solution is not reached, for the successive improvements can be terminated at any stage of the calculations, leaving a better plan than that originally conceived. Another excellent advantage of this technique is that the characteristics of the network model are fully investigated in the zone of project duration of greatest interest. A series of solutions is produced with almost static project duration, all in the correct order by alternately decompressing and compressing, and all in the vicinity of the specified project time. Furthermore, these solutions are very useful later in controlling the project during construction because a variety of plans to suit the contract time (at various costs) is available for consideration.

As an example, consider Figure 6.16. In (*a*) the network shows only the utility data for all the activities; the data for activities 3–5, 3–6, and 4–5 comprise two-stage utility curves. Each cost slope is indicated in relation to the duration at which the slope changes. Thus the network at (*a*) is not in any

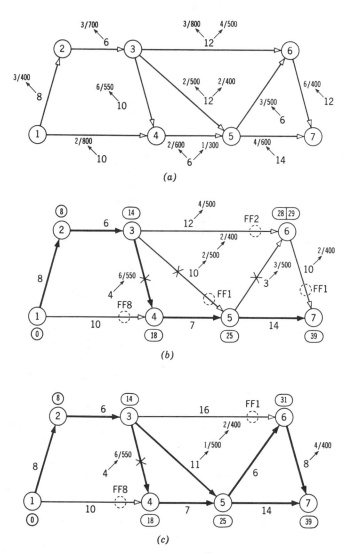

Figure 6.16 Optimization of nonoptimal solutions. (a) Project network showing utility data only. (b) Initial trial (I), $T_P = 39$ days, $C_P = +9400$. (c) Least-cost form of initial trial (II), $T_P = 39$ days, $C_P = +6200$.

way a solution because as yet no durations have been assigned to any activity; it is merely a preliminary layout of the project activities in proper sequence, with utility data indicated.

Assuming now that a project duration of 39 days is specified, a first trial (marked I) is shown in (b). When this solution is examined, and its critical path determined, improvements can be effected, because useless crash is

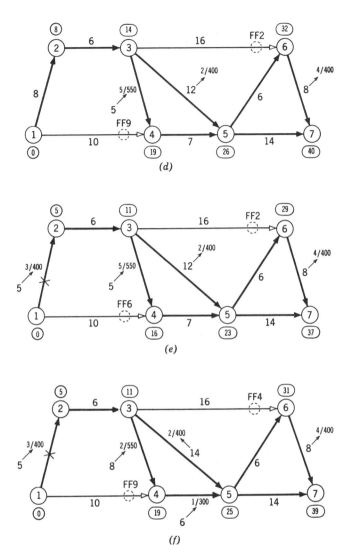

Figure 6.16 Optimization of nonoptimal solutions (*continued*). (*d*) First decompression (III), $T_P = 40$ days, $C_P = +5150$. (*e*) First compression (IV), $T_P = 37$ days, $C_P = +6350$. (*f*) Second decompression (V), $T_P = 39$ days, $T_P = +4200$.

apparent in activities 3–5, 3–6, 5–6, and 6–7. By decompression of those activities with available float, the least-cost form of this initial trial, still at the specified duration, is shown in Figure 6.16c. This is network II, and should naturally be cheaper; indeed the immediate cost saving of this first improvement is 3200 units.

The next three steps (marked III, IV, and V) represent successive solu-

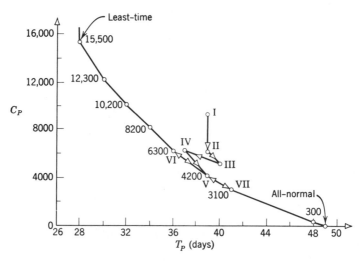

Figure 6.17 Optimization of nonoptimal solutions.

tions, comprising first an improving decompression yielding the greatest cost saving in the network, then a compression giving the cheapest decrease in project duration (to less than that specified), followed by a decompression back to specified duration, and so on; these steps are illustrated in Figure 6.16d, e, and f.[1] In this particular example, step V completes the process, for further alternate decompression and compression will regenerate step V, which indicates that the optimal solution has been achieved.

Figure 6.17 shows the complete direct-cost curve for the project, together with the successive solutions from the initial nonoptimal trial (I) through to the optimal (V) for the specified 39 days duration. The calculations give successive solutions that zigzag about the selected duration and gradually approach the optimal solution (V). The process terminates when a repetitive cycle is produced—in this case, when successive solutions are obtained cycling between the points (V)–(VI)–(V)–(VII)–(V). The nonoptimal solution for least-cost (II) of the initial trial (I) represents a significant cost saving; in some cases this step alone may be sufficient for practical purposes.

Because of the zigzagging of the successive solutions which are produced by alternate decompressions and compressions, this method of optimizing nonoptimal solutions is colloquially known as the "zigzag technique."

Considerable skill and experience are required to determine effectively optimal solutions by this procedure, for general rules cannot be given.

---

[1] The reader should carry out these calculations as an exercise. The entire project direct-cost curve in Figure 6.17 may be generated if so desired from the utility data supplied in Figure 6.16a and the optimal solution for 39 days just derived.

However, the determination of the nonoptimal least-cost solution of the initial network is straightforward and is amply justified for any project where the conventional estimate is based on other than an all-normal duration. After the nonoptimal least-cost solution has been found, the zigzag technique may, in some instances, be initiated by a compression rather than a decompression; also it may be necessary to alter significantly the specified project duration in the intermediate steps to ensure producing correct critical paths and eliminating incorrect ones. Except for the final steps, all solutions obtained by this technique are nonoptimal, and consequently can be improved economically. Hence, for highly competitive bidding, the additional calculations necessary to arrive at the cheapest (optimal) solution for the required duration are usually worthwhile.

Evaluation

a. How accurately should the utility data of important activities be determined?

b. How simple should network calculations be?

PROBLEMS

6.1 Carry out full compression of a simple network of three activities having the following utility data.

Activity	1–2	1–3	2–3
Normal duration	20	32	18
Normal cost	1000	1500	2000
Crash duration	10	20	14
Crash cost	1300	2220	2200
Nature of cost curve	Discrete Point	Continuous	Continuous

6.2 A small project may be represented by a network diagram of three activities. Activity 1–2 may be carried out in either of two ways, where the continuous linear utility data are as follows:

Activity 1–2	Method I	Method II
Normal duration	30	15
Normal cost	1000	1800
Crash duration	12	5
Crash cost	2800	3800

Activities 1–3 and 2–3 have continuous linear data as follows:

Activity	1–3	2–3
Normal duration	38	20
Normal cost	1700	1300
Crash duration	18	10
Crash cost	3700	3800

Derive the coordinates of the direct-cost/time curve for the project.

6.3 Adopting the utility data in Problem 5.3, draw the all-crash network. From this determine the least-cost crash solution, and then by decompression derive the project direct-cost/time curve.

6.4 From the following continuous utility data, determine by the zigzag technique the following information:
(a) The minimum project cost for not more than 40 weeks duration.
(b) The minimum possible project cost and its corresponding duration.
(c) The minimum possible project duration and its corresponding cost.
(d) The project cost-duration curve.

Activity i–j	Present Estimate		Alternative Estimates	
	t	c	t	c
1–2	8	$ 1,000	5	$ 2,200
1–4	10	7,400	8	9,000
2–3	6	1,500	3	3,600
3–4	4	4,300	10	1,000
3–5	10	7,800	12	6,800
"			and 14	6,000
3–6	12	9,000	16	7,000
"			and 9	11,400
4–5	7	2,000	6	2,300
"			and 4	3,500
5–6	3	3,000	6	1,500
5–7	14	10,200	10	12,600
6–7	10	2,500	8	3,300
		$48,700		

7

NETWORK CALCULATIONS IV: SCHEDULING AND RESOURCE LEVELING

7.1 USE OF FLOATS IN SCHEDULING: EARLIEST OR LATEST START

After the arrow network diagram has been analyzed and all the event times established, the scheduling of all the activities may proceed. The typical activity A_{ij}, which has an estimated duration t_{ij}, cannot begin until event (I) is satisfied, and its completion must occur before event (J) can be satisfied. Mathematically, the complete schedule for this activity becomes:

$$
\begin{array}{lll}
\text{Duration} & t_{ij} & \\
\text{Earliest start time (EST)} & T_I^E & \\
\text{Earliest finish time (EFT)} & T_I^E + t_{ij} & \\
\text{Latest start time (LST)} & T_J^L - t_{ij} & (7.1) \\
\text{Latest finish time (LFT)} & T_J^L & \\
\text{Maximum time available} & T_J^L - T_I^E &
\end{array}
$$

It is apparent that this schedule is based on the calculations of the times T^E and T^L for each event. Float times, if applicable to the activity A_{ij}, are included in the maximum time available and may be easily calculated from the equations in Section 4.5. The results of activity schedule calculations of this general type are shown in Table 2.1.

A more specific schedule can be developed if all the events in the network are allotted definite occurrence times. In this case the times T^E and T^L are replaced with the selected event time T. When preparing this specific schedule, a consideration of floats becomes of major importance. The adopted schedule for activity A_{ij} is then

Duration	t_{ij}
Start time	T_I
Earliest completion time	$T_I + t_{ij}$
Latest completion time	T_J
Scheduled float available	$T_J - (T_I + t_{ij})$

$$(7.2)$$

The determination of specific event times depends on many factors, which include the availability of resources, satisfactory manpower requirements, available float times, management decisions, and the general pattern of the project. The derivation of these definite event occurrence times is best illustrated by a simple example, which is now considered in detail.

Figure 7.1a is a small arrow network diagram for a project, showing activity durations, floats, and the critical path, including the times T^E and T^L for each event. The manpower and equipment resources required for each of the project activities, and which determine the activity durations, are shown in Figure 7.1b, where activity 1–2, for instance requires 10 workers plus equipment A (for its duration of 8 days), whereas activity 2–5 needs a work force of 15 men and equipment B, and so on. It is thus seen that each equipment A and B is individually required for two separate activities. (As an alternative to employing a second network to show required resources, some planners prefer to use appropriate notes and symbols, or coded data, on the main network diagram; the method adopted is immaterial.)

An important factor in scheduling is the classification of activities into two groups, continuous or intermittent. When a *continuous operation* activity is started, it must be worked without interruption until finished; an *intermittent operation* activity can proceed piecemeal in isolated sections at irregular periods of time. This division becomes vital when a choice is to be made whether it is advantageous to do part of an activity at a particular time and the balance later. In this example activity 3–6 in Figure 7.1b is assumed to be an intermittent operation, whereas all others require continuous operation from start to finish.

Examination of Figure 7.1 shows that many possibilities exist for scheduling the noncritical activities, within the limits imposed by the network diagram (a) and the resources (b). Each possibility selected implies a definite rate of application of resources (manpower and equipment) and of consumption of the necessary materials.

One obvious schedule is to begin every activity as soon as possible. This

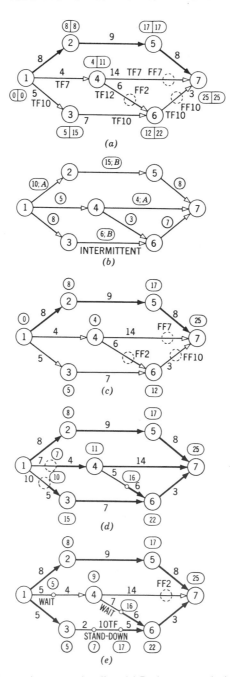

Figure 7.1   Scheduling and resource leveling. (*a*) Project network showing durations and floats. (*b*) Project resources—manpower and equipment. (*c*) Earliest start network. (*d*) Latest start network. (*e*) Network for compromise schedule, after resource leveling.

earliest start schedule is presented in Table 7.1 and shown diagrammatically in Figure 7.1c. In the table a separate row is used for each resource needed by each activity (W for workers, *E* for equipment); free float times appear as the time gap between the last scheduled day for a chain of activities and the terminating event time for that chain. All activities are scheduled for their normal durations; no float is scheduled for any particular activity, and therefore free float becomes a future safety margin only. This schedule requires a definite daily application of resources as indicated at the bottom of the tabulation; the labor force varies from 8 to a maximum of 28 (for 2 days only), with an average of 19 workers; equipment *A* is used continuously for 18 days, two sets being required for 4 days, whereas equipment *B* is required continuously for 12 days, with 2 sets being essential for 4 days. When the running (cumulative) totals are computed—as shown in the last three rows of Table 7.1—the rate of application of resources can be shown graphically by plotting this cumulative total consumption against project duration, as in Figures 7.2 and 7.3. In these diagrams, the lines marked "earliest start" refer to the data listed in Table 7.1. Obviously, the steeper the slope, the more intensive is the use of a resource.

Another schedule is starting each activity as late as possible. This latest start schedule appears in both Table 7.2 and Figure 7.1*d;* the rate of application of resources is plotted as "latest start" in Figures 7.2 and 7.3. The work force varies from 10 workers to a maximum of 28 (for 2 isolated days), the average again being 19. Equipment *A* is used for an 8-day and a 14-day period, separated by a 3-day stand-down; one set of equipment is required. Equipment *B* is required continuously for 13 days, two sets being necessary for 2 consecutive days. It will be seen from the table that all floats are discarded and hence all chains become critical—with this schedule a delay in any activity delays the project completion.

Comparison of the two schedules in Figures 7.2 and 7.3 shows that the earliest start schedule for all resources requires a heavy initial rate of investment (application of resources), which decreases considerably later in the project. The latest start schedule, on the other hand, permits a small initial rate of investment but requires a considerable increase later. It is therefore apparent that the free float available with the former schedule is purchased, in the form of a safety margin, at the cost of a heavy initial rate of expenditure; and that the low initial rate of expenditure for the latter schedule is obtained at the cost of discarding all float and accepting the risk of a completely critical network.

7.2 USE OF FLOATS IN ACTIVITY SHIFTING FOR RESOURCE LEVELING

It will be obvious that a compromise between these two schedules (earliest and latest start) for the project described in the preceding section should

Table 7.1 Earliest start schedule

Activity	W/E	1	2	3	4	5	6	7	8	9	10	11	12	13	14	15	16	17	18	19	20	21	22	23	24	25
1-2	W	10	10	10	10	10	10	10	10																	
1-2	E	A	A	A	A	A	A	A	A																	
2-5	W									15	15	15	15	15	15	15	15	15								
2-5	E									B	B	B	B	B	B	B	B	B								
5-7	W																		8	8	8	8	8	8	8	8
1-4	W	5	5	5	5																					
4-7	W					4	4	4	4	4	4	4	4	4	4	4	4	4								
4-7	E					A	A	A	A	A	A	A	A	A	A	A	A	A								
4-6	W					3	3	3	3	3	3															
1-3	W	8	8	8	8	8																				
3-6	W						6	6	6	6	6	6	6													
3-6	E						B	B	B	B	B	B	B													
6-7	W													7	7	7										
Total Workers		23	23	23	25	23	23	23	28	28	28	25	25	26	26	26	19	12	8	8	8	8	8	8	8	8
Total A		1	1	1	1	0	1	2	2	1	1	1	1	1	1	1	1	1	0	0	0	0	0	0	0	0
Total B		0	0	0	0	0	1	1	2	2	2	2	2	1	1	1	1	1	1	0	0	0	0	0	0	0
Running Totals — Workers		23	46	69	92	117	140	163	186	214	242	267	292	318	344	370	389	418	420	428	436	444	454	460	468	476
Running Totals — Equipment A		1	2	3	4	4	6	8	10	13	14	15	16	17	18	19	20	21	22	22	22	22	22	22	22	22
Running Totals — Equipment B		0	0	0	0	0	1	2	3	5	7	9	11	12	13	14	15	16	16	16	16	16	16	16	16	16

Annotations on the bar chart:
- Critical Path Activities 1-2-5-7 (activities 1-2, 2-5, 5-7); No Floats
- Noncritical Chain 1-4-7 — Free Float of 7 Days
- Free Float of 2 Days in Chain 1-4-6; l_{1-4}
- Noncritical Chain 1-3-6-7 / Chain 1-3-6-7 — Free Float of 10 Days

135

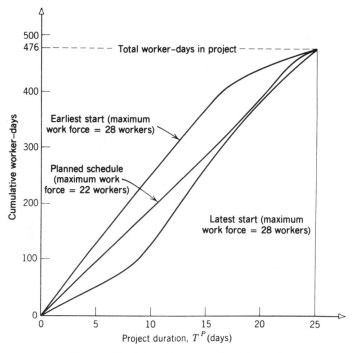

Figure 7.2   Rate of investment of labor only.

result in a more constant work force (and hence a more uniform rate of investment) over the whole project duration and a more efficient usage of equipments *A* and *B*. The development of such a compromise schedule is known as *resource leveling*.

The procedure for resource leveling is shown in Table 7.3. Here the managerial constraint (or strategy) adopted is to give first preference to the leveling out of the equipment requirements and second preference to maintain a fairly constant labor force. The critical activities (having no float) must be scheduled first since their event times are fixed. The tabulation is therefore so presented that chains of activities are considered in order of increasing total float.

The resource requirements of the critical activities 1–2, 2–5, and 5–7 are thus recorded first in the first three rows of Table 7.3, numbered (I.) Next are the noncritical chains, which may occur at any time within the limitations of their available floats; that is, their starting and finishing dates are adjustable in the program anywhere within the range between the timing of their initial and final events provided their durations are not altered. This moving of activity times, within the range of their available floats, is called *activity shifting* and is the essential procedure for resource leveling.

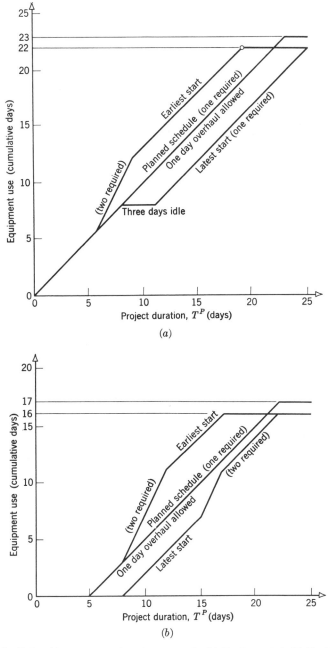

Figure 7.3   Rate of investment of equipment only. (a) Equipment A. (b) Equipment B.

Table 7.2 Latest start schedule

Activity	W/E	1	2	3	4	5	6	7	8	9	10	11	12	13	14	15	16	17	18	19	20	21	22	23	24	25
1-2	W	10	10	10	10	10	10	10	10																	
1-2	E	A	A	A	A	A	A	A	A																	
2-5	W									15	15	15	15	15	15	15	15	15								
2-5	E									B	B	B	B	B	B	B	B	B								
5-7	W																		8	8	8	8	8	8	8	8
1-4	W								5	5	5	5														
4-7	W												4	4	4	4	4	4	4	4	4	4	4	4	4	4
4-7	E												A	A	A	A	A	A	A	A	A	A	A	A	A	A
4-6	W																	3	3	3	3	3	3			
1-3	W											8	8	8	8	8										
3-6	W																6	6	6	6	6	6	6			
3-6	E																B	B	B	B	B	B	B			
6-7	W																							7	7	7
Total Workers		10	10	10	10	10	10	10	15	20	20	28	27	27	27	27	25	28	21	21	21	21	21	19	19	19
Total A		1	1	1	1	1	1	1	1	0	0	0	1	1	1	1	1	1	1	1	1	1	1	1	1	1
Total B		0	0	0	0	0	0	0	0	1	1	1	1	1	1	1	2	2	1	1	1	1	1	0	0	0
Running Totals — Workers		10	20	30	40	50	60	70	85	105	125	153	180	207	234	261	286	314	335	356	377	398	419	438	457	476
Equipment A		1	2	3	4	5	6	7	8	8	8	8	9	10	11	12	13	14	15	16	17	18	19	20	21	22
Equipment B		0	0	0	0	0	0	0	0	1	2	3	4	5	6	7	9	11	12	13	14	15	16	16	16	16

Chart annotations:

- Critical Path Activities 1-2-5-7
- No Floats
- Free Float and Total Float of 7 Days Discarded
- Ind. F = 5 Days Discarded — t_{1-4}
- Noncritical Chain 1-4-7 Now Critical
- Free Float and Total Float of 10 Days Discarded
- Noncritical Chain 1-3-6-7 Now Critical

Table 7.3 Schedule for minimum equipment and then for minimum men.

Activity	W/E	1	2	3	4	5	6	7	8	9	10	11	12	13	14	15	16	17	18	19	20	21	22	23	24	25
1-2	W	10	10	10	10	10	10	10	10																	
1-2	E	A	A	A	A	A	A	A	A																	
2-5	W								15	15	15	15	15	15	15	15	15									
2-5	E								B	B	B	B	B	B	B	B	B									
5-7	W																		8	8	8	8	8	8	8	8
1-4	W						5	5	5	5	5																*(TF 5 Days — Discarded)*
4-7	W																										*(Discarded)*
4-7	E									(A)																	*(Overhaul A)*
4-6	W									A	A	A	A	A	A	A	A	A	A	A	A	A	A	A	A	A	*(Overhaul A / TF 7 Days Discarded)*
										4	4	4	4	4	4	4	4	3	3	3	3	3	3	4	4	4	*(FF 2 Days)*
1-3	W			8	8	8	8	6	6	6	6	6	6	6	6	6	6	6	6								
3-6	W						6	6	(B)	B	B	B	B	B	B	B	B	B	B								*(Stand-down = 10 Days; TF of 10 Days Used)*
3-6	E						B	B	B	B	B	B	B	B	B	B	B	B	B								*(Overhaul B)*
6-7	W																						7	7	7	7	*(No Overhaul B Allowed / B B)*
(I)	Total Workers	10	10	10	10	10	10	10	10	15	15	15	15	15	15	15	15	15	8	8	8	8	8	8	8	8
	Total A	1	1	1	1	1	1	1	1	0	0	0	0	0	0	0	0	1	0	0	0	0	0	0	0	0
	Total B	0	0	0	0	0	0	0	0	1	1	1	1	1	1	1	1	0	0	0	0	0	0	0	0	0
(II)	Total Workers	10	10	10	10	10	10	10	10	15	15	19	19	19	19	19	19	19	12	12	12	12	12	12	8	8
	Total A	1	1	1	1	1	1	1	1	0	0	1	1	1	1	1	1	1	1	1	1	1	1	1	0	0
	Total B	0	0	0	0	0	0	0	0	1	1	1	1	1	1	1	1	0	0	0	0	0	0	0	0	0
(III)	Total Workers	18	18	18	18	18	16	16	10	10	19	19	19	19	19	19	19	19	18	18	18	18	18	19	19	15
	Total A	1	1	1	1	1	1	1	1	1	1	1	1	1	1	1	1	1	1	1	1	1	1	1	1	0
	Total B	0	0	0	0	1	1	1	1	1	1	1	1	1	1	1	1	1	1	1	1	1	1	0	0	0
(IV)	Total Workers	18	18	18	18	18	21	21	15	20	19	19	19	19	19	19	19	22	21	21	21	21	21	19	19	15
	Total A	1	1	1	1	1	1	1	1	1	1	1	1	1	1	1	1	1	1	1	1	1	1	1	1	0
	Total B	0	0	0	0	0	1	1	1	1	1	1	1	1	1	1	1	1	1	1	1	1	1	0	0	0
Running Totals	Workers	18	36	54	72	90	111	132	147	167	186	205	224	243	262	281	300	322	343	364	385	406	427	446	461	476
	Equipment A	1	2	3	4	5	6	7	8	9	10	11	12	13	14	15	16	17	18	19	20	21	22	23	23	23
	Equipment B	0	0	0	0	0	1	2	3	4	5	6	7	8	9	10	11	12	13	14	15	16	17	17	17	17

139

The chain with the smallest total float is 1–4–7. This chain contains activity 4–7, which requires equipment A and 4 workers for 14 continuous days. This activity is therefore scheduled first in this chain to provide continuous use of equipment A. However, since this equipment was used immediately before in the critical activity 1–2 and may need overhauling after its constant use maintaining the timing of event 2, a 1-day overhaul of equipment A is scheduled before beginning activity 4–7. Activity 1–4 (requiring 4 days) is left floating at this stage, since 9 days are available to it and its scheduling may be suited later to smoothing out manpower requirements. Notice that the scheduling of this chain has permitted the continuous use of equipment A, with a 1-day safety factor for overhaul and 2 days of free float still available to offset unforeseen setbacks. The resource situation is now updated, as shown in the three rows of Table 7.3 numbered (II), activity 1–4 is omitted at this stage since it has not been scheduled.

Chain 1–3–6–7 is considered next since its total float is now the smallest. This chain includes activity 3–6, which needs 6 workers and equipment B for a total of 7 days (not necessarily consecutive). Activity 1–3 requires 5 days, which permits only 3 days use of equipment B before the critical activity 2–5 starts (it also uses this equipment). It is therefore better, since activity 3–6 cannot be postponed completely until after day 17, to carry out this activity intermittently. Consequently, there is an allowance of 2 days use of equipment B on activity 3–6, with a 1-day overhaul before starting the critical activity 2–5, proceeding to complete activity 3–6 after finishing 2–5. This decision puts all the available total float into activity 3–6 as a stand-down of 10 days; consequently, since no float is left for activities 1–3 and 6–7, they are now scheduled. The situation then becomes that numbered (III) in Table 7.3. Activity 1–4 is still omitted and a continuous operation for one set of equipment B has been achieved.

The scheduling of activity 1–4 (5 workers for 4 days) and activity 4–6 (3 workers for 6 days) is now determined. Activity 1–4 must be finished by the ninth day; if its start is delayed by 5 days, a smoother total labor demand is obtained for the project. Activity 4–6 could start on the tenth day and must be finished by the twenty-second day; it therefore has a float of 7 days, which can best be used for labor leveling by discarding it, thus delaying the start of activity 4–6 by 7 days. With scheduling finalized in this way, the final resource situation is derived in the rows numbered (IV) in Table 7.3. Figure 7.1*e* shows the network diagram corresponding to this compromise schedule.[1]

---

[1] The deliberate delay to the start of an activity, or to its continuous execution, introduces into the network diagram a number of *artificial "delay" activities* (having time but no cost) within the length of the original activities. These are seen in activities 1–4, 4–6, and 3–6 in Figure 7.1*e*. On completing activity shifting and resource leveling procedures, the final network diagram thus contains a variety of artificial activities and artificial events, and the nodes must be renumbered from start to finish of the project before tabulating the final works schedule.

The rate of application of resources for the compromise schedule, plotted in Figures 7.2 and 7.3 as "planned schedule," shows that reasonably good resource leveling has been obtained using minimum equipment and a fairly constant (and smaller) work force. One set of equipment A is required for 23 days, with a 1-day overhaul after critical activity 1–2, and with a 2-day free float in case trouble is encountered in activity 4–7. Equipment B (one set only) is used continuously for 17 days, with a 1-day overhaul just before starting critical activity 2–5. The labor force varies from 15 to 22 workers, with an average of 19 workers per day. This compromise schedule therefore almost satisfies the ideal situation and still provides for unforeseen delays. The leveling of resources has been well worthwhile.

Activity 6–7 (7 people for 3 days, and now critical) may, if desired, be assigned some extra labor, either to shorten its normal duration or to crash it partially if it is delayed. In other words, a reassessment—on the basis of a constant work force of 22 workers—could be made in the normal durations of some activities to recreate free floats for the chains 1–4–6 and 1–3–6–7, if this is deemed desirable. Whether this is done or not, the ultimate schedule and network diagram now become the construction plan, and actual dates can be assigned to every activity and event. It is then a simple matter to prepare a time-scaled network or a critical path bar chart construction program for the project; indeed the tabulation (Table 7.3) is itself one form of program chart.

With resource leveling of large projects, it is helpful to tabulate the separate labor skills and plot the manpower required for each against time; this may be done when the original schedule is reviewed and amended. A similar plot may be prepared for each item of equipment. The procedure is somewhat complicated because of the interrelations between activities; but the effort is amply rewarded by the production of a detailed works schedule in which problems of labor and equipment and financial fluctuations are overcome as far as practicable.

In chains of activities involving numerous operations, with various labor requirements and skills and different types of equipment, the most satisfactory way of carrying out detailed resource leveling is to begin with the labor craft (or major equipment item) having the greatest fluctuations. Smoothing this out first, *without regard to other skills or equipment,* one then proceeds with a second craft (or piece of equipment), and so on, one skill (or machine) at a time, until a satisfactory overall labor and equipment force is attained. With complex chains the same craft or machine may have to be reviewed several times and relevant activities shifted and reshifted before an acceptable solution appears. As pointed out, the best approach is first to shift those activities with small float and then those with large float.

In drawing up the tabulation, before beginning resource leveling, it is often helpful to sketch in (as on a critical path bar chart) the time ranges available for all the noncritical activities; this will ensure not making any incompatible decisions during the activity shifting procedure, especially when reshifting is found necessary. An alternative aid, preferred by some planners, is to block out all times *not* available for the execution of each activity. Finally, once the ultimate works schedule is determined, financial expenditure and income may also be added to the resource leveling tabulation, as demonstrated in Section 10.2; in this way the estimator can present a complete picture of all the essential resources in strict conformity with the network model.

7.3   SCHEDULING FOR TIME LIMITATIONS

Normally, CPM is used to plan a project so that it will be carried out in the minimum economic time; it has been shown in Chapters 5 and 6 how compression and decompression enable the project duration (and cost) to be varied to determine this time. In many projects, however, certain events must be completed by a specified date or within a stipulated period. To enable the network analysis to be valid under such circumstances, it is necessary to show such time constraints on the diagram by means of artificial activities.

Figure 7.4 shows a project network in which it is specified that event 7 must be achieved within 20 weeks of the commencement of work, event 15 is to be achieved within 12 weeks thereafter, and the entire project must be completed in not more than 48 weeks. The activities 0–7, 7–15, and 15–20 are therefore artificial activities denoting such time constraints, the total durations of these three equaling the specified project duration of 48 weeks. The relevant events 7, 15, and 20 are known as *milestone events,* and are shown as double circles to distinguish them from the ordinary nodes.

This network would first be analyzed in the usual way to determine the most economical solution for the scheduling of events 7, 15, and 20 (ignoring the presence of the three time constraint activities). If this solution conforms with the three time limitations, it is acceptable; if not, then each subnetwork is crashed as appropriate until the artificial time activities are all critical.

It will be clear that this use of artificial time limitation activities enables the duration between any two or more events to be controlled.

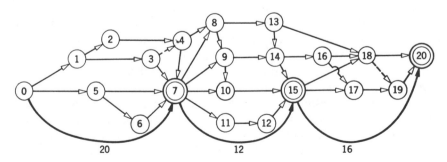

Figure 7.4   Use of artificial activities and milestone events to show time limitations on a project network diagram.

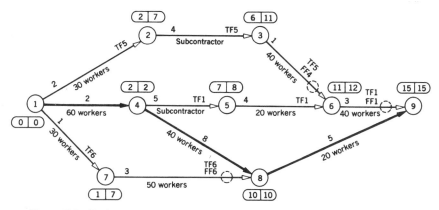

Figure 7.5   Project network and labor requirements, before resource leveling.

7.4   EFFECT OF RESOURCE LIMITATIONS ON PROJECT DURATIONS: RESOURCE SCHEDULING

Inherent in the initial approach to critical path network analysis is the assumption that the resource requirements for the various activities can be met, *whenever required,* from the available project resources, independently of the requirements of other concurrent activities. If this is not the case, and if the network diagram does not properly so indicate in its logic (such as by the use of crew and equipment constraints), then the basis for the determination of the critical path will be invalid. When resources are strictly limited, the situation can arise where chains that are otherwise noncritical must exceed their total float while waiting for their special resources to become available, thus delaying the entire project.

Consider the case shown in Figure 7.5 in which activities 2–3 and 4–5 are to be done by subcontract, with other activities requiring crew sizes as indicated. The minimum project duration is 15 weeks, and all activities are classified as continuous operations. It can be shown (excluding the subcontractor's forces) that the maximum number of workers required at any stage of the project to meet this duration with an earliest start schedule is 140 people; by resource leveling with the available floats this can be reduced to a maximum of 90 workers, using the procedure described in Section 7.2. Suppose, however, that for some reason (for example, a management constraint) it is necessary to restrict the maximum manpower requirement to 70 workers in addition to the subcontractor's forces. The procedure follows that ex- plained in Section 7.2, except that, when float is no longer available, critical and noncritical activities must be further shifted and reshifted until the mini- mum project duration is achieved without exceeding the imposed manpower limitation.[1]

---

[1] In some recent CPM computer packages, such activity shiftings can be handled graphically on the computer screen.

Table 7.4 Schedule for minimum project duration with not more than 70 people

Activity	Duration	Crew Size	1	2	3	4	5	6	7	8	9	10	11	12	13	14	15	16	17	18
1–4	2	60	60	60																
4–8	8	40			40	40	40	40	40	40	40	40								
8–9	5	20											~~20~~	~~20~~ *(b)* ~~20~~	~~20~~	20	20	20	20	20
4–5	5	nil			×	×	×	×	×											
5–6	4	20								20	20	20	20	*(c)*						
6–9	3	40												~~40~~ ~~40~~ *(c)*	~~40~~	40	40	40	40	
1–2	2	30	~~30~~ ~~30~~ *(a)*	~~30~~ ~~30~~	30	30														
2–3	4	nil			×	× *(a)*	×	×	×	×										
3–6	1	40						40 *(a)*	40											
1–7	1	30	~~30~~ *(a)*	30			30						40			40				
7–8	3	50	~~50~~ *(a)*	~~50~~	~~50~~ ~~50~~	50							50 *(b)*	50	50					
(I) after (a)			60	60	70	70	70	40	40	110	110	110	80	60	60	60	20	20	20	
(II) after (b)			60	60	70	70	70	40	40	60	60	60	110	90	90	60	20	20	20	20
(III) after (c)			60	60	70	70	70	40	40	60	60	60	70	50	50	60	60	60	60	20

Table 7.4 shows the solution to this resource scheduling problem. The critical activities are scheduled first, followed by noncritical activities in order of increasing float, all entered in the tabulation for the earliest start schedule in the first instance. The first step, designated (*a*) in Table 7.4, is to shift activities 1–2 (and 2–3), 3–6, 1–7 and 7–8 within the limits of their available float, since critical activity 1–4 needs 60 people and all other activities require more than 10 workers; manpower is now satisfactory to the end of week 7, the position shown in row (I) in the table. The next step (*b*) is to shift activity 7–8 (and hence critical activity 8–9) beyond its available float to reduce the labor requirement below the specified 70 until the end of week 10, the smallest shift of critical activity 8–9 being 3 weeks (that is, the float overrun of activity 7–8). This current situation is shown in row (II) in the table. Since the project duration is now 18 weeks, it is obvious that activities 3–6 and 6–9 have gained 3 weeks additional float, and can be further shifted within this amount as shown in step (*c*). The situation is now summarizerd in row (III) of the table and satisfies the imposed labor limitation. Hence the minimum project duration under these circumstances is 18 weeks, an increase of 3 weeks due solely to the resource limitation. If desired, the subcontract activity 2–3 can be further shifted so that it will follow activity 4–5, for 6 weeks float is still available to it. With this additional subcontractor constraint, the final network for this project will be as seen in Figure 7.6; the nodes must, of course, be renumbered. It is important to note that the network must be reanalyzed for float each time available floats are exceeded.

An alternative resource leveling procedure is to examine the project in successive steps of one time unit at a time, shifting chains of activities as required at each step until a valid solution appears for each successive time unit examined; activity-shifting begins with the chain having the largest total float. To illustrate this method, consider again the problem shown in Figure 7.5. Table 7.5 is prepared with all activities scheduled for EST in their order

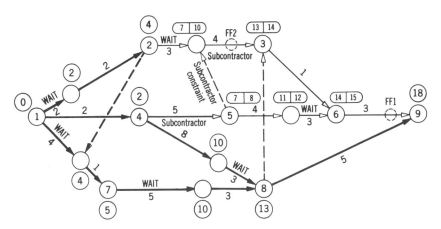

Figure 7.6 Final network for project in Figure 7.5, after imposing the personnel limitation of 70 workers.

Table 7.5 Resource leveling with worker limitation of 70 people by unit time steps

Activity	Duration	Crew Size	1	2	3	4	5	6	7	8	9	10	11	12	13	14	15	16	17	18
1-4	2	60	60	60																
4-8	8	40			40	40	40	40	40	40	40	40								
8-9	5	20											20	20	20	20	20			
4-5	5	nil			X	X	X	X	X											
5-6	4	20								20	20	20	20							
6-9	3	40												40	40	40				
1-2	2	30	~~30~~	~~30~~	30	30														
2-3	4	nil		~~X~~	~~X~~	~~X~~	X	X	X	X	40									
3-6	1	40							~~40~~	~~40~~	40									
1-7	1	30	~~30~~	~~30~~	30															
7-8	3	50	~~50~~	~~50~~	50	50	50	50												
Workers for EST			120	140	90	90	40	40	80	60	60	60	40	60	60	60	20			
Workers after step 1			60	120	120	90	90	40	40	100	60	60	40	60	60	60	20			
Workers after step 2			60	60	100	120	90	90	40	60	100	60	40	60	60	60	20			

Development beyond Step 2 not shown

146

of increasing float (as in Table 7.4), and the EST personnel requirements are listed for each time unit. It is seen immediately that the people limitation of 70 workers is violated from the outset (week 1). Hence the chains 1–7–8 and 1–2–3–6 (in that order) are shifted 1 week to the right to give a valid solution for week 1: the personnel position is then as shown at the bottom of Table 7.5 after step 1. The next step is to examine week 2; chains 1–7–8 and 1–2–3–6 are again shifted one week to give the worker position noted in Table 7.5 after step 2. For steps 3 and 4 (weeks 3 and 4) it is necessary only to shift chain 1–7–8 to reduce the personnel requirement to 70 workers. These and subsequent steps are not shown in the table, it being left to the reader to develop the final solution as an exercise;[1] this will appear as designated in row (III) of Table 7.4.

The initial stages, outlined above, of the superposition of a limited resource logic for the project of Figure 7.5 can be represented on the network diagram. Figure 7.7 shows the situation initially (*a*); and at the end of the second (*b*) and third (*c*) weekly time intervals. Both arrow and circle networks are shown, and it is clear that for changes in logic, circle notation is superior. Note that it may be possible to suppress initial technology logic (now superseded by resource logic) to reduce the number of arrows in the diagrams; however, it must be recognized that if this reduction is effected, some information is lost and may have to be regenerated at a later stage.

The resource scheduling procedure outlined above can be generalized and formulated as follows: Consider two activities *A* and *B* as shown in Figure 7.8 in general CPM bar chart form. Assuming that a resource conflict occurs for *A* and *B* only when both activities are working during any specific time interval and not otherwise, then the conflict can be resolved by either scheduling *B* to follow *A*, or *A* to follow *B*. Obviously, it is better to select the sequence that increases the project duration the least.

If *B* is to follow *A*, then the best strategy is to schedule *A* as soon as possible (that is, at EST_A) and *B* as soon as *A* is finished (that is, at EFT_A). The increase in project duration T_P then becomes

$$\Delta T_P \,(A \text{ then } B) = (EFT_A - LST_B) \quad \text{if} \quad LST_B < EFT_A \atop = 0 \qquad\qquad\qquad \text{if} \quad LST_B \geq EFT_A \right\} \quad (7.3)$$

The second equation means that no project extension results since *B* can be shifted after *A* within the float available to *B*.

---

[1] For week 5 no shifting is required. For weeks 6 and 7 only activity 7–8 is shifted, reaching the end of its float and becoming critical. No shifting within remaining floats will satisfy week 8; consequently, the critical path and project time must be extended. For week 8 shift activity 7–8 (and 8–9, both critical), and for weeks 9 and 10 shift activities 7–8 (with 8–9) and 3–6; notice that the chains 1–2–3–6 and 4–5–6–9 have now gained 3 weeks additional float. Hence for weeks 11, 12, and 13 shift activities 3–6 and 6–9; weeks 14 to 18 will also be satisfactory, and a valid solution has been found. The leveling of subcontract activities 4–5 and 2–3 can follow as before.

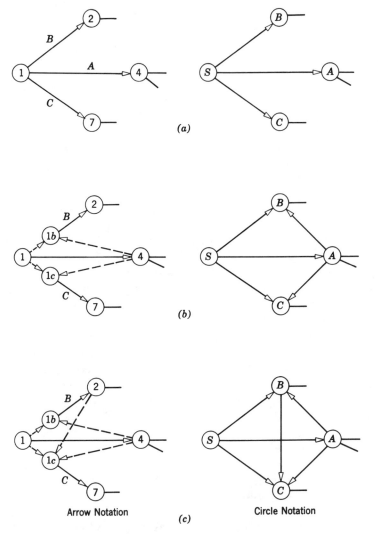

Figure 7.7 The staged superposition of limited resource logic for the project of Figure 7.5.

Similarly, if A is to follow B, then the increase in project duration becomes

$$\left. \begin{array}{ll} \Delta T_P \; (B \text{ then } A) = (\text{EFT}_B - \text{LST}_A) & \text{if} \quad \text{LST}_A < \text{EFT}_B \\ = 0 & \text{if} \quad \text{LST}_A \geq \text{EFT}_B \end{array} \right\} \qquad (7.4)$$

The best scheduling sequence is then readily determined.

If more than two activities are simultaneously involved in a resource conflict, a systematic evaluation, similar to that indicated above, becomes

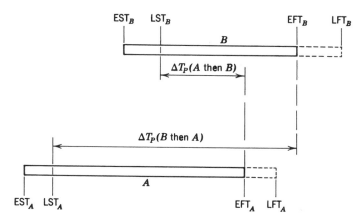

Figure 7.8   Increase in project duration due to sequence scheduling.

necessary for each possible scheduling sequence. Since the result of each search is to order two activities only, the process may have to be repeated again and again until the resource conflict has been removed. Before proceeding to the next time interval, the normal CPM calculations must be repeated on the new network model to obtain updated values of EST, LST, EFT, LFT for the network activities. The method then continues on through all the remaining time intervals in the project.

It is interesting to note that if the resource scheduling strategy is adopted of not shifting "in progress" activities, then equations 7.3 and 7.4 shift resource-conflict activities in the total float order. Thus if activities B and C simultaneously enter in resource conflict with "in progress" activity A, then equations 7.3 and 7.4 yield:

$$\left.\begin{aligned}\Delta T_P \text{ (A then B)} &= (\text{EFT}_A - \text{LST}_B) \\ \Delta T_P \text{ (A then C)} &= (\text{EFT}_A - \text{LST}_C)\end{aligned}\right\} \tag{7.5}$$

and since
$$\text{EST}_B = \text{EST}_C$$

and
$$\text{TF}_B = \text{LST}_B - \text{EST}_B$$
$$\text{TF}_C = \text{LST}_C - \text{EST}_C$$

there results

$$\left.\begin{aligned}\Delta T_P \text{ (A then B)} &= \text{EFT}_A - \text{EST}_B - \text{TF}_B = \text{constant} - \text{TF}_B \\ \Delta T_P \text{ (A then C)} &= \text{EFT}_A - \text{EST}_C - \text{TF}_C = \text{constant} - \text{TF}_C\end{aligned}\right\} \tag{7.6}$$

Thus the activity with largest total float is shifted in sequence first.

Limitations of manpower, finance, equipment, or other resources frequently occur in practice. Sometimes these resource limitations can be incorporated into the network logic by means of dummies or allowed for by skillful resource leveling. When this is not possible, the project duration is increased and the usual critical path (based on unlimited resources) may be radically altered, as can be seen by comparing Figure 7.6 with Figure 7.5. It is therefore obvious that, in some realistic diagrams, the critical path, although still being the most time-consuming path through the network, is not easily determined, for it depends on the availability of a wide variety of resources and on the network characteristics.

For a limitation in maximum allowable financial expenditure during construction, the procedures are similar to those for a maximum personnel limitation, using either of the methods illustrated in Tables 7.4 and 7.5, except that the estimated cost per unit time for each activity is substituted for the crew size in compiling the tabulation. Activity shifting and summation of requirements will follow, until the final totals are found to lie within the specified financial limitation. With equipment limitations either of these procedures can be adopted, and indeed personnel and equipment limitations can often be handled simultaneously (as done in Table 7.3), whether available floats have to be exceeded or not. It is simply a matter of logical activity shifting and reshifting until the desired result appears.

Another method of handling equipment limitations, or a limitation in the availability of specialist technical crews, is illustrated by considering the small network diagram shown in Figure 7.9a. Associated with each activity is a duration t_{ij} and a resource requirement R_{ij} for its duration. Figure 7.9b presents the development of the critical path in the usual way, based on the assumption that unlimited resources are available. The actual resources necessary for this project duration of 32 days are two sets of equipment I and one set of equipment II, as presented in Figure 7.9c. The second set of equipment I is required for only approximately one-fifth of the project time, and this requirement cannot be reduced by activity shifting within the available floats.

Now suppose that only one set of equipment I exists (or can be made available). Consequently, the network diagram in Figure 7.9b is logically incorrect and that of Figure 7.10 must be substituted. Notice that considerable alterations have been made to ensure that correct logic holds for the special resource limitations now imposed on the project. The determination of the critical path then proceeds as usual, giving a project duration of 36 days. This increase of 4 days is attributable directly to the imposed resource limitation. The actual cost of this enforced delay can be determined, if necessary, for comparison with other alternative construction methods (if any are feasible). It is important to appreciate that both network diagrams (Figures 7.9b and 7.10) are logically correct in their respective portrayal of their different resource constraints. In both diagrams the critical paths are validly determined as critical time paths, that of the second network differing from the first solely because of the network characteristics and the resource constraints. The price paid for the imposition of the equipment limitation is thus the increase in project duration, as was the case for the personnel limitation example (Figures 7.5 and 7.6).

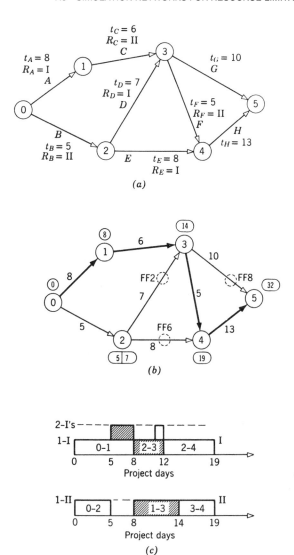

Figure 7.9 Effect of resource limitations. (a) Network diagram. (b) Critical time path (unlimited resources available). (c) $T_P = 32$ days. Actual resources required.

7.5 SIMULATION NETWORKS FOR RESOURCE LIMITATIONS

In projects with highly interconnected networks and varied resources, it is virtually impossible to anticipate and incorporate all the resource constraints that may become necessary. A realistic approach is then to formulate a network that resembles the usual diagram in its presentation of the valid construction logic pertaining to the activities and at the same time providing

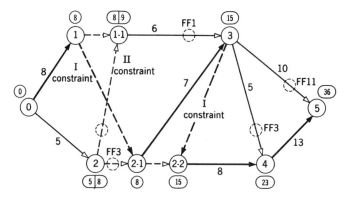

Figure 7.10   Critical time path. One set only of equipment I and II available.

for the control on network timing and critical paths dictated by the imposed resource limitations. Such a network is a special type of arrow diagram incorporating appropriate resource availability and utilization graphs. The analysis of the network proceeds under the constraints dictated by the resource graphs and thus has features related to simulation; consequently, this type of network is referred to as a *simulation network*.

In this network, each separate resource considered in the project has an individual resource availability graph, showing to what extent in time and numbers it is available in the project. During the scheduling of the work, as soon as an activity has been assigned a resource for a specific period, this resource utilization is recorded on the resource availability graph, thus indicating whether and when the particular resource will be available for other activities.

Before any activity using project resources can commence, the necessary resources must be available. Such activities in a simulation network are called *conditional activities*. The arrow representing this kind of activity incorporates near its start a *diamond-shaped "test box"*; its purpose is to question the availability of the essential resources required for this activity. For instance, suppose an activity needs n sets of equipment X in order to finish in its all-normal duration; the question at the test-box is: "Are n sets of the equipment X available *now?*" If the answer is "Yes," the activity can commence; if "No," the activity must wait, or be deferred, until the equipment becomes available. It therefore follows that, with this simulation network concept, the critical path determination proceeds hand in hand with resource scheduling and may have to be revised several times if unwanted characteristics develop.

To illustrate the technique for the problem postulated in Figure 7.10, a simulation network diagram, together with its associated resource availability and utilization graphs for the equipments I and II, is shown in Figure 7.11. The comparison of these two network diagrams demonstrates that the

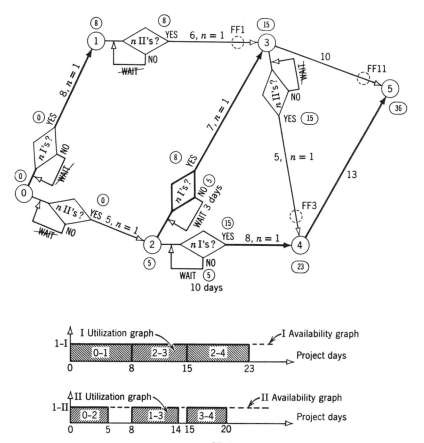

Figure 7.11 Simulation network diagram showing composite critical path. Critical time path activities are 0-1 and 4-5. Critical resource path activities are 2-3 and 2-4.

constraint dummies of Figure 7.10 (which were logically required for the specific resource allocation concerned) are no longer necessary. This simplification of the network is obtained, however, at the expense of continual close correlation of the usual network calculations with the relevant resource graph during the entire network analysis. In this particular problem, various schedules are possible, depending on the preliminary precedence in allocating resources to the activities; that used in Figure 7.11 gives preference to full utilization of equipment I and precedence to activity 2–3 over activity 2–4, both of which require equipment I.[1]

---

[1] Having mastered this technique, the reader should reschedule this project, giving precedence to the full utilization of equipment II as far as possible, and should also consider nonuniform resource availability graphs.

The method of calculation will be clear from the following logical approach. On starting with activity 0–1 (requiring one set of equipment I for 8 days), the answer as to equipment availability at day 0 is "Yes"; hence this activity proceeds and the availability and utilization graph for equipment I is blocked out for 8 days with activity 0–1, as shown. The situation is similar for activity 0–2, requiring 5 days use of equipment II. Proceeding to activity 1–3 (requiring equipment II for 6 days, starting at day 8), the answer from the availability and utilization graph for equipment II is "Yes," and thus it is scheduled as shown; notice that the EFT of event 3 by this chain 0–1–3 is 14 days at this stage of the calculations. When activity 2–3 is considered (requiring equipment I for 7 days), the answer from the graph at day 5 is "No," so the start of this activity must wait for 3 days (to day 8 when the answer will be "Yes"); this is recorded on the simulation network, making the EFT for event 3 now equal to 15 days (5 + 3 + 7). Similarly, for activity 2–4 the answer at day 5 is "No," but since at day 15 it will be "Yes," a 10-day wait is necessary. This procedure is repeated step by step throughout the network, until the final event is reached at day 36. Notice the intermittent use of equipment II and the full employment of equipment I.

The most interesting development is that the critical path is no longer a continuous chain through the simulation network; the usual continuity of Figure 7.10 has become the discontinuous path of Figure 7.11 because the resource constraint dummies are no longer inserted in the simulation network diagram. When resource limitations affect the project duration, the critical sequence becomes a combination of those activities lying on the *critical time path* and those on the *critical resource path* and is thus called the *composite critical path*.

The technique of using simulation networks can be employed for any project where fixed limitations may exist in manpower, equipment, finance, or other resources. The simpler problems may be solved by manual calculations as shown above; the complex situations will need computers.

7.6   COMPRESSION OF SIMULATION NETWORKS WITH RESOURCE LIMITATIONS

An interesting development in network compression techniques is the optimal compression of simulation networks with imposed resource limitations. Once resources are allocated to a project and the utility data computed, optimal compression calculations may proceed; however, with simulation networks, there is an added complication in that, if critical resource paths develop, additional limitations are imposed on available compressions by the temporarily (or permanently) unavailable resources.

An illustration of these compression problems is the simulation network model shown in Figure 7.12a; this is identical with the problem shown in Figures 7.10 and 7.11. It has been assumed in this case that the conditional

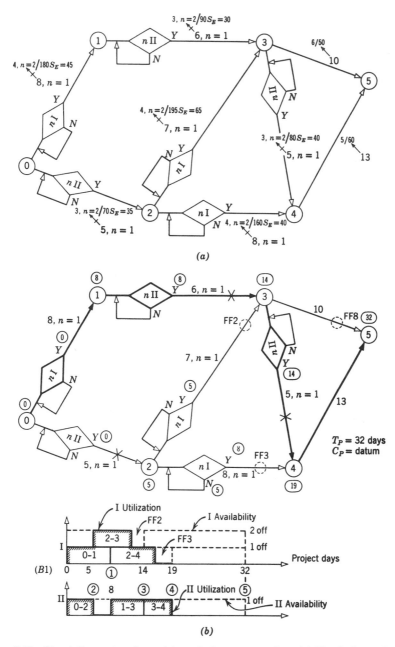

Figure 7.12   Simulation network model ready for compression. (a) Simulation network model—utility data. (b) Resource allocation B: all-normal solution (B1).

activities can be performed only in one of two ways: either in the all-normal manner using one piece of equipment or in a fully crashed manner using two pieces of equipment. Consequently, all conditional activity compressions are "jump" compressions. In the utility data of Figure 7.12*a* are shown the effective cost slopes S_E, together with the total cost of changing from the all-normal to the crashed method of performing each activity.

In this example, three resource allocations, *A, B, C,* for the project have been considered. Allocation *A* has one piece of equipment I and one piece of equipment II; *B* has two pieces of I and one of II; and *C* has two pieces of each equipment. To conserve space here, the network compression calculations are demonstrated for allocation *B* only with equipments I and II; calculations for the other two allocations are similar in principle. No other resources are introduced in this example, but the problem could, of course, be extended to include personnel, finance, etc., if these are subject to management constraints.

By using resource availability graphs for allocation *B*, the all-normal solution (*B*1) is developed in Figure 7.12*b*. This is characterized by each conditional activity using one piece of equipment only (although more are available) to avoid needless cost. The project duration T_P^N of 32 days is identical with that of Figure 7.9, because no resource limitation affects the critical time path. This all-normal (cheapest direct cost) solution for resource allocation *B* is assumed as datum for compression cost comparisons.

In considering this solution (*B*1), it will be noticed that because only one piece of equipment II is available to the project, activities 0–2, 1–3, and 3–4 cannot be speeded up any further and are therefore already crashed from a practical viewpoint. The critical path activities 0–1 and 4–5 only may consequently be considered for compression. Activity 4–5 has a potential compression of 5 days and a cost slope of 60 units per day. No network limitations exist (activity 3–5 has a free float of 8 days), and therefore the project duration can be shortened 5 days. Activity 0–1, on the other hand, has a potential "jump" compression of 4 days, a total cost of 180 units, and an effective cost slope of 45. However, the effect of this "jump" compression on the project duration is limited to 2 days (as a trial calculation will show); this is because the critical path switches to 0–2–3–4–5, leaving activity 0–1 with a 1-day FF. Hence the effective project cost slope from this reduced duration of 2 days is 90 units per day (4 days decrease in activity 0–1 at 45, divided by 2 days decreased in project duration).

Compression stages for this simulation network are therefore:

1. *Compress activity 4–5 by 5 days to full crash at a total cost of 300 units.* This solution (*B*2) is not illustrated, but is similar to *B*1, except that activity 4–5 now has a duration of 8 days.
2. *Compress activity 0–1 by a "jump" of 4 days, with an effective reduction in project duration of 2 days, at a total cost of 180 units.* This solution (*B*3) is shown in Figure 7.13*a*.

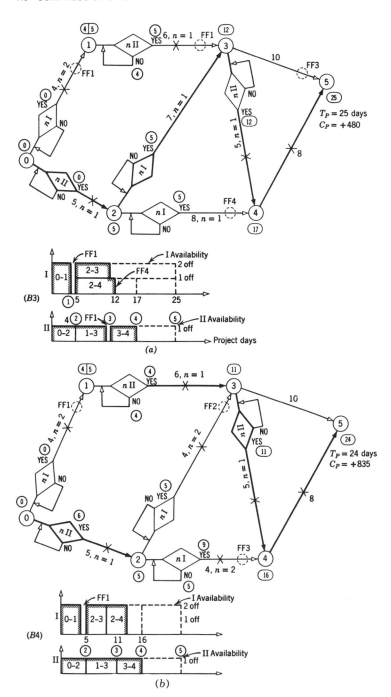

Figure 7.13 Compression of simulation network model. (a) Resource allocation B: second compression (B3). (b) Resource allocation B: final compression (B4).

The switched critical path is apparent and results from the necessity for the start of activity 2–3 to wait for satisfaction of event 2, notwithstanding the fact that all equipment I is idle from day 4 to day 5, as seen in resource graph I in Figure 7.13*a*. The new critical path in *B*3 contains only one uncrashed activity, 2–3. It can be "jump" compressed by 3 days if both sets of equipment I are used; however, a close study will show that, if this activity only is compressed, event 4 still occurs at day 17 and no reduction occurs in project duration. Thus despite the fact that event 3 can occur 1 day earlier (at day 11), activity 2–4 cannot begin until equipment I is free from activity 2–3 at day 9. Hence, unless activity 2–4 is also "jump" compressed, the crashing of activity 2–3 would be useless. Therefore:

3. *"Jump" compress activities 2–3 and 2–4 for a 1-day reduction in project duration, at a total cost of 355 units.* The result of this expensive compression is shown in Figure 7.13*b*. No further compression is possible.

The derivation of the optimal project direct-cost curve for resource allocation *B* is seen plotted in Figure 7.14 as well as those for resource allocations *A* and *C*.[1] The cost curves for *B* and *C* start with the same all-normal duration of 32 days, derived previously in Figure 7.9*b*. In other words, in this particular case, resource allocation *A* is the only one with a limitation preventing the usual CPM all-normal solution; its effect was previously seen in Figure 7.11. This has a most important practical significance in that it shows that the project would be underequipped if resource allocation *A* were employed.

In more general problems, families of curves would develop from the several possible restricted all-normal durations between the points corresponding to *A*1 and *B*1 in Figure 7.14 until the resource allocations exceed the minimum required for the usual CPM all-normal solution; thereafter, another family of curves would radiate from the real all-normal duration as the various resource allocations imposed their limitations on the compressions of the simulation network model. This latter feature is demonstrated by the way in which the *C* curve departs from the *B* curve and offers cheaper optimal solutions (in terms of direct cost) at the expense (indirect cost) of bringing to the site the additional resources provided (in this case, one extra set of equipment II).

The particular problem examined above is, of course, extremely simple. Uniform resource availability graphs were assumed, no activity shifting was required, and consequently no maneuvering was necessary within the resource utilization part of the graph; also the "jump" compressions, although a proper practical consideration, reduced considerably the number of possible compression stages. It is felt, however, that from this indication of the

---

[1] The reader may derive these as an exercise.

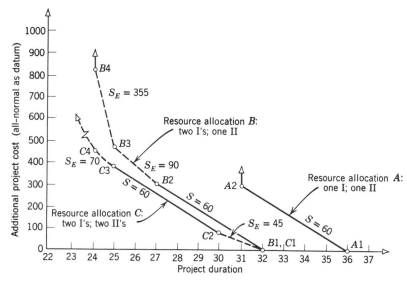

Figure 7.14 Project direct-cost/duration curves for various resource allocations.

nature of compression calculations for simulation networks, more complex problems can be undertaken by the reader.

In conclusion, it is apparent from Figure 7.14 that, in effect, a three-parameter problem is developing, with the variables of project duration, number of pieces of equipment I, and number of pieces of equipment II. This means that a multidimensional cost surface could be evolved, for which more general optimizing compression calculations are feasible, thus enabling computation of the optimal resource allocation for a given project duration. This development leads immediately to the field of digital computers, which will be considered further in Chapter 13.

7.7 OPTIMUM SOLUTION WITH RESOURCE LIMITATIONS

After deriving the project direct-cost curves for various resource allocations and computing the actual indirect costs for providing each of these resource allocations on site, the problem arises of selecting which allocation to adopt for the project. In other words, which resource allocation will give the best overall economical solution? The general method for determining this optimum solution, discussed in Section 5.8, can be readily applied to projects involving resource limitations, as the following exercise will show.

The project direct-cost data for the problem discussed in the preceding section are given in Figure 7.14. Taking the cost of providing resource allocation A as datum, assume that allocations B and C require additional indirect costs of 200 and 400 money units, respectively; assume also that the general indirect costs for the project are 80 money units per day. The project total cost curves can now be derived for each resource allocation and appears as shown in Figure 7.15. The minimum project total cost is for a project duration of 27 days, with resource allocation *B*. Hence this is the optimum solution.

7.8 AN OVERVIEW

The limited resource problem arises naturally in the CPM process because of the positions adopted during the planning and the manner in which the network model is developed. The emphasis on the separation of the planning and scheduling functions ensures an initial concentration on the development of the construction logic and the deferment of project resource considerations. Frequently, the deferment is total and tantamount to a blind acceptance of project resource requirements as dictated by the CPM schedule based on earliest starts. Often this attitude is discovered only the first time a schedule is

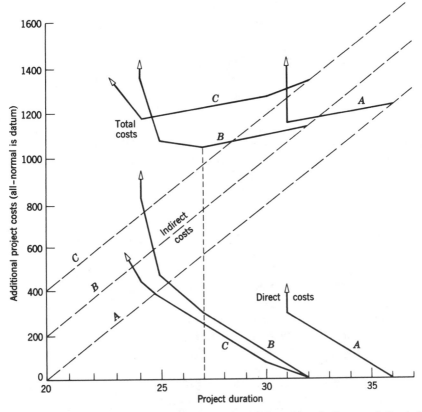

Figure 7.15 Project cost curves for various resource allocations. Optimum solution is 27 days duration with resource allocation *B*.

generated, which leads to a conflict between the resources that can be made available to a project and those apparently needed as determined from the network model.

The prime emphasis on the planning function in the CPM process means that there is a strong tendency to formulate the construction logic independent of project resource requirements. This tendency, coupled with the initial assignment of resources (implied or otherwise) to each individual activity during the activity utility data definition stage, gives little scope for resource considerations during the formative stages of the construction plan and the network model. In fact, the CPM process has been described as being based on the assumption of infinite resource availability.

The CPM process leads to two classical limited-resource problems, the resource leveling problem and the resource scheduling problem. The *resource leveling problem* adopts the network model and project duration based on the assumption of infinite resources and activity utility data, and searches for that schedule of the project activities that removes peak requirements of limited resources, or minimizes the number of certain specific limited resource types. The resulting resource requirements are then often adopted as being a natural attribute of the project, when, in fact, they are really a consequence of the specific construction plan and model developed independently of resource requirements.

The *resource scheduling problem* commences with the network model and a project duration based on the assumption of infinite resources and activity utility data, and seeks that minimum time extension of the project duration that will produce a practical schedule of the activities based on resource requirements that can be met from a defined resource availability while still maintaining the construction logic. This approach does not question the resources originally assigned to the activities nor their durations.

Currently, no formal mathematical model exists that considers the interaction between the various possible resource allocations per activity, the construction logic, and defined acceptable resource availabilities. Many heuristic models exist based on arbitrary criteria, which obtain feasible solutions by considering a limited view of the problem area and taking advantage of a particular structuring of the limited resources problem. None consider tradeoffs between different resource allocations for an activity and its effect on project resource requirements, and all require a new start (human-initiated) to provide feedback if unacceptable solutions result. Possibly the most successful attempts to date are those that incorporate repeated subjective input, in an on-line real-time mode on computers, which permit reevaluation of activity status, construction logic, and acceptable resource availabilities. In this way the user can repeatedly make tradeoffs between planning and scheduling. Although optimal solutions are rarely obtained, at least the solutions so derived are feasible, useful, and understandable to the construction planner.

Recent research has been directed toward eliminating the rigidity implied in a full definition of the construction logic. That is, no attempt is initially made to specify completely the form of the network model, a necessary prerequisite for the CPM process. Instead the construction logic is considered as being made up of three types of specifications: (1) the identification and ordering of collections of activities that must be contiguous, these collections forming rigid subgraphs that must be incorporated into the network model without distortion; (2) the identification and relative ordering of collections of activities that are generally structured relative to each other, but that within limits can be further separated or closed in time, and can therefore be incorporated into the network model in an elastic fashion; and (3) those activities that are independent and can be located almost anywhere in the project. The precise ordering is effected in such a way that either predetermined resource availabilities are not exceeded or resource requirements are kept to a minimum. These methods are heuristic; they require computer processing for execution, but have the advantage of combining the planning and scheduling stages without undue penalty in the formulation of the construction logic.

In conclusion, it is doubtful that a formal mathematical method will ever appear because of the increasing number of sophisticated requirements that the method must meet. Even if it does eventuate, it is doubtful if it will be as computationally efficient as current heuristic methods.

Evaluation

    a. Why bother with resource levelling and/or scheduling?

    b. Do you consider the portrayal of a separate resource network worthwhile?

    c. How are separate resource networks related to the project network model?

PROBLEMS

7.1. Given the following network data, carry out resource leveling within the minimum project duration for the smoothest use of equipment and smallest number of workers, if only one piece of equipment A is available and all activities are continuous.

Activity	Duration	Workers	Equipment (Incl. Operator)
1–2	2	3	B
1–4	2	6	A
1–7	1	4	B
2–3	4	—	A
3–6	1	4	A

Activity	Duration	Workers	Equipment (Incl. Operator)
4–5	5	—	A
4–8	8	4	—
5–6	4	2	B
6–9	3	4	A
7–8	3	5	B
8–9	5	2	B

7.2. Using the information derived in Problem 7.1, determine the shortest project durations for the following cases of resource limitations:

   (a) Not more than 8 workers, one A and one B, where the continuous operation of B is essential, whereas that of A is not.

   (b) Not more than 6 workers, one A and one B, where continuity of equipment operation is not essential, but is more desirable for A than for B.

7.3. Given the following network data, develop a schedule for minimum equipment requirements.

Activity		Duration	Mode	Manpower	Equipment		
I	J				A	B	C
1	2	8	Continuous	4	1	1	0
1	3	7	Continuous	10	0	0	1
1	5	12	Intermittent	5	0	1	0
2	3	4	Continuous	6	0	0	0
2	4	10	Intermittent	7	1	0	0
3	4	3	Continuous	5	0	1	0
3	5	5	Intermittent	8	0	0	0
3	6	10	Intermittent	7	0	1	0
4	6	7	Continuous	11	0	1	1
5	6	4	Continuous	3	1	0	0

8

PRACTICAL PLANNING WITH CRITICAL PATH METHODS

8.1 PRELIMINARY AND DETAILED PLANNING OF PROJECTS

The most important advantage of CPM is that management is compelled to plan and think logically from start to finish of a project. Every works operation, and the relevant constraints, must appear on the network, the development of which forces the planner to consult the various departments connected with the project. Hence all concerned acquire a fuller understanding of the problems, for it is in this preliminary planning phase that most of the vital decisions are made.

Normally, several alternative methods for carrying out the project are possible; these are each shown on separate networks, and associated calculations and schedules are made from approximate utility data. Although the final selection of a method for construction may occasionally depend on detailed calculations, a decision can usually be made from these preliminary networks and from approximate utility data.

With large and complex projects, the preliminary networks may be simplified by grouping together appropriate activities to reduce the size of the diagram until the best solution is determined; for example, formwork, reinforcement, and concreting for a particular part of the works may be conveniently grouped into one activity. Once the best preliminary solution has been obtained, this condensed network diagram may be expanded (if necessary) to show each individual activity in more detail. The necessity for activities showing important materials deliveries and supply of vital drawings should not be overlooked.

From the preliminary networks one or more may be chosen for complete examination. This requires compilation of complete utility data and a detailed

164

network for the scheme. Isolated subnetworks of parts of the works, where alternatives still seem favorable, may be more closely examined. After deciding on a final scheme, a complete schedule for the project is prepared. Equipment, manpower, and other resources are then listed, and resource leveling requirements investigated.

Considerable attention should be paid to the constraints, especially those imposed by manpower and special skill requirements. Often it is helpful to lay out the networks to a time scale (as in Figure 3.9) to facilitate the proper understanding of constraints; and it is usually easier to lay out the original network assuming unlimited resources and then to impose the constraints on it. Special attention must be given to events and activities along the critical and near-critical paths, and a check of activity relationship is worthwhile.

Starting with the critical path, a reexamination of all constraints reveals whether any further speeding-up can be achieved, either by a rearrangement of sequences or of resources, still at the all-normal level. This examination of the critical path also reveals many especially critical events. The activities preceding each of these may then be reviewed to see if any alternatives exist which will achieve any of these events in less time. In any project there is no unique method of construction, and the best plan is obtained only after considering all possible alternatives. The selected method is then represented by a unique network diagram: the cheapest feasible (all-normal) solution.

Following this procedure, crashed solutions may be examined according to the size and nature of the project; ultimately, the final network model emerges. Thus, before site operations begin at all, the selected optimum plan for the project has been carefully chosen and agreed on. This plan depends not on experience and judgment only, but on a mathematical and logical solution to the problems involved.

8.2 PREPARATION OF UTILITY DATA

The proper presentation of activity utility data is of major practical importance. There are several types of utility curves, which may be broadly classified as follows:

1. Smooth continuous concave curves, as in Figure 2.2 *a*, which are ideal and theoretical; these are never encountered in practice.
2. Piecewise linear curves, such as in Figures 2.2 *b*, 3.11, and 6.1, which are known as multistage utility curves and which approximate the ideal. The direct-cost/time relationship is continuous in that any duration between normal and full crash is feasible and valid.
3. Single-stage straight-line linear approximations to the ideal or to multistage curves, when utility data are available only for normal and crashed durations (as shown in Figures 6.14 and 6.1), but where again the direct-cost/time relationship is validly continuous.

4. Noncontinuous discrete point curves of the type referred to for activity 1–2 in Figure 6.4. These may consist of two or more points, depending on the nature of the activity and the method of crashing. Here the direct-cost/time relationship is not continuous, and only specific durations are feasible and valid.

5. Mixed concave-convex curves, as seen in Figure 6.9, where the direct-cost/time relationship may be continuous, but where it is desirable to use only certain durations that are on the concave portions of the curve. The inefficient durations are thus avoided, as was done in Figure 3.14 for the combination of activities *E* and *F*. Here the utility data take the form of a discrete point curve.

In practical planning it will rarely pay the estimator to prepare such detailed utility data that the curve will closely approach the ideal theoretical shape. At the most, piecewise linear approximations are adequate for continuous direct-cost/time relationships: usually, single-stage (that is, constant cost slope) data are sufficiently accurate. The important factor to be recognized and considered is whether the utility curve is, in fact, continuous or whether (as is often true) discrete point utility data are essential; this decision is fundamental if the network compressions are to yield feasible and valid solutions. Discrete point data are precise in their relationship to the actual direct costs and durations; they are not an approximation, as with other types of practical utility curves.

The preparation of the utility data required for the practical planning of construction projects may therefore be taken directly from the conventional cost estimates. In most project planning, all relevant alternative construction methods are examined as the direct-cost estimate for each activity is being prepared; thus the utility data for these are immediately available when the network is being developed. The proper presentation of activity utility data therefore requires little additional effort by the busy estimator and entails merely the listing of all the separate activities with their appropriate durations and corresponding direct costs. Cost slopes (when required) may be computed by calculator and added to the model as the network analysis proceeds.

To illustrate the advantages of this procedure, several examples of CPM planning of actual construction works follow.[1]

8.3   EXAMPLE OF THREE-SPAN BRIDGE

This example is presented first in order to demonstrate, in some detail, the development of a satisfactory feasible solution for executing a simple project

---

[1] To conserve space, the listing of activity sequence data and constraints has been omitted from these examples, and several steps in network development may be taken at one time. As an exercise, and to aid in a full appreciation of these examples, the reader may wish to list the activity and constraint data illustrated on the networks. Other alternative approaches to these projects may then be tried.

on a one-shift basis. Various alternative proposals were investigated in sequence with the object of producing a "near all-normal" solution; as will be seen, only a small amount of compression was used to derive the final network for this bridge. Network diagrams are drawn as sketches, for the busy planner has no time for finished drawings; and essential information only is recorded as the construction plans develop. Once the final network is attained, a proper drawing may then be made, showing all floats and utility data; this is then the network model for the project.

This project comprised the construction of a prestressed concrete girder bridge of three 30-m spans, with a cast-in-situ deck, supported on two river piers and two abutments on level banks. The piers and abutments had precast concrete pile foundations, with pile caps below water level, each therefore requiring cofferdams. The river was reasonably stable, with no flood problem, and pier cofferdams could be withdrawn as soon as the lower half of the pier shaft was completed. A girder casting yard was necessary on the site, and the prestressed girders could easily be transported by trolleys running on rails across the flat banks and over the river on falsework. The concrete piles could either be cast on the site or purchased from a concrete products manufacturer in a neighboring town. Access was available to either bank, the existing road crossing the stream a short distance downstream by a timber bridge, which the new structure would replace. Local supply of labor was adequate, and a 40-hour working week was feasible. From direct costs in the works estimate, utility data for the project were as given in Table 8.1. Indirect costs and overheads were estimated at $360 per working day. The contract time was 16 months (345 working days).

First Network

It was originally assumed that pile-driving would be a critical operation so that the structure could proceed rapidly; it could therefore pay to purchase the piles for this reason alone. It was also assumed that, since access to both banks was easy, it would be advantageous to drive both sets of abutment piles consecutively while the falsework was being constructed; and then to drive the piles in both piers. In this way, no floating plant was required, and sufficient falsework could be erected for access to the piers by the time the two abutments had been driven. The falsework was of timber, on timber piles, constructed entirely by mobile crane (which could drive the timber piles with a double-acting hammer). One set of cofferdam material, one set of pier formwork, and one set of abutment forms were considered adequate. One pile-driving rig could be provided for the concrete piles, and the cofferdam installations and removals could be handled by the mobile crane and double-acting hammer.

Figure 8.1 shows the preliminary network prepared on the above assumptions if piles were cast on the site. For convenience, activities $J2$ and $K1$ were grouped, because they had the same duration and were concurrent; likewise $J3$ and $K2$ (a practical step to keep the network as small as possible).

Table 8.1 Utility data for three-span bridge (40-hr week)

Activity		Normal Duration (Days)	Normal Direct Cost	Alternative Proposals: Time and Cost	
A	Move in and prepare site	30	$ 9,000		
B	Pile and girder casting yard	40	18,600	Girder casting only 30 days	$15,000
$C1$	Cast piles for Abutment A	25	23,400 ⎫	Purchase piles from	
$C2$	Cast piles for Abutment B	25	23,400 ⎪	manufacturer:	
$C3$	Cast piles for Pier No. 1	25	23,400 ⎬	Zero	$105,600
$C4$	Cast piles for Pier No. 2	25	23,400 ⎭	time	delivered
				45-hr week	
$D1$	Drive piles in Abutment A	27	7,800	24	$ 8,100
$D2$	Drive piles in Abutment B	27	7,800	24	8,100
$D3$	Drive piles in Pier No. 1	23	6,000	21	6,300
$D4$	Drive piles in Pier No. 2	23	6,000	21	6,300
$E1$	Cofferdam—install at Abutment A	15	16,000	13	18,150
$E2$	Cofferdam remove; install Pier 1	20	21,000	18	21,210
$E3$	Cofferdam remove; install Pier 2	20	21,000	18	21,210
$E4$	Cofferdam remove; install Abutment B	20	21,000	18	21,210
$E5$	Cofferdam remove from Abutment B	15	3,000	(Not crashed)	
	(Note: Net saving in cost if D and E are both done by pile rig—$24.00)				
$F1$	Erect falsework in Span 1	25	12,000		
$F2$	Erect falsework in Span 2	25	12,000		
$F3$	Erect falsework in Span 3	25	12,000		
$F4$	Remove falsework, all spans	20	6,000		
$G1$	Reinforced concrete, Abutment A	20	15,000		
$G2$	Reinforced concrete, Pier 1	40	33,000 ⎫	To release cofferdams these	
				were treated as 4 activities	
$G3$	Reinforced concrete, Pier 2	40	33,000 ⎭	of 20 days each	
$G4$	Reinforced concrete, Abutment B	20	15,000		
$H1$	Manufacture PC Girders, Span 1	70	96,000		
$H2$	Manufacture PC Girders, Span 2	65	96,000		
$H3$	Manufacture PC Girders, Span 3	65	96,000		
$J1$	Erection of PC Girders, Span 1	15	5,400		
$J2$	Erection of PC Girders, Span 2	15	6,000		

Table 8.1 *(Continued)*

Activity		Normal Duration (Days)	Normal Direct Cost	Alternative Proposals: Time and Cost
J3	Erection of PC Girders, Span 3	15	6,600	
K1	In-situ concrete deck, Span 1	15	9,000	
K2	In-situ concrete deck, Span 2	15	9,000	
K3	In-situ concrete deck, Span 3	15	9,000	
L	Approaches, handrails, etc.	30	21,000	
M	Clean up and move out	10	6,000	
	Total Direct Cost of Bridge		$730,800	

The calculated minimum project duration was 332 days. The initial assumption that it would be advantageous to drive both sets of abutment piles before the piers (gain time for falsework construction) was not substantiated; this is obvious from the total float available for activities $F1$–$F2$–$F3$ at event 20. This initial network therefore was useful in presenting a picture of the job, but was certainly capable of improvement.

Second Network

The effect of driving the piles in order across the river—that is, abutment A, pier 1, pier 2, abutment B—was the next obvious step; piles were still to be cast on the site (cheaper than purchasing). The result is seen in Figure 8.2: a minimum project duration of 317 days. This proposal not only saved 15 days, but still provided considerable float in pile casting and falsework construction at no extra direct cost. Analysis of plant scheduling showed that the mobile crane still had sufficient time to carry out falsework and cofferdam operations in the required order.

Third Network

Retaining the sequence of pile-driving adopted in the second network, the effect of purchasing the piles was investigated next; activities $C1$ to $C4$ were thus eliminated, and both cost and time savings were obtained in a smaller casting yard. The result is shown in Figure 8.3: the minimum project duration was reduced to 295 days. In addition, it was found that, with the time now available, the pile-driving rig could cope with both the concrete piling and the cofferdam work. The mobile crane could therefore proceed uninterrupted with falsework construction and thereafter be released from the job.

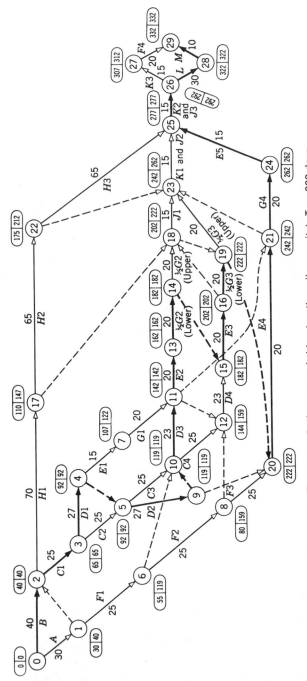

Figure 8.1 First network, three-span bridge (casting piles on site): T_P = 332 days.

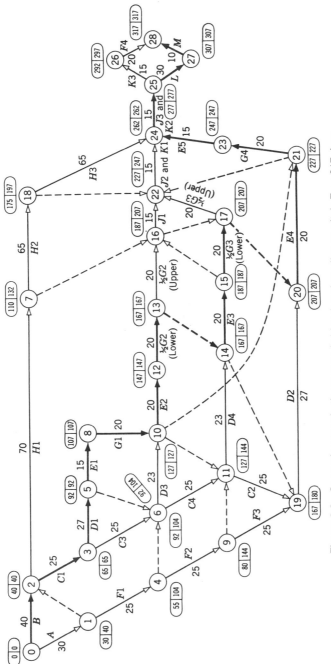

Figure 8.2 Second network, three-span bridge (casting piles on site): $T_P = 317$ days.

171

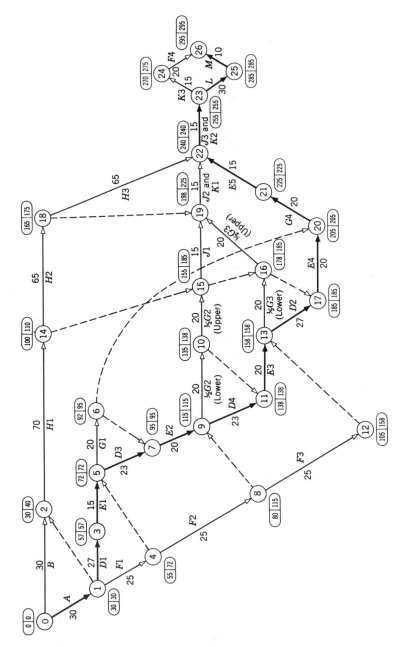

Figure 8.3 Third network, three-span bridge (purchasing precast piles): $T_P = 295$ days.

Compared with the second network plan, this scheme cost $12,000 extra for the piles, but saved $2400 in crame charges and $3600 in casting-yard construction (activity B), a net extra cost of $6000. The indirect cost saving of 22 days at $360 was $7920, so this scheme just managed to pay off. On this network, the critical path followed the pile-driving operations; this led to the suggestion that the pile rig crew should work a 9-hour day to avoid probable delays in the project.

Fourth Network

The utility data for a 45-hour week for activities D and E are given in Table 8.1—the piles were still to be purchased. Figure 8.4a shows the network for this proposal: the project duration was now 285 days. With an increased direct cost of $1980 for overtime plus the additional net $6000 from the third network—*total extra direct cost was $7980,* compared with the second network; *the indirect cost saving of 32 days at $360 = $11,520* therefore made this proposal more attractive than the third network.

In addition, the critical path now lay through the casting yard (where further compression was not feasible), and the pile-driving operations had a float of 8 days throughout. From a practical viewpoint this was a more preferable critical path, because precasting operations are more easily controlled than other site work; and furthermore, pile-driving procedures are such that some float is practically essential.

This network presented a satisfactory plan for a "near-normal" solution and was therefore adopted for bidding purposes. The total cost of the works, compared with the original proposal (first network) was:

	First Network	Fourth Network
Direct Costs	$730,800	$738,780
Indirect Costs	119,520	102,600
Total	$850,320	$841,380

This final network, drawn to a time-scale, is shown as Figure 8.4b.

Conclusion

Although the reduction in cost from the first to the fourth proposal was barely $9000 in this case, the use of CPM enabled a superior project plan to be evolved, from a practical viewpoint, with greater certainty of the relationship between all the operations of the job. Further revision of this project plan, during construction of the bridge, is discussed in Chapter 9.

The complete CPM analysis, from the conclusion of the direct-cost estimate to the completion of the fourth network (including preparation of utility data), took one estimator just over a day. Before the end of the second day,

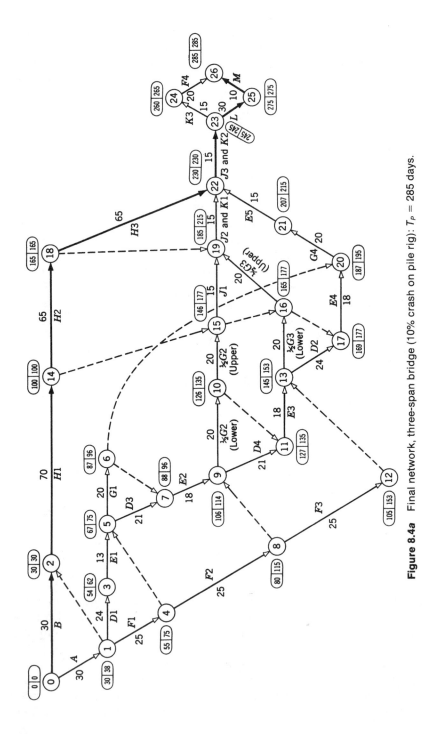

Figure 8.4a Final network, three-span bridge (10% crash on pile rig): $T_P = 285$ days.

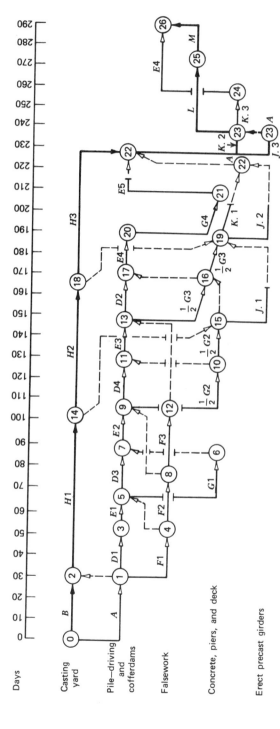

Figure 8.4b Final network, three-span bridge (drawn to time-scale): $T_P = 285$ days.

175

the scheduling and a first draft of the bar graph construction program had been completed.[1] The use of CPM therefore gave an assured works cost reduction of $8940, and a sound plan for doing the job, at the expense of 2 days' extra work by the estimator.

8.4   EXAMPLE OF ROCK-FILL DAM

Many construction works require careful assessment of site hazards. The effects of such hazards, and the constraints which they produce on relevant activities, may be shown on network diagrams by means of artificial "hazard activities." This technique is particularly applicable to the constraints in-duced by river behavior in dam construction works, as the following example of a rock-fill dam project illustrates.

This rock-fill dam, approximately 300 m long and 55 m high, involved the winning, placing, and sluicing of 400,000 cu m of rock in the main bank, with 27,000 cu m of packed rock on the upstream face, supporting a reinforced concrete face slab (7000 cu m). The spillway, located at one end of the dam, required 14,000 cu m of concrete and an additional 1000 cu m would be needed for outlet works, and so on. A good quarry site was available at an average haul of 1 km, and it was estimated that about 85% of quarry rock would be suitable for use. The river had a reasonably regular behavior, with a definite flood season. It was specified that bank construction begin at the end of a flood season and that the rock-fill be completed to El 425 m before the next floods; placing of rock-fill could then continue to El 450 m, leaving a suitable flood gap and thereafter the embankment could be completed to final level (El 460 m).

River diversion was to be handled in four stages: (1) through a temporary diversion culvert under the dam, which would ultimately be part of the outlet works (this could carry normal river flow and minor freshes); (2) after con-struction of the inlet tower and spillway and after the bank construction had reached El 450 m (with flood gap), the stream could be taken into a temporary opening in the base of the inlet tower and then through the culvert; (3) on completion of the main bank and a reasonable proportion of the face slabs, the temporary opening in the inlet tower could be plugged and the reservoir permitted to fill; (4) the outlet pipeline from the inlet tower to the downstream valve house would be installed in the culvert.

Because construction labor was in short supply, it was decided to attract workmen to the site by providing a long working week and hence higher wages. The utility data, based on a working week of six 9-hour days (original network), is shown in Table 8.2; also shown is a revision, based on two 8-hour shifts on excavation and rock-fill, and one 9-hour shift on other work, 6 days

---

[1] The reader may care to carry out the scheduling and prepare the construction program showing relevant float times as an exercise.

Table 8.2 Utility data for rock-fill dam

Activity		Original		Revised	
		Time (Days)	Direct Cost	Time (Days)	Direct Cost
A	Preliminary works	50	$ 447,000	60	$ 474,000
B1	River diversion, stage 1	60	102,000	60	102,000
B2	River diversion, stage 2	30	12,000	30	12,000
B3	River diversion, stage 3 (plug)	10	9,000	10	9,000
B4	River diversion, outlet pipeline	35	66,000	35	66,000
C	Excavation, dam site	130	441,000	75	465,000
D	Excavation, spillway	150	630,000	86	663,000
E	Excavation, quarry; and construction of bank:	(405)		(240)	
E1	Rock-fill to El 425 m	125 ⎫		75 ⎫	
E2	Rock-fill to El 450 m	125 ⎬	3,210,000	75 ⎬	3,384,000
E3	Rock-fill to El 460 m	160 ⎭		95 ⎭	
F	Drill and grout dam site	80	151,200	80	151,200
G	Permanent roadworks	20	63,000	20	63,000
H	Valve house embankment fill	15	18,000	15	18,000
J1	Concrete in spillway	150	960,000	150	960,000
J2	Concrete face-slabs to dam (Note: 25% may precede end of floods by working on sides of valley)	165	1,071,000	165 (Crash to 103)	1,071,000 (1,078,200)
J3	Concrete in inlet tower	70	198,000	70	198,000
J4	Concrete in outlet valve house	60	81,000	60	81,000
J5	Concrete closure, inlet tower	24	13,800	24	13,800
K1	Metal work in inlet tower	20	78,000	20	78,000
K2	Metal work in valve house	20	72,000	20	72,000
L	Clean up and move out	25	12,000	25	12,000
	Total Direct Cost of Dam		$7,635,000		$7,893,000
	Crash J2				($7,900,200)

per week, used in later networks. The construction of camp accommodation was included in activity A in each case.

Indirect costs for the project amounted to $1500 per working day. The construction time allowed was 20 months.

Specified constraints, additional to those already stated, were: (1) activity F must be completed before E2 begins; (2) G must follow E3, and H must follow B3; and J2 cannot begin until E2 is finished.

The flood season lasted for 6 months with a month's uncertainty as to starting and finishing dates. It was decided to assume a flood period of 31 weeks. Adopting a 6-day week for the project (312 working days per annum), it was assumed for estimating purposes that the flood season would cover 190 working days and the clear season 120. By the use of "hazard activities," the influence of floods can be shown in the network diagram as constraints on the activities affected by flood conditions.

First Network: 54-Hour Week

The specified starting date for the bank construction controlled the project starting time. With a 6-day working week, the total construction time permitted was 520 working days. In Figure 8.5a is seen the network diagram developed from the original utility data. The constraints due to flood hazards are indicated along the top, together with the hazard activity dummies and notes on the diagram; these affected all work except activities A, B1, B4, D, E2, E3, G, J1, J4, J5, K2, and L (H could be done if the flood were not severe). For clarity in following separate chains of activities, the start of the project has been divided into three separate events, numbered 0, 1, and 2; these are connected by dummies to retain the logic. The project duration was 505 days by this scheme, which was considered very tight for works of this type. Furthermore, it can be seen that because of the flood constraints, nearly every path in the network is critical; and, in addition, the network still contains flood dangers, notably at events 9, 10, 17, and 23.

A flood contingency was therefore essential, and was computed (on the basis of worst possible damage, plus delays and penalties) at $288,000. The works cost estimated with this original network thus became

Total direct costs	$7,635,000
Indirect costs, 505 days at $1500	757,500
Flood contingency at maximum damage	288,000
	$8,680,500

Even with this contingency, this scheme did not offer a very attractive proposition for the construction of the works.

Revised Network: Partial Shiftwork

It was obvious that a shorter project duration had to be achieved and that the flood hazards should, if possible, be removed. After some trial networks, it appeared that two 8-hour shifts on excavation and rock-fill and one 9-hour shift on other operations would be satisfactory. The utility data were revised, as seen in Table 8.2; after crashing the later 75% of activity J2 to a two-shift operation, the revised network appeared as in Figure 8.5b. Activity K1 was postponed until after the flood season, to be done immediately prior to K2, thus providing continuity for a metal-work crew; and since the hazard activity dummies are now real dummies, the network is logically tied together. Event

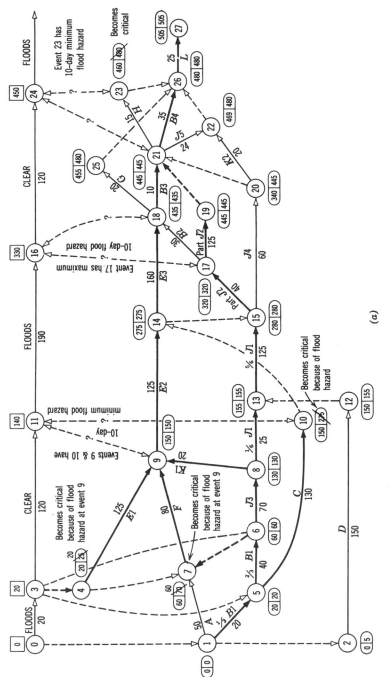

Figure 8.5(a) First network, rock-fill dam: $T_P = 505$ days.

(a)

179

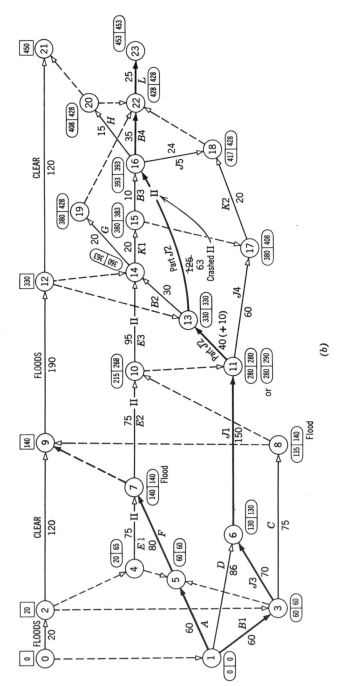

Figure 8.5(b) Revised network, rock-fill dam: $T_P = 453$ days.

13 could be achieved 10 days before the assumed end of the flood season; if the floods finished early, the project time could be reduced. However, for safety, advantage was not taken of this, and a project duration of 453 days was adopted.

The critical paths on this revised network diagram were 1–3–6–11–13–16–22–23, with 1–5–7–9 for the first clear season. Project duration was well within the allowed time, and no flood contingency was required; indeed, a saving of 10 days' indirect cost ($15,000) could be made in all probability if the flood season ended on its predicted time (the flood hazard period of 190 days included for probable lateness). There was a very reasonable float on the two-shift rock-fill operations, and a possibility of slackening off to single shift toward the end of embankment construction; subsequent resource leveling proved that this could be done. In other words, this revised proposal included a small but sufficient hidden contingency.

The works cost estimate for this revised network was the following:

Total direct costs	$7,900,200
Indirect costs, 453 days at $1500	679,500
	$8,579,700

Final Network: Crash Finish

Because of the necessity to crash the last 75% of activity $J2$ to a two-shift operation to attain the solution shown in the revised network (Figure 8.5b), the next obvious step was investigating optimal compression beyond event 13. Figure 8.6a shows this portion of the network as in Figure 8.5b, but with the necessary compression utility data added. No crashing of the critical path was possible before event 13 because of its control by the flood season; nor was it considered advisable to compress the final activity L (22–23). Therefore the two activities 11–13 and 22–23 are marked with a cross (X) to indicate that they are not available for compression. Furthermore, if activity 13–16 (the last 75% of $J2$) was to be crashed from its existing two-shift to a three-shift operation, it would be necessary to complete event 17 *before* event 13 to avoid a clash in concrete requirements between activity $J4$ and the fully crashed $J2$ (due to mixer-house capacity); thus a new constraint would have to be introduced.

The first full compression was of activity 16–22 ($B4$)—the cheapest cost slope available—which immediately introduced a new critical path. Subsequent compressions of activities 16–18 ($J5$) by 6 days and 11–17 ($J4$) by 20 days—with introduction of a new dummy—enabled a full compression of activity 13–16, together with full compression of 15–16 ($B3$) followed by partial compression of activity 13–14 ($B2$). These compressions resulted in two looped critical paths in the final network, as seen in Figure 8.6*b* and reduced the project duration by 37 days to a total of 416 days.[1]

The direct-cost increases due to these compressions were the following.

For activity 16-22, a compression of 17 days at $180 = $ 3,060
For activity 16-18, a compression of 6 days at $ 30   180
For activity 11-11*A*, a compression of 20 days at $300   6,000
For activity 13-16, a compression of 20 days at $360   7,200
For activity 15-16, a compression of 5 days at $ 90   450
For activity 13-14, a compression of 12 days at $120   1,440
                                                        $18,330
Additional shift supervision for 81 days at $75   6,075
Hence the total extra direct cost of this crashing= $24,405

The reduction in indirect costs for 37 days at a rate of $1500 per day was $55,500, giving a net saving for the crash finish of $31,095 and a works cost estimate for the final network of $8,548,605.

Conclusion

Before applying critical path planning to this project an estimate had been taken out by the contractor in the conventional manner, on the basis of the 54-hour week. The conventional bar chart construction program showed that the timing was tight for flood periods and also for the completion date so that a conventional contingency would have to be added. On the erroneous but normal assumption that a faster program would not result in overall de-creased cost, the conventional estimate for bidding purposes was composed as follows:

	Conventional Plan
Direct costs	$7,635,000
Indirect costs (20 months=520 days)	780,000
Contingency, 5%	420,750
Total works cost excluding profit	$8,835,750

The critical path network analysis used the same direct and indirect costs to obtain its utility data and commenced with the 54-hour-week proposal as a preliminary examination of the problems of the project; this immediately

[1] The reader should formulate the logic and carry out the network compression calculations as an exercise. Obviously, this final network model requires redrafting with the events renumbered to accommodate the new constraint introduced as dummy 11*A*-13; the reader may wish to do this, too, and then perhaps to investigate this project for financial resource planning after mastering Section 10.2.

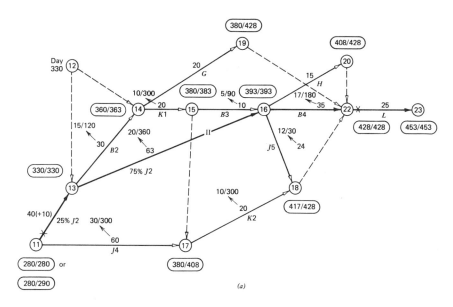

(a)

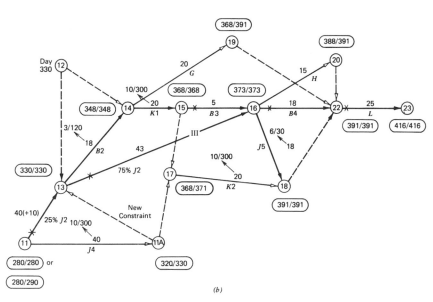

(b)

Figure 8.6   Final network (crash finish), rock-fill dam: $T_P^C = 416$ *days*. (a) Concluding section of network ready for compression. (b) full compression of concluding section of network.

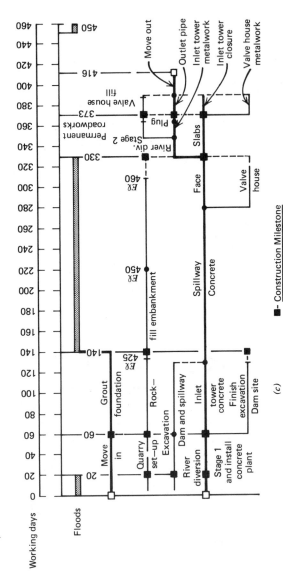

Figure 8.6 (*continued*) Final network (crash finish), rock-fill dam: $T_P^C = 416$ days. (*c*) Time-scaled network, showing construction milestone events.

disclosed the flood risks, as well as showing that they had been overpriced at 5% arbitrary contingency sum. The subsequent critical path planning proved not only that the flood risk could be removed but that the second flood season could be avoided, at the same time gaining considerable reductions in cost, a less rushed move-in time, and a much more satisfactory sequence of operations; even the crash finish pulled over $30,000 off the cost.

The application of critical path methods to this project at the bidding stage resulted in a reduction of the original conventional bid estimate by almost $288,000 and gave the contractor a superior construction plan for the execution of the works. It required a little over 2 days work by the CPM planner and was based entirely on the contractor's own estimates of cost rates and productivity.

Construction Milestones

Having examined resource requirements and made consequential adjustments to the program, the final network is then drawn to a time-scale, as in Figure 8.6c; the schedule at the top may be either in working days (as shown) or by calendar dates. The next step is to determine those *milestone events* (see Section 7.3) that have a time limitation or constrain two or more site operations so that during the construction phase they may be monitored for the purpose of project control (see Chapter 9). In this example, the first construction milestone is shown at day 20; being the end of the flood season, access will be available across the river, so work must begin in the quarry (stripping overburden) as well as in the spillway and river bed (start excavation). River diversion (Stage 1) and the installation of the quarry and the concrete plant should also begin; these, together with all the other preliminary work, must be finished by the second construction milestone at day 60 to permit the foundation grouting, rock-fill embankment, and inlet tower concrete to commence. Damsite excavation proceeds up the valley sides. With adequate float, placing rock-fill in the dam can have work continuity within the imposed constraints but will not be critical at any time. The third construction milestone is at day 140 when the rock-fill must be at least to El. 425, the damsite excavation and grouting must be finished, together with the inlet tower, and the spillway concrete should be in progress. If desired, some intermediate milestones may be scheduled, but the fourth one shown is at day 330 (the end of flood season) when the embankment, the spillway, and the outlet valve house are all finished, and sufficient face slabs (at least 25%) have been concreted to permit river diversion Stage 2. This is followed by other activities to permit diversion Stage 3 to be done before the fifth milestone, at day 373, by which time all the face slabs should be finished. The remaining activities are then completed by day 416.

Construction milestones are valuable indicators on a network program, emphasizing important time limitations or drawing attention to relationships between specific physical activities and resulting consequences or the imposition of constraints, including obligations that may arise with subcontractors and suppliers. On major projects, milestone events are essential to define interaction between two or more independent networks or between different contracts; they are then called *interface events* and are discussed in detail in Section 15.3. The master schedule, as summarized in Figure 8.6c, thus illustrates the essential characteristics of the adopted construction plan and draws attention to important dates along the way. It may be extended later, if desired, to show float details (as in Figure 2.5), or it may be reproduced in greater detail by the provision of subnetworks (fragnets) for each of the major activities in the project; but in the form shown in Figure 8.6c it provides the essential framework for the preparation of a bid, thereby permitting the scheduling of equipment requirements, the determination of crew sizes, and the forecasting of adequate project finance (see Chapter 10).

It should be added, before leaving this example, that when major hazards exist on a construction site, a series of network diagrams may be drawn up (during the preliminary planning) to determine the advantages of one starting date in comparison with another. This will often provide additional insight into the effects of the hazards on the work and of the relative risks involved in the various alternatives investigated. On certain occasions it may prove economical overall to delay the start of a project quite considerably, and then to adopt a crash program, in order to finish the job within the permitted time. Critical path analysis is the only logical way to investigate this aspect of construction planning. Furthermore, it can be employed to assess and compare the costs of risks which have to be tolerated.

8.5 EXAMPLE OF SEWER MAIN

This sewer submain extension, 1.65 km long, was laid partly in trench (1100 m) and partly in tunnel (550 m). The tunnel section presented no unusual difficulty, but sections of the trench passed through very bad ground, requiring considerable support. Access to the tunnel section was by several manhole shafts forming part of the works. This project, located in a developing country, presented a specific problem (after determining the construction method)—that was to obtain the optimum solution for the cheapest overall cost since the work was entirely labor-intensive but not urgent.

Alternative Solutions

There were only three feasible solutions: the all-normal (working one shift per day), a crashed least-time (working as fast as possible from a practical viewpoint), and a partially crashed solution. The utility data for the various activities at different shifts are given in Table 8.3. The relevant networks for the three solutions are seen in Figure 8.7; for convenience some concurrent activities are grouped in the diagrams. Following activities B and C, there are

Table 8.3 Utility data for sewer main

Activity	All-Normal		Crashed Two-Shift		Crashed Three-Shift	
	Time (Days)	Direct Cost	Time	Cost	Time	Cost
A. Move-in	12	$ 3,000	6	$ 3,300	4	$ 3,600
B. Excavation in trench	300	33,700	150	39,200	100	44,700
C.　　　　　in tunnel	305	36,150	155	38,100	105	40,050
D.　　　　　in shafts	72	6,250	36	6,600	24	6,950
E. Backfill (trench only)	50	(Taken)	25	(Taken)	17	(Taken)
F. Pipelaying in trench	50	8,500	25	8,700	17	8,900
G.　　　　　in tunnel	46	5,500	25	5,700	16	6,000
H. Concrete in trench	70	11,400	35	11,650	25	11,950
J.　　　　　in tunnel	60	11,000	30	11,250	20	11,500
K.　　　　　in manholes	60	4,600	30	4,700	20	4,800
L. Metalwork, etc.	60	1,050	30	1,075	20	1,100
M. Clean up and move out	6	(Taken)	3	(Taken)	2	(Taken)
Total Direct Costs	$121,150 All-normal, one shift		$130,275 All-crashed, two shifts		$139,550 All-crashed, three shifts	
Total Indirect Costs	$140 per day		$160 per day		$200 per day	

artificial activities introduced to cover waiting periods; these are labeled "wait" and represent an assessed delay, after finishing excavation, to permit the completion of other activities.

The all-normal single-shift solution required 353 days and involved fluctuations in labor requirements. The practical least-time solution has a number of critical paths and could not economically be crashed further, except to work shifts on activities A and M, which was not considered feasible; the project duration was 145 days (see Figure 8.7c). The optimum solution must lie between these two.

Examination of the network in Figure 8.7a showed if activity C were crashed that activity B must also be crashed for a satisfactory network; this was then developed as the first-crash solution and is seen at (b). After a few other trials, it proved to be the only practicable alternative; its duration was 203 days.

Optimum Solution

The time-cost curves were plotted in Figure 8.8, where it is clear that network (b) is the most economical method. The most interesting feature of this analysis was that, with the purpose of finishing the job as fast as possible, the practical least-time method of working would have been selected except for

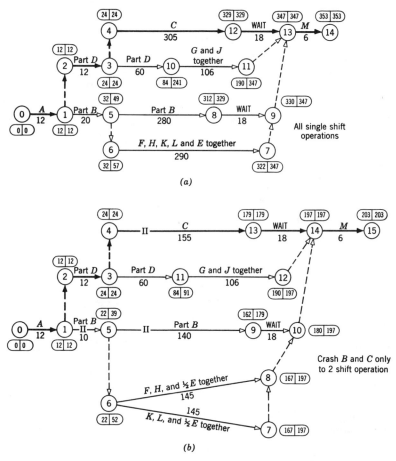

Figure 8.7 Normal and crashed networks, sewer main. (a) All-normal solution: $T_P^N = 353$ days. (b) First crash solution: $T_P^1 = 203$ days.

CPM; had this been done, it would have needlessly cost an additional $4570. It took barely a half day's work with CPM to obtain the optimum solution.

The scheduling and financial budget for this project are developed later in Figures 10.1 and 10.2.

8.6   ALLOWANCES FOR NORMAL LOST TIME

Every construction job will normally lose some time because of snow or rain. This should be recognized in preparing the network model for a project so the project duration and the scheduled event dates are consequently more realistic. Determination of inclement weather cost allowances is made by the estimator when computing the labor rates for the direct-cost estimate. These

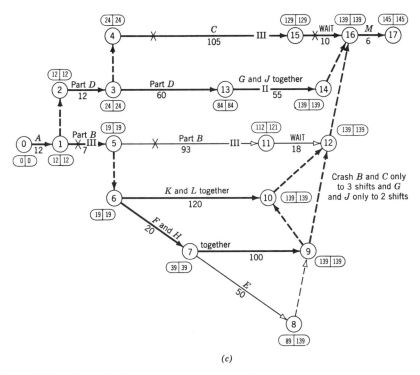

Figure 8.7 (*continued*)   Normal and crashed networks, sewer main. (*c*) Practical least-time solution: $T_P{}^C = 145$ days.

allowances are then included in the manhour rates as part of the labor on-cost; for a realistic network diagram these weather allowances should be translated into time when considering the duration of each activity. For example, if 1 day per week is expected to be lost because of inclement weather, then those activities affected by these conditions should have their theoretical durations increased by about 20%. Activities that may proceed irrespective of the weather do not require such adjusted durations. By notation on the network, it can be made clear to the field staff which of the activity durations include this allowance for lost time: one method, for instance, is to show duration as ''10 (+2)'' to indicate that the computed no-delay duration of 10 days has, in addition, an allowance of 2 days for expected delays; this activity is then assumed to require 12 days' duration in the network calculations.

Other normal lost time will occur at weekends and public holidays. Network durations show only working days and for programming, therefore, these must be converted to calendar dates. This, however, requires no adjustment on the diagram, but rather an attached tabulation showing the actual dates corresponding to specific days on the network timing. Depending on the hours to be worked on the project, weekends and public holidays may be

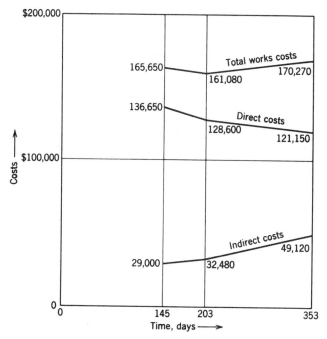

Figure 8.8   Optimum solution, sewer main: $T_P = 203$ days.

counted as working days or not, as appropriate. If the project is delayed during construction, a new date tabulation is easily prepared. This is much more preferable than trying to show dates on the network diagram itself. In preparing date tabulations, and indeed in the compilation of network diagrams, it must be remembered that some activities and constraints are specified by calendar dates and not by working days; simple examples are concrete curing periods and formwork stripping times; these lose no time through weekends, public holidays, or inclement weather.

For a simpler explanation and for clarity in presenting the preceding networks and examples, no allowances were shown for normal lost times. However (to be realistic in practice), every network model and every project schedule should provide for predictable lost time that may be regarded as normal, as well as including as constraints all the essential time delays pertinent to the job. This has been done in Figure 8.9, using one of the many conventions adopted for indicating lost time allowances; in this connection, the authors recommend the following.

Essential time delays should be shown as artificial activities, having time but no cost. If, for instance, it is specified that the formwork of a concrete beam cannot be removed for a week after casting, then the stripping of this beam must be removed from the concreting by an artificial activity having a duration of 5 for a 5-day working week (or 6 for a 6-day working week), as

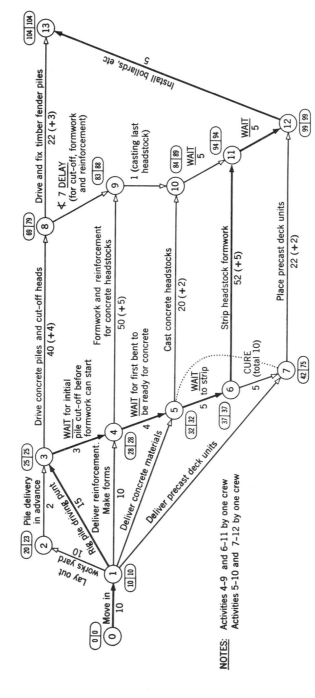

Figure 8.9 First draft of network diagram for jetty construction showing allowances for normal lost time and essential time delays required.

191

shown with activities 5–6 and 10–11 in Figure 8.9. Similarly, if a 2-week curing period is specified before any loading may be placed on a member, an artificial (delay) activity is introduced between casting and loading; this was done in Figure 8.9 with activity 5–7 (at the same time incorporating the shorter delay of 5 days, for stripping formwork, as activity 5–6).

Expected time delays, which may *not* occur, are shown as increases to the normal or crashed durations of an activity, and should be in parentheses to indicate that they are not factual. Expected time delays include the normal lost times which generally occur, and are predictable, but not precisely computable—such as inclement weather; the method of showing these is seen in Figure 8.9 for activities 3–8, 8–13, 4–9, 5–10, and so on.

Thus in a network there are two ways of providing for lost time that is expected to occur on the site. If the delay is specific or if it is assessed as essential for any valid reason, it is represented as an artificial activity within a finite duration and zero cost (as was done with the "wait" periods in Figure 8.7). If, on the other hand, it is expected to occur, but may not, then it is shown as an addition to the durations of the activities that it will affect (as was done with activity 11–13 in Figure 8.5*b* and Figure 8.6). The addition, however, is written separately in brackets so that the field staff may differentiate between an efficient speeding-up of an activity and a gain in time because a probable delay did not occur. Thus, both in planning and in project control, the network diagram presents a logical and practicable delineation of the project, resulting in a plan that is reasonably capable of achievement on the site.

8.7 PLANNING FOR ANNUAL WORKS PROGRAMS

Many organizations carry out construction and maintenance works with their own day-labor forces and equipment, operating on an annual financial budget; in addition, some of them may apportion part of their work on contract. When possible, they endeavor to provide regular employment for their labor and plant forces (which preferably should remain fairly steady) so as to carry out efficiently a maximum amount of work each year within their fixed annual budget. Examples of these organizations include public utility authorities, local government councils, certain government departments, and some large industrial undertakings. All of them will derive considerable benefit from using CPM in planning their annual works programs.

The planning procedure is based on the principles already discussed. First, detailed direct-cost estimates are prepared for all the works awaiting attention in the following year and in the subsequent near future; the list of works should be then arranged in order of preference for construction. Usually the list is larger than can be accomplished with available funds, so it is useful to know at least the more urgent works. Where jobs are of sufficient magnitude, they may be individually submitted to network analysis to determine their durations and costs. Whether or not this is done, it is essential that accurate predictions of time and cost are made for every item on the list of works. Second, the manpower demand and a plant schedule are derived for each item; where necessary for the major jobs, resource leveling may be undertaken to minimize fluctuations in their requirements.

The individual works are then coordinated into a combined network diagram, accommodating first the urgent and important works, adjustments being made as expedient to ensure continuity of labor and plant utilization. The floats are then calculated and recorded; if desired, a CPM bar chart may be drawn up to visualize the problem at this stage.

During analysis of the individual job networks, and in the combined network diagram, every endeavor should be made to retain all-normal operation. If some activities must be crashed, the additional costs must be derived and added to the utility data, for the annual budget is usually not very flexible. At the conclusion of this stage, the major works program and costs are established, manpower and plant schedules for the combined operations are computed, and the available floats and gaps tabulated. The minor jobs are then fitted into the gaps or into the available floats, thus completing the year's program within the budget.

As a simple example of this method of planning for annual fixed expenditure, consider a small local government body, whose principal equipment comprises two motor graders, one bulldozer, a front-end loader, and a small portable concreting plant. With these it carries out general maintenance and improvement works within its area, all major construction (and works beyond its capacity) being let out on contract. This council owns and operates service vehicles only, general haulage jobs being done by hired trucks when required. The annual funds available for direct-cost works expenditure amount to $980,000. Table 8.4 shows the works awaiting attention, in the preferred order of precedence, together with the utility data pertaining to each. It is impossible to accomplish this program completely since it exceeds available funds. The problem is to find the optimum works schedule that takes full advantage of resources, bearing in mind that the rainy season (of 2 months' duration) occurs from weeks 37 to 46.

After computing the utility data from the detailed cost estimates and small networks for individual jobs, the combined network diagram for the year was begun with items *A, B, C, D, E, F, G, H,* and *I* (the principal essential works); these were rearranged as required until a reasonable sequence appeared, and then the other works were fitted into relevant spaces. After further rearrangement, the final network looked like Figure 8.10.

Examination of this diagram shows that the specified items were programmed to avoid the rainy season, that there are reasonable allowances for normal lost time, as well as a general float of one week, and that the equipment is kept reasonably well occupied. The total expenditure amounts to $967,200, which is very satisfactory in comparison with the annual funds available, allowing for the one-week float. (Some money must be available to cover expenditure for this week's work if this float is not consumed.) There is, of course, no critical path in this network, because of the deliberate

Table 8.4   Utility data for shire works program

Works Item	Duration (Weeks)	Estimated Cost
A. Construction of new bridge	26	$ 240,000
B. Approach roadworks to A	8	41,200
C. Deviation of Smith Street	4	25,600
D. Maintenance of gravel roads	76	91,200
E. Resheeting of gravel roads	12	96,000
F. Bitumen sealing to B, C, and part E	3	104,000
G. Flood protection levees on river bank	11	60,000
H. Town drainage works extension	7	28,900
I. New culverts on Sydney Road	14	52,000
J. Reconstruction of Sydney Road	19	163,200
K. Reconstruction of Jones Road	7	36,000
L. New culverts on Browns Road	5	28,000
M. Reconstruction of Browns Road	12	84,000
N. Kerb and gutter town streets	15	62,400
O. Repairs to White Creek Bridge	3	14,400
P. Sewage treatment works extension	10	54,400
Q. Bitumen sealing to Sydney Road and Browns Road	3	128,000
R. Reform White Creek Road	1	1,200
S. Reform and resheet River Road	8	52,000
T. Renew timber decking, Jones Bridge	3	16,000
Total Direct Cost of Works Pending		$1,378,400

provision of float. The EFT and LFT of events 20, 21, 22, and 23 demonstrate this point. Activity J (23–24) may start either at week 50 or 51 depending upon float conditions at that time; its timing is not vital, because it is planned to continue into the next financial year. Activity M can begin at week 51 if the one-week float is still available then.

In a large organization, there may have to be considerable activity shifting within the proposed program period to accommodate all the work in a smooth-flowing sequence and within the total allowable expenditure while still retaining a reasonably fixed labor force and avoiding idle plant time. Normal lost times must be provided for and some margin of float preserved for unforeseen delays. Such planning will ensure that the Council members, or members of the governing Board, fully understand why certain works are scheduled at their specific times and why they cannot be done economically at other times.

It is obvious that adequate project control must be exercised during execu-

Plant Required	Remarks and Constraints
By Contract	Urgent before rainy season
Dozer; grader, *F/E* loader and hire trucks, 4 weeks	To follow *A,* before rainy season
Dozer, *F/E* loader and hire trucks; grader 3 weeks	Any time this year before *F*
Graders (one 50, one 26 wks)	
Grader and hire trucks; *F/E* and dozer at gravel pit	Any time this year, before *F*
By Contract	After rainy season, this year
Dozer, *F/E* loader and hire trucks	Before rainy season, this year
Concrete plant	Before rainy season, this year
Concrete plant and *F/E* for 10 weeks; dozer 2 weeks each, before and after	Before rainy season, this year
Dozer, grader, *F/E* loader and hire trucks	Before rainy season next year
Dozer; grader, *F/E* loader and hire trucks, 4 weeks	Any time before *T*
Concrete plant	Any time; avoid rainy season
Dozer, grader, *F/E* loader, and hire trucks	With or after *L;* avoid rainy season
Concrete plant	After *H;* avoid rainy season
Concrete plant and *F/E* loader	
Concrete plant	
By Contract	After *J* and *M;* after rainy season
Grader	After *O*
Grader, *F/E* loader, and hire trucks	
By Contract	After *K*

tion of the program and that revisions may have to be made for unforeseen conditions; hence the necessity to preserve some float for possible activity adjustment during the year and some funds for the resulting cost variations. Activity shifting was discussed in Chapter 7. The procedures for project control follow in the next chapter.

8.8 PLANNING FOR REPETITIVE OPERATIONS

Frequently, in construction work a series of activities will recur in a specific order, either continuously or from time to time. A simple example is the construction of a high, reinforced concrete retaining wall in which the sequence of activities between events 10 and 15 in Figure 8.11 will be repeated for every lift of concrete in the wall. Obviously, with such a series, the crew size and duration for each activity will be arranged to provide, as far as possible, continuity of work between events 10 and 15; or, alternatively,

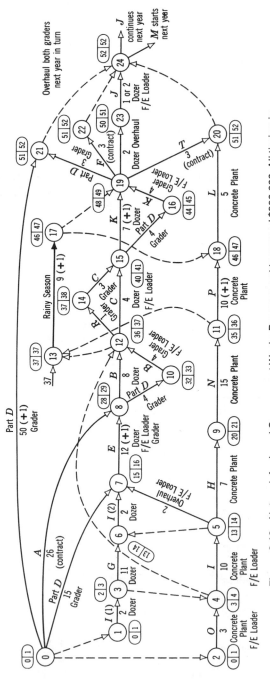

Figure 8.10 Network for Local Government Works Program not to exceed $980,000. All timing in weeks; general float—1 week; total expenditure—$967,200.

other work on the project will be planned to enable the activities between events 10 and 15 to be done by appropriate crews without loss of time.

In the simple case of Figure 8.11, if the various operations were all carried out by one crew, this group of activities could be combined coveniently into a single activity 10–15, comprising the six detailed tasks involved from the finish of placing concrete in *lift x* to the finish of placing concrete in *lift y*. In the network diagram of the project this single activity would then have a single appropriate duration and would be repeated for each lift of the wall. If, however, the various tasks are carried out by different crews, they cannot be combined into a single activity, because it is not then possible to show the individual durations and tasks affecting the work of each crew.

In practice, there are many cases of repetitive operations in which the series of activities cannot be logically combined. In these cases the repeating subnetwork must be shown each time it occurs, and some care will be necessary to ensure that the correct network logic of interdependency of activities is maintained; this can only be achieved by the judicious use of dummies. A typical example is seen in Figure 8.12 in which details of successive lifts of the concrete retaining wall are shown employing three separate crews: one for concrete placing, another for preparing the construction joint and installing the reinforcement, and a third for the formwork. Between lifts in this wall each crew carries out work on other parts of the project.

A similar but more complex situation occurs in the construction of multistory buildings where the finishing trades are scheduled to follow one another in a predetermined order from floor to floor. Figure 8.13 demonstrates in arrow notation the necessity for dummies to preserve the logic; for instance, on any floor the carpenter and plumber cannot start until the bricklayer has finished, and yet they are not dependent on one another—and the plasterer cannot begin until the carpenter and plumber are both finished; however, the finish carpenter (joiner) and electrician, who follow the plasterer, are dependent on one another and must work each floor together, after which the painter can follow. Obviously, the double events are required to differentiate between the end of work by a trade on one floor and the start of the same trade on the next floor; if only one node were used, the logic of the network would be incorrect, since it would show, for example, that painting on the third floor

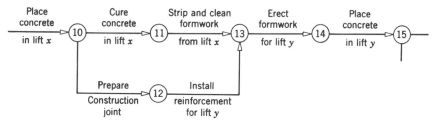

Figure 8.11 Sequence of activities for each lift of a high concrete wall.

198

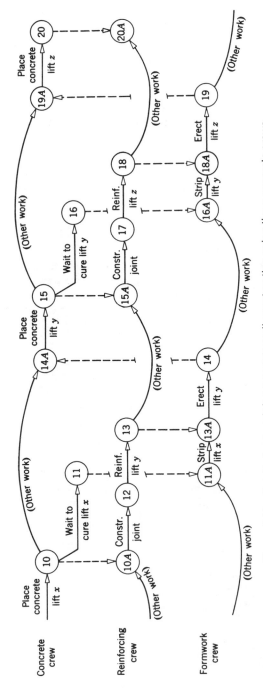

Figure 8.12 Detailed network for concrete wall construction using three separate crews.

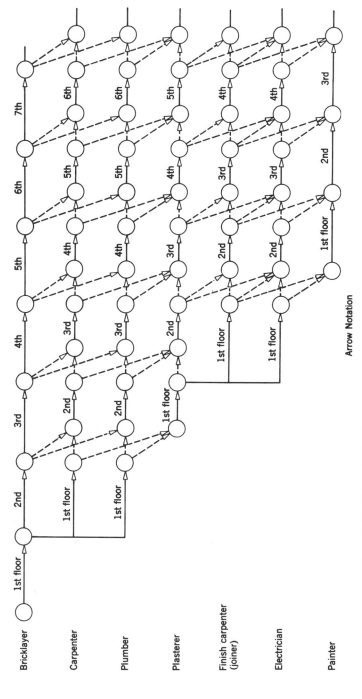

Arrow Notation

Figure 8.13 Network showing repetitive operations by the finishing trades on a multistory building project.

199

depended on completing plastering on the fourth, plumbing on the fifth, and so on.

Figure 8.14 models the same situation in circle notation and illustrates the general advantage a precedence diagram has over arrow notation in quickly representing complex logical situations.

It is interesting to note the subnetworks in both representations for a typical floor (say, the third). Figure 8.15 shows these subnetworks in arrow and circle notation side by side. It is possible to convert one network into the other by making simple changes in the modeling of the activities. If this is done, it will be seen that the logic for both networks is, in fact, identical. Detailed fragnets may also be prepared for each trade throughout the building if desired.

In many repetitive situations, such as that indicated in Figures 8.13 and 8.14, it is not necessary, and in many cases not desirable, for field management personnel to develop a network plan in detail because in such repetitive situations they intrinsically know the inherent construction logic involved and may feel constrained in any attempt to follow such a detailed network plan. In these situations the scheduling of the various subcontractors and trades to each floor may be accomplished by a simple tabulation sheet with a complete list of floor activities laid out across the top and the various floors listed one below the other down the left-hand side of the sheet. Using this layout, the work schedules for each floor activity can be portrayed and simply updated by entering the relevant start and occupancy dates for each subcontractor and trade crew involved. In essence, this corresponds to a compact bar chart schedule in which no attempt is made to portray the details of construction logic which have, however, been prepared from CPM planning by the project manager.

8.9   IMPORTANCE OF RESOURCE PLANNING

The development of critical resource paths, arising from the limitation of resources on a project, has been discussed in Chapter 7. Very rarely will unlimited resources be available in practice, and, consequently, the critical path(s) in actual construction projects are usually composed of a mixture of those activities lying on a critical time path and those on a critical resource path; in other words, a valid and realistic project network model may be expected to contain a composite critical path or paths.

This may not be apparent, however, unless the resources necessary to execute the various activities within the stipulated durations are indicated on the diagram. Consider, for example, the time-scaled network shown in Figure 8.16, being part of a typical CPM program submitted by a contractor at the beginning of a project. As drawn, Figure 8.16 shows correctly the planned sequence of operations and the estimated duration of each activity, but it does not indicate the resource requirements necessary to achieve the indi-

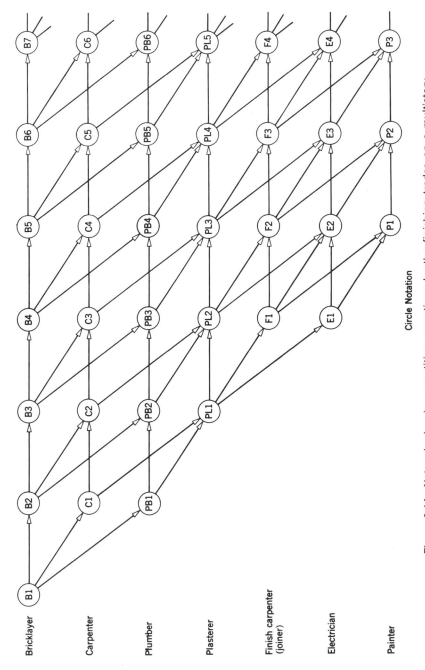

Circle Notation

Figure 8.14 Network showing repetitive operations by the finishing trades on a multistory building project.

201

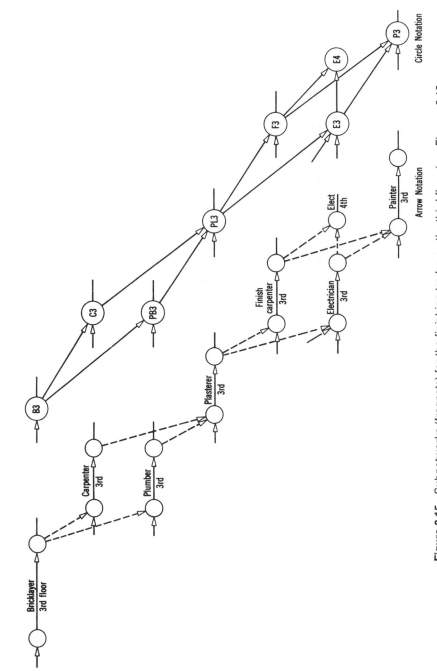

Figure 8.15 Subnetworks (fragnets) for the finishing trades on the third floor (see Figures 8.13 and 8.14).

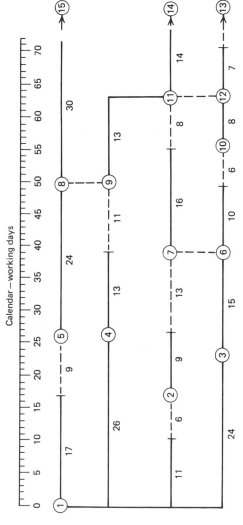

Figure 8.16 Time-scaled project network showing activity durations but not the resources required.

cated durations (for which the contractor will, in fact, need three different items of equipment, which may be designated A, B, and C). It would appear that there is no critical path, every chain having apparent float. If, therefore, extra work requiring 5 days was ordered in activity 1-5, the engineer would hardly feel disposed to agree to any extension of time, since activity 1-5 would not be considered critical. The contractor, on the other hand, would claim that the construction plan had been disrupted because of interference with the resource allocations, and would then demand not only time but monetary recompense for idle equipment as well. The reason for this impasse is that the network is not realistic: to be valid and practical it should appear as in Figure 8.17, showing how the contractor planned the work around the equipment usage. There is now no doubt that there are two critical paths—that of resource A and that of resource B—so that interference by extra work on either of these paths will prolong the project duration. (Resource C is not critical, having true float available.) Needless to say, the network in Figure 8.17 must be redrawn to introduce new event nodes, which will then be renumbered, and the resource paths would in practice be shown in colors.

Then if 5 extra days of work is required in activity 1-5, it is immediately clear that activity 2-7 must wait to start, thereby delaying the chains 5-8-9-11-14 (equipment A) and 5-8-15 (equipment B), and that B will be idle for 5 days between events 4 and 5, awaiting the arrival of A from activity 2-7. The contractor now has a realistic explanation for the claim for both time and money, which the engineer can see. Alternatively, to avoid the extension of time and any payment for idle equipment B, the engineer may elect to pay for the temporary introduction of another equipment A (to be rented by the contractor) plus any necessary additional manpower, to carry out the extra 5 days of work during the "wait" period in activity 1-5, thereby permitting the original equipments A and B to work as planned without any prolongation of the project duration. The engineer's decision is simply based on the compara-

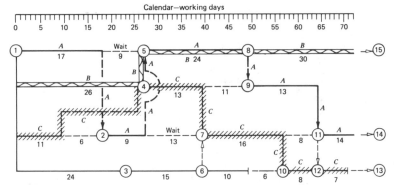

Figure 8.17 The network of Figure 8.16 with resource requirements added.

tive costs of the two alternatives clearly available from a study of the network.

It is consequently of importance in projects planned around available resources that the network show clearly the resource usage and the critical path(s) arising from them. The construction of a multistory building, or any structure where tower cranes or cableways are the principal equipment, are common examples of this problem. The entire construction program may well be developed around the continuous employment of the crane or cableway hook, and any interference with its planned usage may lead to a major disruption of the work.

Evaluation

a. Carefully examine several professionally prepared project networks for:
  1. method of portrayal
  2. level of detail
  3. typical activity size
  4. identification of critical paths
  5. identification of floats
  6. whether manually or computer prepared

  What conclusions can you make?

PROBLEMS

8.1. Given the following network, draw the diagrams for
  (a) the earliest start schedule and
  (b) the latest start schedule.

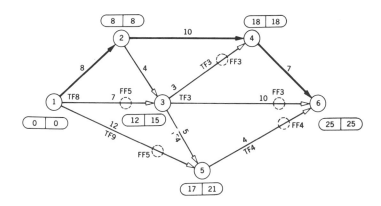

8.2. It is desired to purchase a new crane. Prepare a network showing the proper logic for the following activities:

Obtain quotations
Place the order
Delivery
Installation
Notify the authorities regarding inspection
Appoint an operator
Train the operator
Have the crane inspected
Test the crane

8.3. In the construction of a cross-country welded steel pipeline the basic sequence of operations is: survey, clearing, trench excavation, lay out pipes, weld pipes, test pipes, and backfill; delivery of pipes along the line is concurrent with trench excavation, and all other operations are to be spread out along the line at 1-km intervals. The separate crews for each construction operation are to be organized to progress at the rate of 4 km per week, the pipeline being 40 km long.

   Draw sufficient network diagram for this project to indicate logically the construction planning and determine the minimum project duration required in the field. (Construct the network diagram roughly to scale, with each horizontal chain representing the progress of one construction operation, such as survey, clearing, excavation, or delivery.)

8.4. A new road deviation with concrete pavement, shown below in longitudinal section, is 11,600 m long. It is to be constructed in accordance with the following conditions:

   (a) The balanced earthworks from Ch.00 to Ch.58(00) may be done at

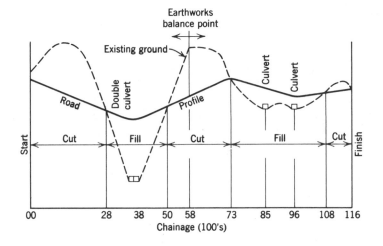

the same time as the balanced earthworks from Ch.58(00) to Ch.116(00) using two separate independent crews.

(b) The double box culvert will be built by one crew, and another crew will build the two small culverts. Concrete may be supplied either from the Paving Batch Plant or from small independent mixers at the culvert sites, whichever is expedient.

(c) One small slip-form paver will do all the concrete paving work, and all the shouldering will then follow with one crew after the concrete pavement is cured.

(d) Seeding the embankments with grass must be left as late as possible.

(1) Prepare a network diagram and determine the minimum possible project duration.

(2) If independent concrete mixers are used for the culverts, what is the latest day for delivery of the Paving Batch Plant to the site, so that the paving crew may have continuity of work (no idle time at all)?

Activity Description	Duration
Del. Rebars—Double Box Culvert	10
Move-In Equipment	3
Del. Rebars—Small Culverts	10
Set up Paving Batch Plant	8
Order and Del. Paving Mesh	10
Build and Cure Double Box Culvert Ch.38	40
Clear and Grub Ch.00-58	10
Clear and Grub Ch.58-116	8
Build Small Culvert Ch.85	14
Move Dirt Ch.00-58	27
Move Part Dirt Ch.58-116	15
Build Small Culvert Ch.96	14
Cure Small Culvert Ch.85	10
Cure Small Culvert Ch.96	10
Move Balance Dirt Ch.58-116	5
Place Sub-Base Ch.00-58	4
Place Sub-Base Ch.58-116	4
Order and Stockpile Paving Materials	7
Pave Ch.58-116	5
Cure Pavement Ch.58-116	10
Pave Ch.00-58	5
Cure Pavement Ch.00-58	10
Shoulders Ch.00-58	2
Shoulders Ch.58-116	2
Guardrail on Curves	3
Seeding Embankments with Grass	4
Move out and Open Road	3

8.5. If, on the project in Problem 8.4, it were decided to do all the "Move Dirt" activities consecutively, using one crew only, what would be the minimum possible project time? Draw the network diagram for this case and state the latest date on which the earthmoving plant could start work.

8.6. A major bridge project comprises 12 steel approach spans on the left bank, 7 steel truss spans in the main channel, and 15 steel approach spans on the right bank. The substructure consists of concrete piers and abutments. On the left bank the abutment and 5 piers are on dry land and also the abutment and 8 piers on the right bank. All piers and abutments are supported on precast concrete piles to be cast on the site and not to be driven until they are at least 28 days old. It is planned that the casting yard will manufacture piles at a rate of one pier or abutment every $1\frac{1}{2}$ weeks.

On dry land pile driving is estimated to require 2 weeks per pier or abutment; on the river, 3 weeks. Construction of the pile caps at each land pier will take 2 weeks, and at each river pier 3 weeks; each abutment or pier shaft above the pile caps can be built in $2\frac{1}{2}$ weeks at any location, and must cure for 2 weeks before steel is erected on it.

Assembly and erection of approach span steelwork over land will take 1 week per span, and over water 2 weeks; assembly and erection of truss steelwork will require 3 weeks per span. The concrete deck slabs in each span will require 2 weeks to form and cast in place at each location and will be constructed alternatively from each end of the bridge to the center, thus averaging 1 week per span.

A period of 3 months should be allowed to move in, procure materials, and set up equipment. Clean up and move out will take $4\frac{1}{2}$ weeks. It is planned to use two pile-driving crews, two pier construction crews, two steelwork crews, and one concrete deck crew with two sets of forms.

Prepare a network diagram for this project, showing the broad scheduling requirements.

9

PROJECT CONTROL WITH CRITICAL PATH METHODS

9.1 NECESSITY FOR PROJECT CONTROL

The construction industry has special features that are not encountered in other industries. First, there is a wide range of operations and processes, from excavation to dam construction, from pile-driving to multistory buildings, from tunneling to bridge erection, from marine work to pavements, and so on—all requiring different construction methods, equipment, and labor skills. Second, the work sites are always temporary and often remote; full-scale production on any one site may last only a few months or at best a few years. Third, the local site management rarely has full control of policy and finance and can never be self-sufficient. Finally, construction personnel are divided into two groups, the more-or-less permanent executives and the transitory operatives. Thus the field organization must be adaptable to the varied conditions from project to project, and must be flexible enough to control adequately the works being executed under a multiplicity of site conditions. Furthermore, the planning and estimating of construction works must take into account these characteristics.

Construction projects are therefore carefully planned and estimated so that they may be successfully completed with regard to quality, time, and cost. Planning helps in selecting the most economical construction method, determining the plant, meeting manpower and financial requirements, proper ordering and delivery of materials, establishing the necessary supervisory staff, engaging competent subcontractors at the proper time, and executing the job within the anticipated cost.

No paper plan, however, will work continually in practice, even if it is perfect theoretically. Unforeseen delays, unpredictable constraints, and unknown factors will all affect the smooth operation, as depicted on the network or shown on a bar chart program. Hence it is essential that management is kept continually and accurately informed of the progress of the works and that precise predictions are made of the effect of each site occurrence on the available resources and future operations.

To review the current procedures and forecast the future requirements of the job, so that the works may be successfully completed, is the primary purpose of project control. To work effectively, there must be some means of determining rapidly sound solutions to the day-to-day problems, so that the essential requirements for remedial measures may be promptly initiated. To do this, it may be advisable to reestimate the cost of the uncompleted portion of the works and to revise the utility data in the light of current costs of operations; it may be necessary to rearrange available resources or to obtain new ones; it may indeed be essential to revise the entire remainder of the program in order to finish the job within the specified time at the least possible cost. By employing critical path methods trouble can be foreseen in sufficient time to determine the required logical revisions for any new situations. The former method, that of speeding up everything on the job running behind schedule, has no place whatever in the offices of managers who have CPM at their elbows.

The more accurate and logical the planning, the easier it will be for the work to be performed in accordance with the program. Highly detailed planning, however, takes time and costs money. Consequently, project planning at the bidding stage may not proceed far enough to provide all the necessary particulars for project control. For this reason, it is essential that, before beginning operations on the site, the project schedule (and its accompanying networks and charts) be reviewed for special details.

Although this reviewing procedure is actually the final phase of the detailed planning, it is also the first step toward the actual site control of the project and should be done with considerable care.

9.2 REVIEWING THE SCHEDULE AND NETWORKS

With the main network diagram completed and doublechecked for inaccuracies or omissions, and with subsidiary networks showing more intricate details of parts of the project worthy of more personal attention, the estimator has a master plan for the scheduling of labor, materials, and plant. Very often in construction planning, however, the network (and its accompanying schedule) may be left at the development stage reached in the examples given in Chapter 8, for this is often adequate for bidding purposes: a range of starting and finishing dates for each activity has been determined and a CPM bar chart (as in Figure 2.5) or time-scaled network (Figure 3.9*b*) drawn up.

When a contract is obtained, however, the fullest details are required. It then becomes necessary to know the optimum dates for the actual start of each operation, having in mind the smooth flow of all the required resources. For the project milestones and critical activities the dates are already fixed, but for noncritical activities these may now be reviewed, leaving some float available for unpredictable delays. Originally, the time estimates for all activities in the project were based on obtaining the optimum duration for minimum total cost. Within this limited duration, noncritical activities can be shifted across the program within their available float; or it may be more practical to use smaller crews (taking a longer time) at no change—or at a saving—in cost, thus helping to smooth out labor and equipment requirements. Alternatively, two activities using the same plant or labor skills, originally planned for simultaneous performance, may now be made consecutive, if one or both of them have enough float. In short, resource leveling is carried out so that noncritical activity durations and starting dates are so assigned that they smooth out avoidable labor, equipment, and budget fluctuations. The procedure for the leveling of labor and equipment was discussed in Chapter 7 and that for financial planning occurs in Chapter 10. In seeking these desirable objectives, however, care must be taken that the float on noncritical chains is not completely consumed; otherwise no means for maneuver will remain during construction.

In some cases, in order not to be overwhelmed with detail, it is desirable to postpone the detailed network planning of major portions of the project rather than include them in a significant form in the overall project network model. Using this planning and management approach, only that portion of the overall project model that falls within a planning horizon is drawn up in detail and used as the basis for resource leveling and scheduling. Then, provided the overall project plan, critical path, and milestones are maintained, the zones of network detail move progressively across the project as it proceeds, so that previously deferred detailed network planning of major portions of the project is resumed as soon as the portions impact the planning horizon of the project management team. This type of project networking is discussed further in Chapter 15, and, when related to milestone planning and control, enables a project team to manage their own limited planning and scheduling time in a way that does not compromise the project as a whole. This approach leads to a technique known as "fast tracking."

The changes resulting in the schedule from the above review will, of course, necessitate changes in the timing on the network diagram itself. Obviously it is preferable for the final network model for the project and the final schedules to be fully developed in the office, if possible at the bidding stage. Even if done later, it is nevertheless essential that all major activity starting dates should be determined for the field staff before beginning work on the site. The tedious detailed scheduling in the field is thus avoided, the field staff can concentrate their efforts toward the construction operations from the start, and the construction manager is presented at the outset with a feasible plan.

On the larger projects, especially where rapid construction is required, this final network model is usually detailed enough to disclose all the essential features of the construction planning yet broad enough to act as the master program for the project. From it there is prepared regularly, as work proceeds on the site, such detailed subnetworks (fragnets) and work schedules as may be required to show, for the short-term future, all the essential operations necessary for each trade, or each section of the project. These fully detailed fragnets should not be prepared for more than a few weeks ahead, since otherwise they require revision in the light of actual performance, delays, and so on; a common practice is to issue them weekly or fortnightly for the ensuing two or three weeks. This is done in conjunction with the regular project reviewing discussed in the next section.

9.3 REGULAR PROJECT REVIEWING DURING CONSTRUCTION

Basically, project control with CPM is the need for regular periodic reviews of the work completed to date, together with revisions of the network model as necessity demands. This technique has been compared to that for guided missiles: formerly one aimed at a target, applying all known corrections before firing the shot; with CPM the trajectory may be continually changed, so that the project is consistently steered to its completion date at least overall cost.

The general procedure is to revise the project network periodically, replacing the original time predictions with actual facts as time elapses. Whenever the activity durations are revised, the network is analyzed to determine whether the critical path and the project duration have been affected. If the job is found to be running behind time, the network may be amended and the appropriate future activities crashed to restore the position. Such crashing may take the form of overtime work, introduction of additional plant and/or labor, resort to shift work, etc. The actual cost of these remedial measures may be forecast and alternative proposals compared with one another in order to determine the optimum overall solution. Sometimes it may even prove economical to accept the delayed completion date.

Delayed noncritical activities may be permitted to consume their available float without affecting project duration. If the delay is of sufficient magnitude to overrun available float, the critical path will change and the uncompleted portion of the network must be reanalyzed. Sometimes it will be advantageous to manipulate deliberately the critical path or to change the sequence of activities when one is faced with site delays and unforeseen problems. Whatever one does, the consequences may be quickly analyzed and costed, which enables a quick comparison with other proposals and with the original estimate.

After the remedial measures have been decided, the network, with its schedules and bar charts, are appropriately revised, and thus a new plan is available for the uncompleted portion of the project. In this way the construction plan may be "updated" whenever necessary.

The usual causes of time delay on construction works include:

1. Incorrect estimates of activity durations.
2. Unforeseen weather conditions or site hazards.
3. Unpredictable delays in delivery of materials.
4. Strikes or other labor troubles
5. Unexpected site conditions.
6. Extras or deductions in works quantities.

The criterion for network revision is the magnitude of the delay to the completion date. For small delays (say, a few days only), which can be tolerated, the tabulation of calendar dates may be revised and a note made on the network diagram. Extensive delays, the necessity to change the sequence of events, or the introduction of new activities to cope with (5) and (6) will demand full revision of the network model in order to depict properly the new plan.

Regular reviewing of site operations may be made at any appropriate time. Usually weekly or two-weekly periods will suffice; but on large, fast-moving jobs (working on a three-shift basis) daily review of the major operations is warranted. Alternatively, reviewing may depend on the occurrence of specific control events. The system is quite flexible and may be suited to the desire of the construction manager. Project reviewing may be confined to examining critical and near-critical activities or it may consider the status of the entire works. Similarly, network revision may cover the whole diagram or may be confined to a particular portion. The degree of uncertainty, the magnitude of the project, the time to completion, and the troubles encountered are all factors influencing the frequency of periodic project reviewing.

It would appear at first sight that it is only necessary to save time on critical activities in order to bring a late project back on schedule. In the periodic reviews of works progress, therefore, there is a tendency to concentrate on activities lying on the critical path. If this is carried too far, however, noncritical activities may slip behind time to such an extent that they become critical. The two methods of guarding against this are either to review all activities in full or to use the current rate of works expenditure as a warning. The first method may necessitate a considerable amount of unprofitable review work, unless there has been serious slippage of the noncritical activities; it is nevertheless the only positive way to determine the matter. In the second method, the current rate of expenditure is compared with the scheduled works expenditure rate. If the critical activities are on time or ahead of schedule, but the expenditure rate is behind scheduled spending, it is likely that the noncritical activities are slipping; on the other hand, under similar circumstances, if expenditure is in accord with or ahead of schedule, no serious slippage has occurred unless the critical path timing has been maintained by the use of resources taken from the noncritical activities (and this latter will be apparent from the costing control—see Chapter 10).

With a view to avoiding unnecessary paper work, and yet ensuring that any slippage of noncritical activities is readily detected, the simplest practical way on a large project is to review critical and near-critical activities at regularly short intervals as well as the works expenditure rate. When the expenditure rate indicates the possibility of a slippage, and in any case at longer regular periods (or at the occurrence of selected control events), a complete activity review is undertaken. If the project is equipped for cost control with CPM (see Chapter 10), the periodic review of the entire works may well coincide with (say) the regular cost periods: if the project is using the factual network technique (see Chapter 11), reviews will follow network updating.

Finally, if it is found at any time that the critical activities are behind time, or barely on schedule, while at the same time the works expenditure is higher than expected, then either critical items are being neglected in favor of noncritical activities or the cost of meeting critical dates was underestimated; the detailed costing of each activity will disclose the actual cause.

9.4 REVISING THE NETWORK DURING CONSTRUCTION

Information for the periodic reviewing of site operations is collected in activity status reports; the simplest form is illustrated in Figure 9.1. Each report should cover every relevant activity in progress, due to start or due to finish, during the review period. This information may be transferred to a copy of the network diagram, which thereby becomes a permanent record of works progress. This transfer is accomplished by, first, coloring each arrow to show the proportion completed (thus finished activities are indicated by a completely colored arrow); second, crossing out the estimated durations of completed activities and inserting the actual duration (in color); third, by crossing out the original earliest and latest finish times in the "time-box" of all completed events and super-imposing (in color) the actual completion time; and finally, indicating in pencil the expected completion time of all activities in progress. The network is now up to date, and may be so revised at every review period.

If all the activities are on schedule and no difficulties are anticipated, there is no more to be done until the next review period. If, however, some activities are not on time, the earliest and latest finish times of all future events are then computed from the new data and written in above the "time-box" at each event; in this way, the currently estimated project completion time (and cost, if desired), as well as the current critical path, are determined for all to see. If this estimate is acceptable, the project may continue as

Critical Path Program—Report on Activity Status

Works Section: _____

Date: _____

Reported by _____

Activity Arrow Numbers	Activity Description	Started ? Not Started ? Finished ?	Scheduled Start Date	Scheduled Finish Date	Status (% complete on above date)	Expected or Actual Finish Date	Reasons For Delay, If Any.

Figure 9.1 Activity status report.

scheduled; this simply means that the current status of the project is tolerated until the next review period.

On the other hand, should the current period duration prove unacceptable, two remedial courses are available. The first course is to recover the position by manipulation of available resources within the present network concept. Redistribution of manpower and equipment will introduce new characteristics in the diagram and may produce new critical paths; any changes will be apparent as soon as the network model is analyzed. The proposed remedial scheme is rescheduled and costed. If this is satisfactory, the project may proceed on this new schedule with new control events; if not, further resource redistribution may be tried. It may be possible to recover the position entirely without recourse to compression. If not, crashing of critical activities is the obvious first step.

The second course is to design a completely new network model from the current position to the project completion event, introducing new construction methods and/or equipment, together with other additional resources, in order to maintain the desired completion date. The new total cost is then determined. If this position is acceptable, the project is rescheduled and proceeds according to this new plan; if not, further alternative plans may be devised until finally the cheapest overall solution is obtained.

The greatest advantage of CPM is the logical determination of the proper activity for attention in order to bring a late program back on schedule. Overtime may be necessary if the works must be speeded up; but instead of accelerating the whole project, CPM indicates the key operation for overtime working, thus saving unnecessary expenditure on unimportant activities.

To assist in the visual appreciation of the network diagram, and particularly in considering the status of the job at any given date, it may be helpful if the diagram is drawn as a time-scaled network, as in Figure 3.9b; this is particularly helpful to supervisors and gangers accustomed to the bar chart type of program—or it may be presented as in Figure 2.5. Alternatively, "time contours" may be added to the diagram, as in Figure 9.2. Various other refinements, warranted by the type and magnitude of the project, may be devised as visual aids, such as codification of each resource pertaining to an activity or the use of symbols I, II, and III, to denote shift work. In short, every endeavor should be made to keep all personnel informed of the current status of the project.

9.5   EXAMPLE OF NETWORK REVISION DURING CONSTRUCTION: THREE-SPAN BRIDGE

The principles of periodic network revision presented above are illustrated in Figures 9.2 and 9.3, which demonstrate the simple techniques of project control as applied during the construction of the three-span bridge discussed in Chapter 8. The final plan for this project was shown in Figure 8.4 and

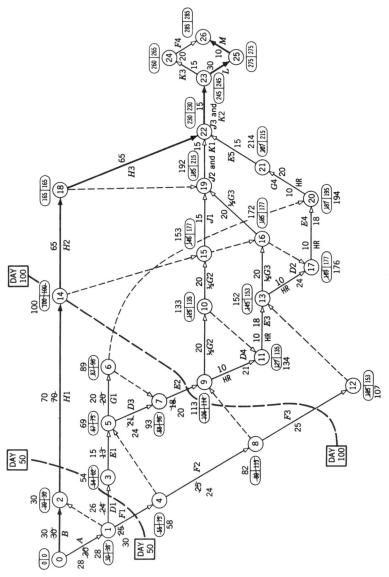

Figure 9.2 Revision of three-span bridge network (Fig. 8.4): status at day 100.

217

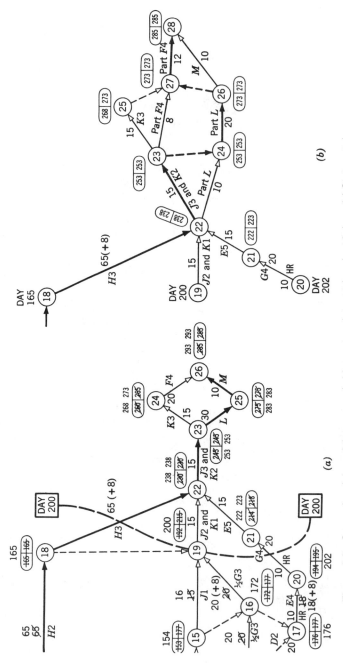

Figure 9.3 Second revision of three-span bridge network to retrieve lost 8 days. (a) Status at day 200 (8 days late). (b) Remedial network from day 200.

entailed the pile-rig crew working a 45-hour week. It was decided at the beginning of the work that the first two major review periods would be at day 50 and day 100, and the "time contours" for these dates were drawn on the network (see Figure 9.2) when work began on the site.

From the status reports at day 50, it appeared that no serious problems had as yet arisen. By day 100, however, it was obvious that the pile crew could not maintain the original estimated durations; event 9 had already lost all but one day of its available float. The corrected durations of completed activities and the forecast finishing times for future activities are clearly seen on the revised network at day 100 (see Figure 9.2).

Although precasting, in situ concreting, and falsework were satisfactory, it was essential to prevent any further slippage of the pile-rig operations (activities D and E); only 1 day's float was now available in this chain. The slippage to date was about 12%: 85 days had been consumed, instead of the 76 days originally allotted. The future timetable could only be maintained, therefore, if the pile crew's working time was increased from 45 to 50 hours per week (i.e., a 10-hour day); this increased productivity would allow the originally allotted durations for activities on the chain 9–11–13–17–20–21–22 to be retained. This chain, having only 1 day's float, was now virtually critical, but further slippage (if any) could still be retrieved by crashing activity 21–22 ($E5$, not yet crashed at all, see Table 8.1).

After these remedial measures were taken at da ' 100, work proceeded reasonably on schedule, with regular project reviewing every 10 days, until unprecedented cyclonic weather caused a general loss of 8 days by day 200; the status of the work at this date is seen in Figure 9.3a.

Logical investigation clearly showed that the position could be retrieved without any crashing of unfinished activities; in fact, the current network should continue to run 8 days late (Figure 9.3a shows the position) until event 22 was reached on day 238. Thereafter some rearrangement of resources would allow part of activity L to begin earlier than originally planned, and similarly some of activity $F4$ could be done before K was finished. The revised network shown in Figure 9.3b demonstrates how the 8 lost days were fully recovered without any necessity for remedial action until 38 days after the slippage was discovered. Without CPM it is certain that immediate speeding-up would have been introduced in activities $H3$, $G4$, and $E5$ in order to bring the works back on time; such immediate acceleration was not only uneconomical, but entirely unnecessary.

9.6 RESOURCE CONTROL—RESTRICTED RESOURCES

It was shown in Chapter 7 that, by means of the resource leveling technique, labor and equipment, materials, and other resources can be allotted to or distributed between the various activities of a project at the final planning stage, in accordance with their respective availability, so as to present the construction manager with a feasible plan and schedule with which to start the job. It goes without saying that, whenever a change occurs in the original plan during the construction of the works (due to delays or any other cause), the revised plan must be reviewed with regard to resource distribution in order for the new schedule to present the optimum use of available, or newly introduced, resources. This resource reviewing is not difficult, requiring merely resource leveling calculations based on the revised network model. Project control of these common vital resources is therefore linked with the control of the individual activities to which they pertain, using techniques already fully described.

There is another type of resource, however, which is extremely vital, but which cannot be controlled directly on site in the normal way, based on consideration of the current activity status of the project. These are known as *restricted resources*. One example is the amount of project finance that has been budgeted as coming from progress payments; another is the amount of storage capacity still unused. Both of these may be vital and must be kept under control, yet neither can appear in a network diagram, because, in each case, the quantity of restricted resource to be controlled is the *difference* between two other resources or quantities of work. On most projects there will be some restricted resources which must be watched intently, *for the completion of particular activities or events earlier than scheduled may have to be prohibited, because of these restricted resources.*

It should be clear that, for job finance from the progress payments, the restricted resource is the difference between the actual amount of expenditure on the works and the actual amount of revenue earned (and received) by the works. The dictum for this type of restricted resource is that it shall always exceed a specified minimum. Consequently, since progress measurements may be expected at certain fixed dates, costly activities may have to be programmed for completion as late as practicable just before these measurement dates; in this way, the expenditure is delayed as much as possible, although ensuring that the revenue derived from such costly activities is earned as soon as possible after the expenditure was incurred. Provided that the requirements are clearly defined, the technique of resource leveling may be used (on a-trial-and-error basis) to make certain, both during the planning and later during the site control of the project, that the terms of this restricted resource are met.

Most construction sites require site storage facilities because, on the one hand, it is difficult to exactly match delivery schedules with site production rates and many construction operations require a certain threshold level of resources before an efficient start can be made on the relevant activity. In addition, some construction processes are interrelated so that progress in one is dependent on progress in the other. Thus the manufacture and site delivery of special items (such as precast units) and the subsequent site handling and erection are interrelated and often tie up heavy handling equipment and specialist crews, which must be fully utilized at all times. In these cases, and especially when such precast units are manufactured on site, it is essential that adequate storage be made available at the site to act as a buffer to absorb the impact of differential manufacture, delivery, site handling, and erection rates. If such site activities dominate field operations, then the provision of adequate storage space and its effective management are critical management problems.

With site storage problems the dictum is again the provision of a minimum restricted resource at all times; this can be very important on a restricted construction site or when delay appears likely to one of the relevant activities and not the other, for this restricted resource is the difference between the production or delivery of an item and its use or erection into the works. As an example, consider the field casting and erection processes for the girders in the three-span bridge project planned in Figure 8.4b, and reproduced here as Figure 9.4a. From the chain 2–14–15 it is clear that, for the rate of production envisaged for this casting yard, at least 116 days (and at most 147 days) precasting production must be stored on the site ($T_2^E = 30$; $T_{15}^E = 146$ and $T_{15}^L = 177$). If activity 14–18 can start earlier than day 100 by faster production in activity 2–14, or if event 15 is delayed beyond day 177, then in each case the storage requirement increases. With this type of restricted resource, therefore, the dictum is not only that a specified minimum must be provided, but (perhaps) that a certain maximum must not be exceeded.

Suppose now that each span is made up of 15 P.C. girders for a total of 45 units. Suppose also that the casting yard requires a 5-day working up period before casting commences and that the first 5 units require 6 working days each for manufacture, the second 5 units 5 days each, all subsequent units 4 days each. Then, assuming that the casting yard is available on project day 30, the first unit enters storage on project day 41, the second on project day 47, and so on until the last is ready on project day 230. A graph showing casting yard production is given in Figure 9.4b.

If the erection rate is one girder unit per day and the erection activities $J1$, $J2$, and $J3$ all start on their respective EST times, then the rate of erection is as shown in Figure 9.4b. The difference between the scheduling of the casting yard operations and the erection operations establishes the required storage yard capacity as a maximum of 24 units (see Figure 9.4b).

Suppose now that the bridge site had ample space for the storage of 24 units (i.e., 116 days production from the casting yard) but only enough to cope with a maximum of 26 units (i.e., 125 days output at the estimated production rate). Activity 15–19 must then be scheduled with its EST of 146 and its LST at 155 *provided that* event 14 is prohibited from occurring before day 100. (Notice that $T_{15}^L = 155$ is quite satisfactory for $T_{10}^L = 135$ and $t_{10-15} = 20$.) The leveling of all other resources would then have to be dependent on retaining these restrictions on event 14 and activity 15–19. The graphic impact of this situation in Figure 9.4b is left to the reader as an exercise. Notice, however, that if the casting yard production and erection rates change, or are rescheduled, the curves of Figure 9.4b change slope and

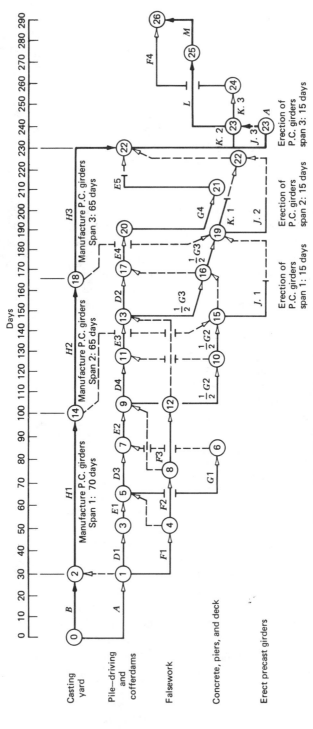

Figure 9.4a Final network, three-span bridge, as in Figure 8.4b.

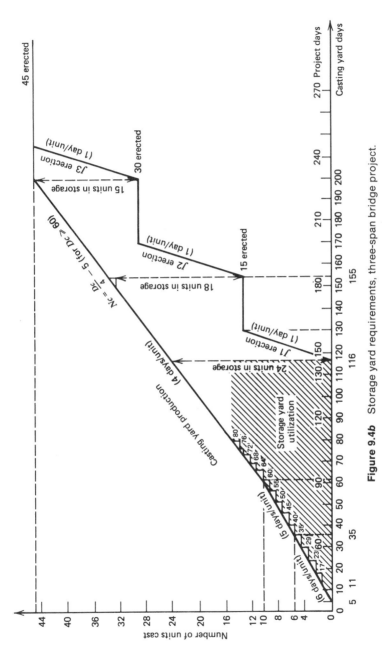

Figure 9.4b Storage yard requirements, three-span bridge project.

223

are moved along the time axis with consequential changes in storage yard requirements or even a complete stoppage of either the casting yard or the erection process.

When this project was reviewed during construction at day 100 (see Figure 9.2), T_{15}^E had become 153 and would therefore be just tolerable (T_{15}^L would still be 155); in the event, T_{15} became 154, as shown in Figure 9.3a, so that a restrictive situation was barely avoided. Had this occurred, it would be necessary either to obtain additional storage space for temporarily housing the bridge units or to close down production; in both cases extra costs would be incurred.

Restricted resources, therefore, are constraints and must be specified as definite minimum (and also perhaps definite maximum) requirements around which resource constraints the network can be planned. Having done this, the wise planner then prepares an alternative scheme for the construction manager, showing what is the most economical remedial course to take if the restricted resource constraint is violated; thus, for the duration of the project, the position can be reviewed regularly and preparations initiated for remedial action as soon as the first signs of trouble appear.

Although restricted resource constraints cannot be directly delineated on the network model, at least their requirements may be designated indirectly at the events and activities concerned. For instance, for the three-span bridge project, the following information could be added to the network diagram (Figure 8.4).

1. On the chain 2–14–18–22, the maximum rate of casting to be permitted
2. On the chain 15–19–22–23, the minimum rate of erection required
3. Near event 15, the maximum storage available.

Project planning and control with restricted resource constraints may be easily handled by manual methods during resource leveling; it is necessary, however, that the events and activities affected by the restricted resources shall be scheduled immediately after the critical path activities, so that their remaining floats are known, before any other activity shifting is undertaken. If this procedure is to be handled by computers, however, it will be necessary to resort to project simulation. In this technique the machine simulates a group of activities linked by the restricted resource. A new technique for the modeling and analysis of construction operations called CYCLONE[1] has been developed for just such repetitive situations. The CYCLONE approach is discussed in Chapter 13 together with the casting yard example introduced above. A series of results is presented, enabling a decision to be made

[1] The CYCLONE (CYCLic Operation NEtwork) modeling technique was developed by D. W. Halpin in a doctoral dissertation at the University of Illinois, Urbana, in 1973. It has been fully described by D. W. Halpin and R. W. Woodhead in *The Design of Construction and Process Operations,* Wiley, New York, 1976.

expeditiously as to the best combination to meet the restricted resource constraint at minimum cost. In the example of the three-span bridge, the CYCLONE method simulates the casting yard and erection processes, and produces the vital information of storage space required with different combinations of production and erection rates on the site. Simulation by computers is not actually a feature of CPM, but it is nevertheless briefly mentioned here because of its adoption in connection with complex projects constrained by restricted resources.

9.7 PROJECT CONTROL IN BUILDING CONSTRUCTION

Because of the multiplicity of operations involved, the building industry is a fruitful field for using CPM in project control. Indeed, it has been acknowledged by many experienced construction managers that only since the introduction of CPM have they really known where the project stands at any desired time. To obtain this result, however, a formal system of progress reporting is essential—and this is particularly true when a building construction company is carrying out more than one major project at the same time. Although the simple status report of Figure 9.1 may be quite satisfactory for general purposes, something more is required when intensive control is desired on complex building operations.

Various systems of progress reporting can be devised, each suited to the problems of the particular construction work and of the building company involved. One very effective system, designed for use where a company possesses planning offices covering large operational areas, is currently used on many building sites; the following description of this project control system is by courtesy of P. E. Consulting Group (Australia) Pty Ltd.

For adequate control at all levels, two documents are used. The first is a daily job progress return, illustrated in Figure 9.5a, which is made out on the site by the immediate supervisor, or perhaps by the project superintendent, depending on the size and scope of the job. It should be filled in no later than the evening of the day to which it refers so that it is available to the office the next morning. Its circulation will depend on circumstances, but it should certainly go to the project manager and to the planning office (where the project networks were originally prepared).

This daily report is simple to use and requires very little time to compile. It is divided into three sections. The first shows the activities completed today, and the second the activities started today, whereas the third section lists those operations that are now five days before their LST. For each of these activities identification is by their network node numbers; the labor skill coding, the activity description, its total float, its latest start or finish day number, the days before its LST or LFT, and the critical days lost (if any) are all shown in the appropriate columns. The first two sections, which merely list the activities finished or begun, enable keeping up to date the records in

Daily Job Progress Return

Job Name .... Contract Number ....

Date .... Project Day Number ....

Node		Nos.	Skill Code	Operation Description	Total Float	Latest Start/Finish Days No.	Days Before Latest Start/ Finish Day No.	Critical Days Lost
I	J							

Operations Completed Today

Operations Started Today

Operations 5 Days Before Latest Start Date

Signature ....

Figure 9.5a Daily job progress return.

the planning office. It is obvious that, where the days before LST or LFT (as appropriate) is a positive number and there are no critical days lost, this part of the project is well under control and indeed ahead of schedule. The third section of the report has two purposes. First, it serves as a warning that activities are due to start shortly, for as soon as an operation first appears here it has only 5 days to go before its LST will expire; thus 5 days' warning of possible trouble is given. Second, it forces the site personnel to look ahead at least 5 days, and hence focuses their attention on preparations for starting an activity, as well as encouraging them to always look ahead in conducting the job.

Daily Critical Operation Summary–Weekly Statement

Contract .... Contract No. .... Project Day Number .... Date............

Daily
Circulation

Project Manager
Area Planning Dept.

Additional
Weekly
Circulation

Area Manager
Group Planning Manager
Area Estimating Dept.

Additional
Monthly
Circulation

Managing Director

Operations Overdue for Start or Within 5 Days of Latest Start Date

Nodes		Operation Description	Days Before Latest Start Date	Critical Days Lost	Skill Code	Critical Man/Machine Days Lost	Reason for Being Behind Schedule	Action Being Taken To Recover Lost Critical Days	Agreed Action
I	J								

Project Manager's
Signature
....

Area Manager's
Signature
....

Figure 9.5b Critical operation summary—daily or weekly return.

227

The second control document is shown in Figure 9.5*b* and is a combined daily critical operation summary and weekly statement. It is made out by the supervisor immediately after the first report and lists the activities overdue for start or within 5 days of their LST. It therefore takes the items from the third section of the daily job progress return and adds to them any activities now overdue for starting. Needless to say, if the project is entirely on time, this document becomes a "nil" return; if, however, any critical activity is late (or a near-critical one is becoming late), this second report provides both information to that effect and a directive for taking action. The activity node numbers, operation description, days before LST, critical days lost, labor skill coding, critical man or machine days lost, and the reason for being behind schedule are all shown on this return, which is forwarded to the project manager and the planning office every day. Again, this daily return forces the site personnel to look at every trouble spot in the project, to state how serious the trouble is (by giving the critical days lost), and then to state the reason for being late.

Once a week these second daily returns are summarized on an identical form, and the area manager's approval (or other instruction) obtained for the action taken or proposed. Thus the weekly statement then becomes the remedial directive. The project manager has the responsibility of seeing that action is taken to recover the lost time. This action will, of course, depend on the circumstances of the case. It may be necessary only to increase the resources being used or to make some slight change in the methods of working; he therefore completes the ninth column of the summary. Where the delay is more serious, however, it may be necessary for the planning department (or the purchasing department) to take specific action and to reschedule part of the project; or it may be necessary to investigate in detail why some supplier or some subcontractor is not producing material or services on schedule.

A meeting is held weekly by the area manager where the weekly statement of the daily summaries is considered and discussed and any further action agreed on. The tenth column of the form is then completed, the whole form photocopied, and sent out by the area manager for immediate action by the appropriate officers. If desired, the same form can be used for a monthly summary to the general manager, or to the Board of Directors; with this monthly summary from each project, a complete picture is presented of all the trouble spots in the operations of the company and of the action being taken to overcome them. It is also abundantly clear which projects are under control and not behind schedule in any way.

9.8 USE OF COMPUTERS IN PROJECT CONTROL

In complex construction works, comprising over 300 or so principal activities, the planning and scheduling are usually done with computers to save the planner much tedious and repetitive manual computation. Even if the network is reduced (by suitable grouping of activities) to less than 300 activities in the planning stage so that manual methods can be adopted, proper project control will be possible only during construction if the formerly grouped activities are now returned to their individual identities. As has been said, a concrete wall may be treated as one activity during project planning, but should be expanded into its constituent formwork, reinforcement, and concreting for adequate project control and costing. For this reason alone, computers may be required during site construction where manual methods were adequate during the original planning.

Furthermore, project control includes not only the processing of activity status reports and the consequent network revisions, but also the costing and accounting procedures discussed in Chapter 10, which are not necessarily part of the reviewing of day-to-day job status. With a computer-oriented control system these phases are combined, and much laborious updating of networks and schedules by manual methods is avoided.

The principal advantage of the computer is that the project schedule can be stored away in the machine to be amended and reprinted as a new schedule (completely replacing the old) whenever necessary. The basis of the schedule is, of course, the network diagram—which must always be formulated manually—from which activities are identified by their node numbers for storage in the machine. Formulated instructions for program changes may then be prescribed to the computer to be executed when desired; hence the human effort is directed toward determining feasible changes and not to the laborious recalculation of the new schedule.

Many computer programs are available, reporting the time status of every activity (on schedule, behind schedule, or overdue in starting, whether critical or not, etc.), providing a new schedule, as well as printing out the critical path together with the three or four next most critical chains through the network. Although this same information can be calculated by manual methods, computers will be essential (and economical) with detailed networks comprising more than 200 to 300 activities.

When project control is expanded to include integrated cost and accounting control, the computer provides both time and cost status reports for each current activity in the network, together with periodic costs, costs to date, overall financial status, future commitments, value of work in progress, etc. A system of this type was devised for the site control of the construction of the $105,000,000 Sydney Opera House and is outlined in Section 9.9 by courtesy of the subcontractors, M. R. Hornibrook (N.S.W.) Pty. Ltd, and the computer company, Australian General Electric Pty. Ltd.

9.9 PROJECT CONTROL AT SYDNEY OPERA HOUSE, AUSTRALIA

The Sydney Opera House is a complex structure, shown in Figure 9.6. It comprises a massive base of cast-in-situ reinforced concrete (in which the various theaters, stages, amenities, etc., are housed) surmounted by fourteen

Figure 9.6   Model of the Sydney Opera House, Australia.

upright shells of precast concrete, stressed down to the foundations. These shells are composed of segments of very complex shape, the total number in the complete structure being nearly 2400, plus over 4000 precast tiled panels. Some idea of the vastness of the building and of the complexity of the project will be gained from Figure 9.7, which shows the work in progress. Its overall

Figure 9.7   Sydney Opera House under construction, June 1964. The outlines of the roof shells are shown. The casting yard in the immediate foreground is not visible.

dimensions are: 195 m long, 135 m wide, and 75 m high from ground level to the top of the largest shell.

All the segments were precast on the site and erected (after curing) by three Weitz tower cranes, each of 300 meter-tonne capacity. Areas available for casting and stacking of segments were extremely restricted, since the site is a promontory adjoining the city proper. Hence it was essential that a coordinated manufacturing and erection program be instituted from the start.

To this end, two distinct networks were initially developed, one to cover the casting yard and the other to cover the erection and in situ work; these networks were respectively composed of 1400 and 1700 activities. As the project progressed and the number of activities was reduced (by their completion), it was possible to refine both networks and then to combine them into one composite diagram for the remainder of the project; even then, the combined network contained 1750 activities and 940 events. This network was updated at regular monthly periods or whenever special circumstances required additional review, using the principles outlined previously in this chapter. Naturally the entire work was done by computers; thus the time status of the project was available to management immediately after the close of a review period.

Time status, however, was not sufficient. The main problem in proper construction control is to have accurate costs available as soon as possible after the close of a review period. With conventional costing systems it may be possible to produce labor costs quickly, but materials costs inevitably lag behind by at least a month; on this project such a situation was intolerable, and consequently computer cost control was deemed essential. Although the initial costing and accounting procedures introduced on the Sydney Opera House project were not directly linked to the critical path network control, the entire system was devised so as to require only simple amendments in order to incorporate the necessary features of such a linkage.

In this system, the daily labor data were punched on cards, showing manhours against activities and cost dissections. Similarly, all materials orders, invoices, and site deliveries were also converted to punch-card records. Purchase orders showed their serial number, cost dissections, suppliers' names, and the estimated value of the goods. Site delivery dockets showed initially the order numbers of the items delivered and the percentage of the total order on this delivery; later other data were added for costing purposes when the invoice was not available. Invoices showed the order number, the invoiced amount, and whether the order was completed.

General details of project cost control with CPM are discussed in Chapter 10. It is sufficient to say here that the principal output from the Sydney Opera House control system included a monthly cost report showing the cumulative and period costs for each cost dissection; estimated materials costs were used when the invoices were not then available so that this report could be ready within 3 days of the close of a cost period. In addition, the system reported on future financial commitments, outstanding deliveries, and any variance between estimated and invoiced amounts on each materials order. Further-more, the monthly purchase journal and payment advice slips (with the supplier's name and address, ready for mailing) were also produced.

9.10 PROJECT CONTROL AT THE SYDNEY ENTERTAINMENT CENTRE, AUSTRALIA

The Sydney Entertainment Centre is a large auditorium seating more than 12,000 people, designed for mass audience enjoyment of popular entertain-ment, indoor sports, and spectacles. The most significant feature of the structure is the massive reinforced concrete hollow-box horseshoe-shaped ring-beam which supports the roof and permits an audience view unobstruc-ted by columns, and to which the sloping auditorium decks are posttensioned. Figure 9.8 shows the model and Figure 9.9 is a longitudinal section of the structure. The auditorium steel-truss roof has a span of 100 m, and the area under the roof is 9000 sq m. The estimated cost was about $40,000,000 and the total time for design and construction was 3 years.

John Holland (Constructions) Pty. Ltd. were the management contractors for the design and construction of the entire project, which was controlled by the "fast tracking" project management technique, requiring very close coordination between the architects, structural, mechanical, and electrical design engineers, the construction team, the various suppliers and the off-site fabricators. Because of this it was necessary to develop a network that contained separate design, construction, procurement, off-site fabrication,

Figure 9.8 Model of Sydney Entertainment Centre, Australia.

and commissioning activities so as to permit rational monitoring of progress and regular updating against each of these components. This network contained about 1000 activities and was computer-analyzed and updated every month, producing the usual early start and total float reports, together with critical activity and short-term planning reports tailored to the project's needs. These reports were subdivided into design, construction, procurement, and administration, permitting close analysis of progress and consideration of alternatives by the team member responsible for each area.

Two basic principles were adopted in all these reports: the first was a simple and meaningful presentation devoid of data processing terms; the second was that scheduled dates were always the earliest dates. In addition to the regular updating, parallel simulation of alternatives were carried out when necessary, to examine possible changes in construction method, timing, and so on. The project completion date was always inserted as a milestone date based on the contract completion date plus authorized time extensions; this enabled real progress to be determined regularly and constructive action to be planned. The short-term planning report was in the form of a bar chart, and progress was measured against this report and used as input for the regular updating.

Resource and cost analysis based on the network was also carried out regularly in order to minimize the effects of labor shortages and increasing costs. This system has allowed the project management team to keep the various critical paths (and alternatives) clearly in view, despite weather and such industrial problems as strikes. Linked to the network was a cost-reporting system that monitored the cumulative and period costs against the budget, and forecast such data as cash flows and likely expenditure based on the logic and duration of the various activities.

9.11 PROJECT CONTROL FOR AUSTRALIAN SUBMARINE REFITS

For many years CPM has been used for major project planning and the control of contracts carried out by Vickers Cockatoo Dockyard Pty. Ltd. at its island dockyard (two dry docks) in Sydney Harbour, Australia. In the past 17 years the company developed its own interesting computer-based network system and applied it to the work of refitting submarines for the Royal Australian Navy.[1] Ship construction and repair operations in a major dockyard require great flexibility in planning because of the ever-changing demands for labor, materials, and information. This is especially so with a submarine refit, which involves a vast number of closely interrelated tasks, particularly during the on-board phases of the work.

---

[1] See "Submarine Refitting in Australia" by J. Jeremy, *Journal of the Royal Institute of Naval Architects,* March 1981.

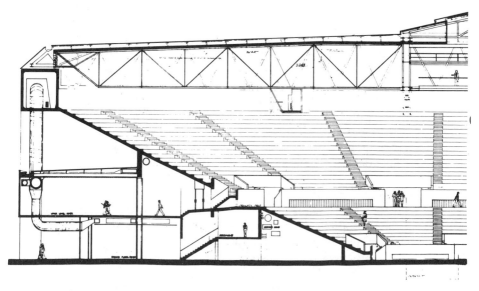

Figure 9.9   Sydney Entertainment Centre, Australia—longitudinal section.

Preparation of a "refit network" for the first submarine began 2 years before the refit start date, and comprised some 8000 activities. It proved too cumbersome, so for the second refit it was slashed to 3000 activities. This also was unsatisfactory, being much too broad to adequately reflect the complex interrelationships involved. By the end of the third refit, sufficient experience had been gained by the planning staff, and a standard computer-based refit network had emerged. This is essentially activity-oriented, with some 6000 activities and 35 milestone events for progress monitoring purposes. The advanced planning is now much simpler as the standard network can be read from the file and adapted to suit the particular project contemplated. The computer is installed in the dockyard and provides a multiproject planning system tailored to the dockyard's ship building and repair requirements; the planning team numbers six people, who handle all the detailed planning and progress control.

As standardized work packages are developed, the number of activities in the standard refit network are being reduced; network detailing is now more in the area of alterations, additions, and modifications approved for the refit. The networks are updated monthly, with full resource analysis involving 13 resources and averaging two resources per activity. Revisions are made monthly, and computer printout schedules are issued to the dockyard foremen. The use of planning staff for progress control has resulted in disciplined and complete information feedback and has been an essential component in the success of the system—as has the use of the company's own staff as

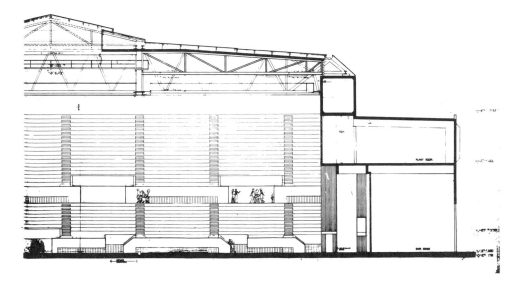

systems analysts and programmers. As a consequence, the reporting process is kept adequately updated so as to produce realistic schedules without last-minute revisions.

Labor costs (recorded daily on each workman's time docket) are input to the labor costing system, from which charges against each work order number are transferred automatically to the planning system where physical progress reporting is also centered. For submarine refits a progress review is made monthly, and a monthly report is made to the Navy. Following the analysis of the refit network on the basis of reported progress, the planning department prepares a progress report which is circulated to senior management. This report is the catalyst for an internal review of progress and an assessment of any logic changes that may be necessary to maintain program. A final network review after these changes provides the basis for the next issue of work schedules. During the later stages of the refit, the tests and trials program becomes the principal control document. Based on a trials sub-network, this program is reviewed and updated on a weekly and, finally, a day-to-day basis.

Evaluation

a. How useful do you consider the manual reports in Figures 9.1, 9.5*a*, and 9.5*b?*
b. Is network revision a worthwhile effort? Why?

PROBLEMS

9.1 A works report for the project in Problem 4.2 at time 20 gives the following:

Activity	Status	Time worked	Time to complete
1–2	Finished	8	0
1–3	Finished	9	0
1–4	Finished	12	0
2–5	Working	12	6
3–4	Finished	5	0
3–6	Not started. 3 to go.	0	13
4–5	Working	6	1
4–7	Working	5	8

All other activities as before.

Determine (*a*) the current status of the project and (*b*) the necessary steps to retrieve the position.

9.2 On the project in Problem 8.5, it was decided to complete all clearing before beginning earth works in order to use the same crew, who would not be available to start until day 12. This would still permit completion in 104 days. On day 35 status reports showed that all work was on time except the culvert at Ch.96, which required 9 days for completion before curing.

(*a*) Draw the original network and update it to show project status at day 35. What is the total delay?

(*b*) What remedial measures can be used to retrieve the position?

10

FINANCIAL PLANNING AND COST CONTROL

10.1 NECESSITY FOR FINANCIAL PLANNING AND COST CONTROL

There is undoubtedly room for improvement in financial management and cost accounting in the construction industry. Competition is essential to healthy business, but so are the profits that enable private enterprise to finance contracts, maintain an efficient staff, and make a reasonable return on the money invested by the shareholders. In construction companies working on relatively small profit margins because of keen competition, the financial planning and budgeting must therefore walk hand in hand with first class technical management, construction planning, estimating, accounting and costing; until this objective is attained, the failure rate in construction companies will remain a serious problem.

Financial planning includes correct calculations of the profit margins necessary to provide for fixed administrative expenses, equipment replacement, dividends, and project financing. The maximum economic volume of work which can be handled with a given financial structure must also be known. Essential prerequisites for fulfilling these intentions are, first, an efficient system of budgeting and costing, and second, a reliable and accurate estimating department.

Since estimating data are based on production performance and costs, it follows that construction accountants should find CPM a powerful, quick-acting tool for cost control and an aid in producing more accurate estimating data. Critical path analysis divides projects into activities of short duration so that each activity may be budgeted and scheduled logically. In the same way,

ity may be budgeted for and costed by the construction accountant, riations from the estimate can be determined as soon as they occur dial steps taken—just as construction managers detect and remedy site variations from their original estimated activity durations.

By comparing estimated with actual costs as each activity is finished, the financial planning becomes more accurate. Furthermore, it may be adjusted as necessary while the work is in progress, just as the actual construction work is adjusted to meet remedial changes in the network model. "Costs to complete" become more reliable and are repeatedly tested as the various activities are concluded. In addition, the cost accountant conversant with critical path procedures will be able to introduce techniques for evaluating quickly the effects of either slackening or crashing of any operation on both activity and total project costs.

By breaking down direct costs into labor, materials, plant charges, and so on, for each activity, the actual cost of each completed operation can be quickly analyzed and compared with the estimate, just as the physical progress of the work is periodically reviewed. Finally, the estimating department can be notified of all significant variations in cost performance and all costs may be annotated to show whether they originated from normal, crashed, or partially crashed operation. The estimator thus has access to more detailed and meaningful cost histories on which to base forthcoming contract bids.

With reliable estimating and accurate cost control, the required profit margin may be properly assessed. Furthermore, a close working relationship will develop between the estimator, the accountant, and the manager because of benefits accruing to each from adopting CPM as a logical solution to their problems of planning and controlling work and resources.

10.2 USE OF CRITICAL PATH METHODS IN FINANCIAL PLANNING

In preparing the overall financial plan for a project, it is necessary first to examine closely the network model, and from it to schedule the essential site costs that must be expended in executing the works; this is readily computed from the activity timing and activity direct costs, to which is added the indirect expenditure at its appropriate rate of occurrence. A plot of these total cumulative site costs against time will show the financial expenditure required.

The next step is to fix the gross profit margin to be added to this particular bid. This depends on the annual gross profit plan for the company and the ratio that the value of this particular contract bears to the total contracts (gross turnover) expected to be obtained. From past records and normal accountancy practice this profit can be determined as a lump sum or (as some companies prefer) as a particular amount per manweek of the average labor force annually employed. In the latter case, for example, if the average overall work force employed by the company in the immediate past has been, say, 10,000 manweeks per annum, and if this particular project has an esti-mated labor content of 3300 manweeks per annum, then it would be marked up with a profit sum equal to 33% of the total gross profit per annum which the company requires in order to stay in business.[1] The major advantage of CPM in this connection is that it provides more accurate estimates of project duration and labor content than is available by former conventional estimat-ing procedures.

Having added this profit margin to the total site costs, and priced out the contract bill to equal the total bid amount required, the network diagram (or its bar chart schedule) now provides an accurate program from which to calculate the rate of cumulative income-earning capacity of the project, in accordance with the progress payment provisions of the contract. This is then superimposed in the usual way on the plot of cumulative site expenditure, to give the financial investment required by the project. Again the advantage of CPM is the accuracy attainable in these calculations, so that the maximum and average overdrafts, and the period for which they are required, are accepted with confidence.

A further direct advantage of CPM is that the gross profit plan of the company has been founded on a system that, in practice, will enable the work to be done within closer budgetary limits than is otherwise attainable, and hence the profits earned are close to those anticipated (except in very unusual circumstances which could not possibly be predicted). It is therefore a simple matter to watch the volume of business obtained, and to interpret the profit consequences with reasonable precision.

The most important feature of CPM in construction finance, however, is the use of compression and/or decompression to determine logically and mathematically the optimum duration of a project for least overall cost. By this feature alone one determines with certainty the most economical plan for doing the whole contract. This, in turn, ensures obtaining a reasonable pro-portion of available work, and hence meeting the financial requirements and profit planning of the company. The accountant will therefore be quick to appreciate the potentialities of CPM network analysis for more efficient administration of financial affairs.[2]

In the development of a fully detailed construction plan for a project, the first use of critical path scheduling was to obtain the optimum project dura-tion; then, by means of resource leveling (see Chapter 7), to take advantage of the float times of noncritical activities to obtain a more even distribution of

---

[1] Gross profit required is the sum of head office costs, business promotion costs, planning and estimating costs, capital retention for future growth, taxation, etc., plus the desirable amount of net profit to be paid as dividends to the company's shareholders.

[2] Some accountants comment that precise utility data for crashing (as well as the compression calculations) are difficult to estimate (and execute). It is well to remember that about two-thirds of all activities never deviate from normality; and that, in the other one-third, few eventually reach full crash. Optimal solutions, therefore, may be satisfactorily derived from reasonably approximate data (see Section 8.2).

labor and other resources, and even to vary the sequence of operations. Next, the final network model was examined in order to determine the overall project financing and profit margin, and thus the total contract price was derived conformably with the adopted project plan. The detailed budgeting for material, labor, plant or other resources required, and for the funds necessary to finance the project, may now be developed.

Budgeting (or forecasting) for the plant, materials, and labor is relatively simple, once the project network has been established; and future variations necessitated during construction can be easily dealt with, because the altered requirements are known well in advance. Finally, from the costed network the total project expenditure and income can be budgeted accurately. This financial resource budget is arranged so as to fit in with the general overall budgeting for the company's operations; by further resource leveling (if necessary) the requirements for, say, cash flow can be suited to the availability of funds. This can obviously be very important with a large company having a number of projects at the one time. Cash requirements can be accurately foreseen, and the peaks and hollows which normally occur can, to a certain extent, be smoothed out.

Consider, for example, the sewer main extension project discussed in Section 8.5, where the network for the optimum solution was found to be that shown in Figure 8.7*b,* and the total works cost was determined as $161,080 (see Figure 8.8). After resource leveling, the final schedule appears as shown in Figure 10.1, with financial expenditure indicated at *A;* from this the works expenditure curve is plotted in Figure 10.2. The gross profit markup was fixed by management at $18,000, giving a contract price of $179,080. It was specified that progress measurements would be made every 4 weeks (20 working days), with payments (less 5% retention) due 20 working days later; on this basis anticipated income was calculated and added to the bottom of the schedule, as shown at *B,* and the works income curve then added to Figure 10.2. As will be seen, the maximum financial accommodation required is $36,380 midway through the job, with a general average of about $25,000. Management thus has an accurate forecast on which to arrange the necessary finance.

It must not be forgotten, however, that the accuracy of any budget depends on the accuracy of the data on which it is based. This maxim is important, because not only does financial planning include the cash resources required to support the required expenditure, but also the income to be obtained from the client during the execution and at the conclusion of the project. Although the expenditure and income-earning capacity may be speedily planned with reasonable accuracy, it is essential to realize that the client may not be so meticulous in payment of claims. Although one cannot perhaps provide against such a contingency, one can nevertheless use the critical path network to determine (from available float) which activities may be postponed until the cash position is rectified; or, alternatively, if additional cash resources must be obtained elsewhere in order to carry on, one can

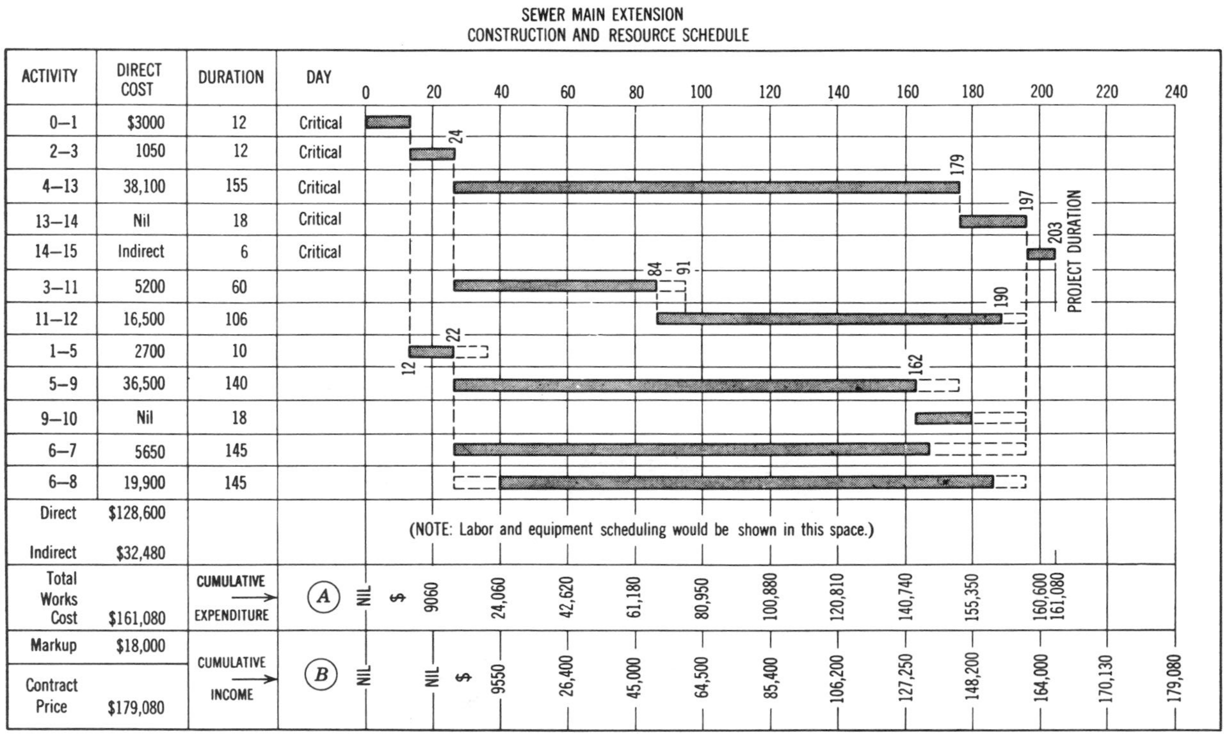
SEWER MAIN EXTENSION
CONSTRUCTION AND RESOURCE SCHEDULE
ACTIVITY
DIRECT COST
DURATION
DAY
0 20 40 60 80 100 120 140 160 180 200 220 240
0–1 $3000 12 Critical
2–3 1050 12 Critical
4–13 38,100 155 Critical
13–14 Nil 18 Critical
14–15 Indirect 6 Critical
3–11 5200 60
11–12 16,500 106
1–5 2700 10
5–9 36,500 140
9–10 Nil 18
6–7 5650 145
6–8 19,900 145
12
22
24
84
91
162
179
190
197
203
PROJECT DURATION
Direct $128,600
Indirect $32,480
(NOTE: Labor and equipment scheduling would be shown in this space.)
Total Works Cost $161,080
CUMULATIVE EXPENDITURE
A
NIL $ 9060 24,060 42,620 61,180 80,950 100,880 120,810 140,740 155,350 160,600 161,080
Markup $18,000
Contract Price $179,080
CUMULATIVE INCOME
B
NIL NIL $ 9550 26,400 45,000 64,500 85,400 106,200 127,250 148,200 164,000 170,130 179,080

Figure 10.1 Construction and finance schedule for sewer main project. (See network diagram, Fig. 8.7*b*).

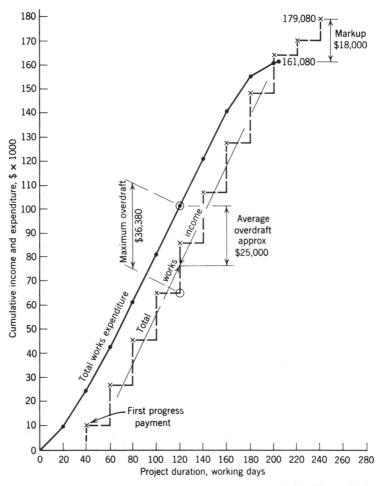

Figure 10.2 Finance chart for sewer main project. (Scheduled in Figure 10.1).

assuredly determine from the network schedule the most economical sequence of activities to pursue.

Finally, in the development of the financial plan, one also determines the major monetary control points for the project. These subsequently become the main network events, around which is built the structure of the cost control system to be initiated as soon as work commences on the site.

10.3 USE OF CRITICAL PATH METHODS IN COST CONTROL

It is not proposed to expound here on details of the conventional accountancy methods adopted in construction. It is, however, important to point out that, with the introduction of critical path procedures for the planning and site control of construction operations, it has become essential for construction engineers to learn something of works accountancy and for cost accountants to familiarize themselves with this new mathematical technique. This applies to both vocations, whether they are engaged as professional practitioners, or as salaried officers in construction companies or in management advisory services. Furthermore, by this new management tool, improved communication is achieved between the construction engineer and the cost accountant: both can now speak the same language, both can now see the project from the same viewpoint, each can readily learn and understand the procedures and the objectives of the other in their respective daily tasks.

It will be realized at the outset that an estimate is intended to be a reliable forecast of probable costs; as such, no matter how accurate, it can only be an opinion. Costs, on the other hand, are the actual historical records of expenditure incurred; as such, they are a statement of facts. When these facts are fully recorded for all the activities in various types of projects, they provide invaluable data for the calculation of future estimates. This accurate correlation between estimated and actual costs has only been practicable since the introduction of CPM drew attention to the great significance of *the activity* in construction.

The concept of the activity as a costable unit is the basis of a new accounting philosophy. The works items customarily provided in a bill of quantities are only remotely connected with the actual detailed operations to be performed, and a costing system based on these items or similar work subdivisions cannot provide proper works control. With the new philosophy, however, each activity is subdivided to suit costing periods with an estimated duration and cost, and it is then a very simple matter to compare these with actual performance in time and money as each subactivity is completed. Estimates of the percentage of work completed to date are eliminated; the actual accomplishments for each subactivity are a much more precise measure of the progress of the works. Herein lies not only the advantage of greater accuracy, but also of greater speed in providing the required answers for both estimators and management.

The use of CPM, therefore, simplifies the manager's task of controlling the costs of a project and provides more accurate information from which to prevent losses occurring. It can also simplify the overall accounting function through the integration of costing with the general accounts. There are many ways of achieving the degree of control and integration required, and it is therefore sufficient to present here only the main points and principles involved.

In industrial operations, one compares the actual cost with the "standard cost," which is (as its name implies) a standardized cost for the operation or work under consideration, and is determined by some analytical process such as work study. In construction this is not feasible, because of the wide variations in the nature of the operations; control must therefore be based upon the estimated cost. This is indeed a very satisfactory way of achieving control, for the profits of any construction project are also based on estimated costs. If adequate control is therefore exercised against the estimate, then the profitable operation of the work can be ensured.

Where cost control systems are based on critical path methods, the requirements of accuracy and speed, in the production of the results, will make the network somewhat more complex. It is necessary to break up major operations of the job into more detailed activities and to bear in mind the objects of control when preparing the diagram. For instance, it is preferable to have, wherever possible, only one trade or one crew in each activity. Furthermore, activities should, as far as practicable, be subdivided into durations not exceeding 1 or 2 weeks; those which would normally take longer should (except for operations like delivery of materials, approval of plans, etc.) be split into two or more activities. This will allow effective control to be exercised over continuing activities by drawing attention to variations from the estimate as soon as they occur during the execution of prolonged site operations.

The scheduling of the network includes the duration of the activity, the number of workers required, their labor skill classification (suitably coded), and the estimated direct costs of labor, plant, and materials. For activities that will be subcontracted, only the duration and total direct cost needed be recorded, and these should be as agreed on with the respective subcontractors. The total of these estimated direct costs, plus the preliminaries and other indirect costs, together with the contractor's overhead and profit, equals the contract price. The estimated direct costs are recorded on a cost card or ledger sheet; each activity is identified by its node numbers in the network. Thus the cost classifications match the defined work items shown in the network model as specific activities; this is a most important feature of this method of costing, as will be understood shortly.

In order to avoid unnecessary duplication, it is preferable to integrate the cost accounts with the general financial accounts as the work proceeds. At the beginning of the project, each activity account, an overhead account, and a preliminary account would be credited with the estimated costs, and the total of these would be debited to a cost-to-complete account. Adjustments would be made to these figures for any variations as received, and also for variations occasioned by provisional items.

Labor Costing

The immediate or project supervisor records the hours spent by site labor on each activity in the usual way. The total gross payroll is then debited to a labor variance account weekly, and the weekly summary of worker-hours spent on each activity is used to debit the labor costs to each activity account; the total of these debits is then credited to the labor variance account.

It goes without saying that basic timekeeping information must be reasonably accurate (as has always been the case for construction costing), for it will be obvious that *time* is the unit by which all productive and unproductive labor is controlled against the estimate. The introduction of CPM has emphasized the significance of time, and it has also added two other important factors: rate of work (normal or crashed) and crew size. It is therefore a waste of clerical effort to compute money costs for each activity or each labor skill; it is the analysis of hours worked (plus observations of crew size, overtime or other premium payments, and variations in the average overall project labor cost rates) which enable the actual performance to be measured against the estimated performance or its site revisions. When required, the gross worker-hours spent during a cost period may be readily converted to money at the average overall project labor cost rate currently applicable.

From the start of a project right through to its completion, one is thus able to compare the actual with the estimated worker-hours for each activity, or for common services or indirect operations, and in total. On appropriate occasions one can check the efficiency of the crew sizes by applying two or more different teams to the same operation and comparing their productivity per worker-hour.

The former detailed periodical labor cost statements are replaced by worker-hour expenditure statements, showing, for each activity:

1. The estimated time required.
2. The time spent to date.
3. The time remaining in the program.
4. The time required to complete the operation, at the current rate of progress.
5. The anticipated gain or loss on the time allowed.
6. On completion, the actual time gained or lost, and its percentage effect on the total labor content of the project.

Figure 10.3 illustrates a statement of this type; these worker-hour reports place the proper perspective on labor costing, both for estimating and for project control.

One other aspect of labor costing deserves further consideration: the unit of work to which labor costs should be equated. Traditionally, all labor charges in the past have been charged against some type of material unit, such as cubic meters of concrete, square meters of formwork, tonnes of steel, and so on. It is now realized that this is, in many cases, most unsatisfactory. One classical example is the conventional method of measuring labor for concrete stair soffit forms on a unit area basis; labor costs associated with this construction are about the same for a 1-m-wide stair as for a 1 ½-m-wide stair, even though the latter contains 50% more area per unit length. Hence, as in many similar instances, it is preferable to adopt a labor cost per activity rather than to use a labor unit cost.

Labor Progress Summary — **Contract No.** .... — **Location** .... — **Week Ending** .... 196....

Operation		Activity i–j	Worker–hours planned	Total worker hrs. to last week	Worker–hours this week	Cumulative total worker–hours	% Complete	Worker–hours needed to complete	Worker–hours left to complete	Loss Expected	Loss Actual	Gain Expected	Gain Actual	Comment
Formwork	FF	14 17	500	220	180	400	70	170	100	70				
Concrete	FF	17 18	180	–	85	85	50	85	95			10		
Formwork	SF	19 23	500	400	210	610	100	–	–		110			
Reinforcement	SF	23 24	200	90	70	160	100	–	–				40	
Concrete	SF	24 27	300	–	285	285	100	–	–				15	
1.		2.		3.	4.	5.	6.	7.	8.	9.		10.		11.

Total Worker-hours Planned for Completed Operations (i.e. 100%)

Total Worker-hours spent on only 100% Completed Operations

Total for Operations in progress

Total for Completed Operations

Delay Analysis:
$$\frac{\text{Worker-hours to date (Col.5)}}{\text{\% Complete (Col.6)}} \times \text{\% Incompleted} = \text{Worker/hours needed (Col.7)}$$

Net = G/L

Net = G/L

Net {Gain / Loss

Discuss Progress:
(a) Efficiency of Workers
(b) Difficulties of Work Methods
(c) Effect on other Work

Figure 10.3 Typical weekly worker-hour expenditure statement.

All construction estimators know that the prediction of labor costs is the most difficult part of their work. This is the field in which they require the maximum amount of help from the cost accountant. The real cost of labor is the number of worker-hours required to perform a specific work task. With CPM, the recording of reliable labor costs has become possible, for each relevant work task may be shown on the network model as a separate and unique activity.

Activities representing specific work tasks must be devised around suitable crew sizes and labor skills. The site costing of these individual work task activities, over a wide range of projects, will eventually enable the compiling of labor task costs. This will then provide the estimators with the proper figures on which to base their future predictions, for they will then have real values, based on factual and costable units of labor.

Costing of individual work processes is not new, as such; its importance has, however, been emphasized by the introduction of CPM, and its development into "activity costing" has resulted. What is new is the concept that labor costing and worker productivity records must show the size (and skills) of the crew and especially whether the activity was completed at normal or crashed speed; in crashed speed, the nature of the speeding up—overtime, shift work, and so on—and the percentage of full available crash should be indicated. Only when such cost data are available will estimators be able to prepare really practical utility data for their compression calculations.

Materials and Services Costing

The system adopted for ordering materials and services, including subcontractors and for collating the commitments incurred, must follow well-proven methods, designed to ensure that purchases do not exceed the planned distribution of budgeted expenditure for the project. When the construction is planned by CPM, the purchase orders are allocated to individual or groups of activities by the appropriate activity numbers. To reduce the volume of clerical work, the quantity of orders issued should be kept to a minimum, but separate orders must be directed to each service, commodity, or collection of materials purchased from one supplier; and separate orders must be likewise issued to each subcontractor.

From the network model of the project, the materials schedule will show the total quantities (and prices) required, the delivery dates to be met, the allocation of quantities and expenditure to each activity, the estimated rate of expenditure, and the total materials fund for the project. Every purchase order should stipulate the first three of these, so that the supplier or subcontractor may be well aware of the requirements expected to be met.

On receipt of materials at the site, the delivery notes are marked with the activities in which the material or service will be used, showing the quantity for each where the delivery covers more than one activity. In some cases this system may lead to a larger number of smaller deliveries, but this in itself provides a very effective control. The delivery notes are immediately checked against the appropriate purchase orders, and are then priced and extended, thus obviating delays arising from waiting for the invoices. These extended delivery notes are used to debit the activity accounts, and the corresponding credits are made to a materials variance account. The debits to the materials variance account are made from the purchase journal, and the balance transferred to a project profit and loss account. Cumulative figures of actual and estimated costs of certain materials (such as cement and bricks) may be separately maintained, if desired, to provide a check against spillage or theft; serious discrepancies disclosed by these figures may then be investigated.

Subcontract performance and costing follows a similar pattern. It is essential in this case that subcontractors, in submitting their prices, be made fully aware of the requirements of the project network and of the effect on the project duration of their own activities. In addition, specific events should be stipulated for assessment of each subcontractor's performance; that subcontractor's progress payments will be computed at each of these control events. During the construction it will be necessary for all parties to cooperate in any desirable replanning, and to define and value variations promptly, so that activity modifications may be adopted without delay when network revision is to be undertaken.

Plant and Equipment Costing

As with labor, the number of productive hours worked (in this instance, by the machine) is vitally important from a costs viewpoint. Plant operation, like labor, is therefore recorded as the time spent on each activity; and from this it follows that the weekly summary of machine-hours consumed (for each class of machine, if necessary) on each activity in the project is the basis from which to begin the costing by debiting plant costs to each activity account and crediting the total to a plant variance account. This principle applies, whether the plant is hired from an outside source or whether it is owned by the contractor (irrespective of how he charges it to the project, which the contractor may do by an internal hire system or by a specific expense write-off); the same processes are used for costing plant and equipment as were used for labor and materials.

Unlike labor, plant productivity may often be estimated with reasonable certainty. Costs are therefore generally estimated against the traditional units (cubic yards of excavation, lineal feet of trench, areas of finished surface, etc.). Nevertheless, there are still a number of machine tasks where the conventional unit cost is not a true measure; it is in these cases that only CPM provides a satisfactory alternative by enabling plant costs to be evaluated for individual tasks represented by unique activities in the same way as labor costs may be determined for special work tasks.

To be of value as a control criterion, performance is required in machine-hours, and the periodical plant costs should be presented as plant-hour expenditure statements, covering each activity, as was described for the presentation of labor costs (see Figure 10.3).

Costing Overheads and Indirect Expenses

Critical path networks can be extended to include activities for unproductive work items, preliminary operations, project services, final clean-up and various other necessary tasks which are usually included (for estimating and costing purposes) in the indirect costs. In this way these important—and sometimes costly—items are portrayed as unique individual activities on the network model, and hence in the costing system; thus they may be separately recorded and reviewed in the project cost returns.

There is indeed no limit to the multiplicity of overhead and indirect charges which may be separately portrayed in this way, except that of practicality and economy. A detailed breakdown should not be incorporated in a network merely to accumulate the costs of trivialities, but only if this will serve some useful purpose either in project control or for future estimating. Where, however, they are of importance, their delineation as network activities will enable regular and accurate evaluation of their magnitude and seriousness; they may thus be watched by management as site construction proceeds.

The manner and extent to which indirect expenses are thus portrayed as additional activities will vary from one single arrow for the duration of the project to a whole complex series of short duration each. Many items of site expense and overheads are worthy of attention in this way on a major project. Indeed, there has been a scarcity in the past of reliable information concerning indirect costs; with the use of CPM this lack can be rectified.

Periodic Cost Reports and Accounts

For cost reports to be of real value, they must be up to date, accurate, and regular. The use of the activity as a costed unit enables all these objectives to be achieved. As was mentioned in Section 9.9, the monthly detailed cost report for a complex project comprising over 3000 activities can, with the aid of computer processing, be presented to management within 3 days of the close of the costing period. On smaller projects, without resorting to computers, comparable speed may be attained.

These cost reports should present not only the costs to date plus forecasts of costs to complete each unfinished activity, but also cumulative valuations compared with total costs to the same date in order to indicate the gross profit so far attained and the final profit now envisaged. A suitable cost report of this nature is shown in Figure 10.4. In such reports, the current and future predictions may be progressively adjusted to follow the vagaries of the work

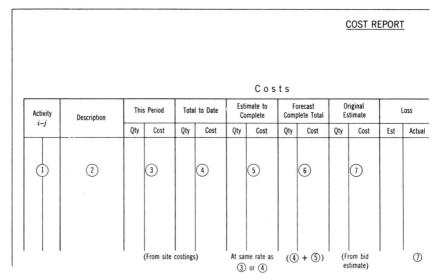

COST REPORT

Figure 10.4   Typical construction cost report.

in hand. At the same time, economies or excesses may be readily assessed, on similar forms to that of Figure 10.3, with respect to labor, materials, plant, site overheads, etc., as may be expedient.

In every cost period the main accounts are brought up to date. It will be remembered that the daily job status reports described in Chapter 9 show which activities have been completed. At the end of each week (or other suitable period) the accounts for the activities finished that week are closed off, and the balance transferred to the project profit and loss account. The journal entry shows, for the information of site management, any significant variations from the estimates that have occurred; a copy of this information can also be made available to the estimating department for use in preparing future estimates. The cost-to-complete account is credited each week with the total estimated costs of the activities finished that week, and this is in turn debited to the client's account. Each month the client's account is also debited with the overheads and other indirect costs incurred that month and the cost-to-complete credited. The client's account is also debited with the profit earned, and this amount is credited to the project profit and loss account. The accounts for overheads and other indirect costs are debited with these items as they occur, and a monthly and cost-to-date report and budget prepared in the usual way for controlling these items.

Where the cost and financial accounts are integrated, clerical procedures are oriented to the rapid preparation of periodical "Profit and Loss" and "Income and Expenditure" statements, as well as balance sheets, overhead expense analyses, and statements showing the source and distribution of funds. With these regular up-to-date reports, management is in a position to

Contract No. ....

Period Ending ....

Valuations

Gain		Total Valuation		% Now Complete	Valuation to Date	Result to Date		Final Forecast When Completed		Planned Profit	Expected Final Result	
Est	Actual	Qty	Amount		Amount	Profit	Loss	Profit	Loss	Amount	Loss	Gain
⑧		⑨		⑩	⑪	⑫		⑬		⑭	⑮	
− ⑥)		(From contract prices)		$\left(\frac{Qty④}{Qty⑥}\%\right)$	⑨× $\frac{⑩}{100}$	⑪ − ④		⑨ − ⑥		⑨ − ⑦	⑭ − ⑬	

review and properly control its construction operations, irrespective of how widespread these may be. The Board of Directors and the company Secretary will therefore appreciate the benefits deriving from the speedy presentation of financial accounts through their integration with a project costing system based on CPM techniques.

10.4 PROFIT CONTROL

The cost report illustrated in Figure 10.4 presents not only costs, but valuations as well, thereby enabling the gross profit position of a project to be constantly watched. For a company with several projects in hand simultaneously, a similar form may be used to summarize for top management the overall operations of the organization, so that the periodical changes in total gross margins are regularly under review. For rapid assimilation, actual profits may be plotted against planned profits, both for individual projects and for the entire operations.

Based on project networks, these figures are not only accurate and currently up to date, but remedial proposals (when required) may be rapidly analyzed on the appropriate network models. Where an element of uncertainty is present in forecasting future trends, the three time predictions (and corresponding costs) of the PERT technique may be introduced, thus providing probable evaluations grounded on sound logic and precise mathematics. In this way the element of risk in taking important decisions is calculated, if not entirely removed, and the relative effects on profit evaluated.

As a further point, consider a single project in which cash is becoming critical (because progress payments are late or for any other reason); an examination of the network model will at once disclose where advantage may be taken of available float to defer purchases or other expenditure, thus temporarily reducing the finance required (and the interest payable thereon) until the cash situation can be rectified. This technique of resource leveling is a most useful adjunct to the planning and control of finance and profits.

The incorrect prediction of profits through inaccurate estimating is immediately apparent when costs and valuations are related to network activities. When this position arises, true forecasts of future trends may be predicted from accurate current site costs to date, applied to the unfinished portions of the network. It must be conceded that this is much better than the sometimes current hope that the position will improve with time! No estimation can be expected to be perfect, and from time to time remedial actions will be necessary; with CPM their relative effects may be logically analyzed before the situation becomes serious and alternative strategies may be given ample consideration before a final decision is taken.

Profit planning, like project planning, is capable of analysis on a network model. It follows that profit control, like project control, is more accurately achieved when based on CPM.

10.5 PERIODICAL CLAIMS FOR PROGRESS PAYMENTS

The adoption of a network as the basis for the cost coding of a project can be put to a further use in valuing the work done for progress payment claims. The treatment required will depend on the nature of the construction and on whether the bill of quantities allows for preliminaries and temporary works as separate items or not. If the preliminary and temporary works are separately itemized and paid for, the costed network can be used for progress claims as it stands; if the contract is one in which the cost of all nonpermanent items is spread over the specified permanent items in the bill (say on a percentage basis), then this spread percentage must be added to the estimated activity costs to arrive at the price for each activity. In either case the network, with suitable variations, still forms the basis for arriving at the amount of each periodical progress payment claim.

There is, however, one very important point which must be appreciated when the costed network is used for this purpose. In the planning, and in the cost control system discussed previously, the cost figures used in the network model for each activity are the estimated direct costs; indirect costs and site overheads are regarded as extra activities, and are so costed, but these values on the network are still estimated costs. Monthly progress claims, however, must be calculated against the contract prices shown in the contract bill of quantities.

Clearly, the difference between the estimated direct cost of an activity representing a billed item and the contract price for this item is the amount of estimated indirect cost and the profit pertaining to this item. It is not difficult to convert the estimated direct cost shown on the activity into the appropriate price shown in the bill by applying a standard percentage correction; this approximation is usually reasonable enough for the purpose of progress payments, from which a retention amount will doubtless be withheld in any case. When the overheads and profit markup are spread evenly throughout the direct cost estimate, this method is precise. However, when the overheads and profit are varying from activity to activity, or from one section of the project to another (as may be the case on some projects to suit particular circumstances, or to suit the cash flow required), some complexities will arise—although the costed network cannot then be adopted directly, a system can be devised to cope with the situation.[1]

One of the great advantages, in the restriction of the duration of each activity to a maximum of a week for costing purposes, is the ease with which sections of the project may be assessed for the proportion completed at the close of each costing period. It will be obvious that the internal costing periods, and the periods specified for progress measurements, must coincide for two reasons: first, so that costs, valuations, and progress claims may be compatible and second, so that duplication of clerical work is avoided. There is no reason, however, why costing periods should not be fractionally shorter than payment periods, if this is advantageous to project control.

The critical path network, each activity of which has a predetermined income payable by the client and an estimated expenditure incurred by the contractor, has thus provided the construction manager with the simplest means of preparing cost records, works valuations, and claims for progress payments. It also provides an accurate assessment of costs, both direct and indirect, involved in works variations that may be ordered by the engineer or architct during the construction period, thereby substantiating legitimate claims for extras (or deductions) logically and mathematically. Furthermore, the effects of delayed performance by suppliers or subcontractors and of enforced delays occasioned by the client (or the client's engineer or architect) can also be determined with precision, so that recompense for such delays may be claimed with reason and justice. The method of evaluation of the costs of variations and delays follows in Chapter 11.

10.6 CONTROL OF LIQUIDITY

In Section 10.2 the finance chart for the sewer main project was developed for the first-crash solution. This plan leads to a maximum overdraft of approxi-

---

[1] The reader may appreciate this situation better by studying the sewer main project schedule shown in Figure 10.1. In the financial planning of this job the indirect costs are to be recovered at a uniform rate with time (irrespective of the direct cost value of the work done each 20-day period), whereas the gross profit markup was spread as a percentage onto the total works cost. The contract bill is then priced so that progress payments may be claimed every 20 working days directly in proportion to the total works cost. The uniformity of the rates of expenditure and income is apparent from Figure 10.2.

mately $36,400 and an average overdraft of $25,000. The adoption of other plans leads to different overdraft requirements. Figure 10.5 indicates the financial requirements of the three plans developed for the sewer main project in Section 8.5. It is obvious that as the project duration decreases, the rate of total works expenditure increases, and both the average and maximum overdraft similarly increase. If the overdraft requirements are plotted against project duration, the graphs of Figure 10.6 are obtained. Using these graphs, it appears that a project duration of about 300 days would require a maximum overdraft of $25,000 and an average overdraft of about $18,000. Clearly, in certain circumstances, it may be necessary to schedule a project to suit available overdraft limitations rather than to adopt the all-normal solution or the least total cost schedule.

In addition to project finance requirements, the estimated demand on cash resources for a project can be determined once a schedule has been evolved. Figure 10.7, for example, indicates the cash demand for the first-crash solution for the sewer main project. The inclined lines represent cash outflow and the vertical lines correspond to the monthly progress payments (income). The

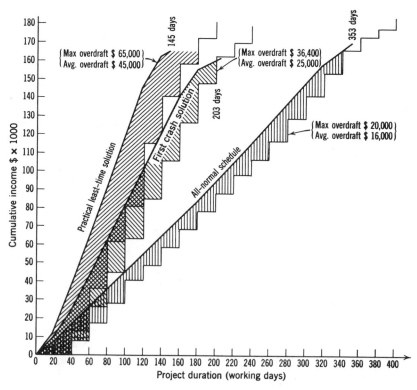

Figure 10.5 Finance charts for sewer main project (see Figure 8.7).

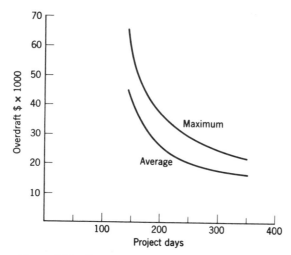

Figure 10.6   Sewer main project overdraft schedule.

actual cash demand is, of course, complicated by many factors, such as the time delay in account billing, the manner in which indirect costs are assessed and paid, the precise time when materials are landed on the site, and equipment hire charges (especially if the equipment is owned by the contractor). In these situations the overdraft schedule shown in Figure 10.7 may be excessive; it should be possible to carry the project on a lower overdraft, because

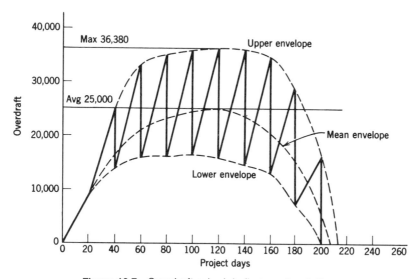

Figure 10.7   Overdraft schedule first-crash solution.

the actual cash flow may be lower than indicated. Nevertheless, the general shape and characteristics of Figure 10.7 will still pertain and should be studied carefully. Then adjustments can be made toward more realistic monetary budgets.

Obviously, the cash requirements of Figures 10.5, 10.6, and 10.7 are estimates only, and the actual amounts needed during construction will in most cases be different. Indeed, by management decision during the course of the work, the cash flow can be varied by activity shifting within available float, or by changing from one construction plan to another. This may become essential in order to eliminate or alleviate a critical cash deficiency. If another construction plan is used, the project duration may change; if only activity shifting within available float is satisfactory, the duration of the project will not alter.

If a broader view is taken, it is possible to relate all existing and planned projects to the overall liquidity situation (or working capital) for the entire organization. In this way CPM can help formulate realistic policies, especially regarding the commencement dates for new projects and construction rates to suit available liquidity. Figure 10.8, for example, may indicate the estimated company cash commitment or working capital status for a specific budget period, say for the next $2\frac{1}{2}$ to 3 years. Committed expenditures for

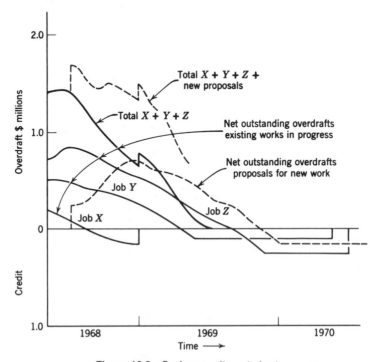

Figure 10.8   Cash commitment chart.

current and planned projects are drawn on this chart to indicate the total cash demands on the working capital during this period of time; incidentally, similar techniques can be used to indicate other resource's requirements. In this way attention can be focused on problems such as the decision to bid for new projects, determination of project duration, and optimum start times, so that financial crises in the company can be planned for, if not eliminated.

In this view new projects can be considered as activities on a large scale that can be shifted within float to suit the resource requirements of the entire company. Thus the project duration for the new project, and its rate of works expenditure, can be considered in the same way as shown previously for activities, using resource leveling concepts and network compression techniques. At the same time, allowances can be made in the working capital budget for investments such as new equipment purchases and other capital expenditures.

It must always be remembered that these cash requirements are estimates, and must be updated with actual facts as the data become known. Clearly, then, in estimating for a budget period, cash flow requirements must be regarded in a probabilistic sense, and there must be sufficient float, when several projects are added together in time, to allow for unexpected cash flows arising from unforeseen situations, change orders, and so on. Whatever the situation, however, it will be clearly realized that cash flow requirements computed from the network give a more reliable and realistic view than other traditional methods.

10.7 A CPM-BASED ACCOUNTING SYSTEM

The availability of large-scale computers and associated data processing and retrieval facilities enabled progress and cost control of the type outlined in Section 9.9 to be adopted at the Sydney Opera House in 1962–1963. Since then, further developments in computer programs have permitted network-controlled accounting procedures (as described in Sections 9.10 and 9.11) to be introduced to the construction industry; thus management may have readily available reliable and regular information concerning the financial status of the entire organization, as well as the detailed time and cost status of every project in progress.

Accounting data comprise first, the necessary entries for each individual financial account, whether debit or credit; second, a series of financial accounts subdivided into assets, liabilities, expenditure, income, accounts payable, accounts receivable, and so on; and third, the information (derived from these financial accounts) to be entered in each line of the financial statement of the enterprise. Hence, if all the financial accounts of the subdivisions are totaled, the result is the entry for that subdivision in the financial statement, and from these statement entries the balance sheet and profit and loss account are derived.

The entries for each individual financial account originate in one of the projects in progress or in the head office, and are suitably referenced to a particular set of cost accounts. Each of these cost accounts, whether they are for a particular project or for central administration expenditure, represents an activity on some CPM network, so that the entire system of accounts is based on activity costing, as described at the beginning of Section 10.3. These cost accounts begin with the estimated quantities and costs appropriate to the account. For a project these will be the estimated direct costs of labor, plant, materials, and services, or the estimated indirect cost if appropriate, whereas for head office accounts they will begin with the budgeted amount for the period. As work progresses, these cost accounts accumulate data comprising actual quantities and costs as these occur, and thus they form the basis of the periodical project cost reports referred to at the end of Section 10.3.

In addition, payroll processing is included in the accounting system, so that not only are labor costs distributed into the appropriate cost accounts, but also wage data are collected into separate employee record files, where each individual's earnings and deductions are accumulated.

Control of works in progress is obtained by simultaneous updating of the project cost accounts and the project network. The resulting information provides the basis for the regular revision with respect to duration and cost of all activities not yet completed. From this updating of each project follows the ability to update the financial records of the complete organization. The result of this combination of CPM with project cost reports and overall company financial status provides management with a dynamic and realistic tool for reporting on current conditions, predicting future results, and proposing changes in methods and procedures to improve performance.

Evaluation

a. Are critical path methods in financial planning worthwhile? Why?

b. How often should cost information be updated, and made available to (i) site personnel, (ii) head-office managers?

PROBLEMS

10.1 Consider a simple project consisting of two parallel and concurrent activities A and B. Contract conditions specify that progress payments are to be made every t days with payments (less $r\%$ retention) due t days later.

Given the following data:

Markup: $\$c_P$ per day
Indirect costs: $\$c_I$ per day

Direct costs: for activity A, $\$c_A$ per day for $3t$ days
for activity B, $\$c_B$ per day for t days

Draw the project finance chart showing the effects on overdraft requirements of shifting activity B within its float of $2t$ days.

10.2 In the sewer main extension project considered in Section 10.2 in the text, the finance chart of Figure 10.2 is derived from the bar chart schedule of Figure 10.1, assuming uniform distributions for direct costs over each activity duration.

Determine the effect on the overdraft requirements for the project if:

(a) (i) $6000 can be considered as an initial landed direct cost for activity 4–13 on its first scheduled day with the remaining $32,100 distributed uniformly over its duration of 154 days.

and   (ii) activity 5–9 similarly has an initial landed direct cost of $8,000 with the remaining $28,500 distributed uniformly over its duration of 139 days.

(b) the contractor owns the excavation equipment used in activities 4–13 and 5–9 and the two landed direct costs represent internal equipment rental payments.

11

EVALUATION OF WORK CHANGES AND DELAYS

11.1 LEGAL CONSIDERATIONS

In the construction of nearly all civil engineering and building projects there will usually be a number of changes in the volume, nature, and order or duration of work to be performed from that envisaged at the start. The term "work changes" is used to refer to all alterations, variations, deductions, extras, or omissions of this nature, whether the works are executed by day labor or contract. The magnitude of these changes depends on a number of factors, not the least important of which include the thoroughness of site investigations undertaken prior to the design, the completeness of the working drawings available at the time of preparation of the estimate or proposal, and unpredictable circumstances during construction. Some of these work changes may result in a variation in the actual cost of the project; others may not. Again, some may require additional time to perform, others may require less, whereas some may have no effect at all on the project duration. Although these effects are important on any project, they present a vital financial problem on works carried out by contract.

In addition to work changes, delays to various parts of a project may also occur, due to eventualities such as late deliveries of materials, procrastination on the part of the contractor or the subcontractors, tardy provision of detailed drawings, protracted inspections and approvals, shortage of labor, strikes and industrial troubles, unusually inclement weather, enforced retardation of work to suit the owner or other contractors on the site, and financial problems. Some of these delays may result in late completion of the entire project, whereas others may have no effect on the total construction time. All these delays usually cost money.

The problems of work changes and delays occur in all contracts, whether firm price or prime cost, but are more difficult in firm price for obvious reasons. With prime cost, the cost is always paid by the owner, but if the contractor is to be treated reasonably, there should be adjustment of the fee and the contract time.

In the preparation of all general conditions of contract, it is therefore necessary to anticipate and provide for such work changes and delays, and to place a legal liability on one of the parties to accept financial responsibility for them. Usually the contract conditions provide that the owner can make changes to the work without vitiating the contract, and that they shall be valued by mutual agreement between the parties, or, failing that, by the owner's engineer (or architect), and the contract price (or fee) adjusted accordingly. In addition, if the work is delayed, the engineer (or architect) is empowered to extend the contract time under certain circumstances.

Delays can be divided into three categories:

1. Those over which neither party to the contract has any control.
2. Those over which the owner (or the engineer) has control.
3. Those over which the contractor (or any subcontractor) has control.

It is generally recognized in modern contracts that delays of type (1) are part of a contractor's normal and legitimate monetary risk, and hence should give neither party grounds for monetary recompense, but that the engineer should extend the contract time in order to protect the contractor from any resulting liquidated damages for late completion. It is also recognized that for delays of type (2) the contractor should receive fair and reasonable recompense, for both cost and time, whereas for type (3) delays the contractor should get no recompense at all.

In law, claims by a contractor for time and cost recompense for work changes and delays due to the acts of the owner are validly reimbursable except when expressly provided otherwise in the contract conditions; but the law also requires the parties to take all reasonable steps to minimize the time and the cost. The method of evaluation of valid claims, for both time and cost, depends on the terms of the particular contract. Usually such determinations are made the responsibility of the engineer (or architect), but no guidance nor any basis on which to make a valuation is given. In some cases it may be stipulated that changes shall be measured and the value determined from the unit rates stated in the contract; but this is of no help in determining a valid time extension or in computing a new rate of work of a different nature from that in the contract.

Work changes include variations in the form of additions and deductions where the volume of part of a work item is changed; they consequently result in more or less cost and time to execute the varied item, and they usually require some rearrangements of resources. Alterations and extras both change the nature of the work and may involve resource replanning as well as changes in costs, whereas the time to do the work may or may not alter. Omissions usually mean less cost, but not necessarily less time, and may result in wasting resources. Frequently, a work change is a mixture of types, as in the omission of one item and its substitution by a variation, alteration, or extra; or there may be a double variation with a deduction from one item and an addition to another. Whether the work change is a mixture of types or not, there is no simple path to its proper evaluation. In practice, when cost and time are not agreed on before the changed work begins, the contractor submits a price for the change, and probably also a request for an extension of time; if these are not acceptable to the engineer (or architect) two serious problems have been created: determination of a fair cost and evaluation of a reasonable time.

With respect to compensation for delays, contract conditions presently in use vary greatly in their provisions. Even if they were all in agreement that time extensions only will apply for type (1) delays, and that time and cost may be recovered in full for type (2), the problem remains of assessing the extra time and/or cost attributable to each delay and of applying this to the project as a whole. It may first appear that only the evaluation of delays of types (1) and (2) is of prime interest in contract administration, since reimbursable delays fall within these two categories. This is, however, fallacious because if two delays are concurrent or overlapping, that is, one of type (1) or (2) and the other of type (3), it becomes essential to evaluate both in order to determine the resulting overall effect on the project. As with work changes, the two problems of fair cost and reasonable time have been created.

Acting for the owner, the engineer (or architect) tends to minimize the value of such work changes and delays, whereas the contractor's natural reaction is to maximize them. Often much unpleasant dispute occurs.

This is a continuing problem in contract administration, because no principles for computing time extensions or costs are provided in a contract. Furthermore, in practice, most construction contract rates include not only the direct cost of the item, but also a proportion of the indirect costs, overheads, and profit for the entire project. Hence any change in the quantity of work, or in its nature or order will affect the return to the contractor of some or all of these components of the contract rates. In addition, when a contractor carries on some work while experiencing delay in other items, labor and plant inefficiencies will be encountered due to the discontinuity of the program and the necessity to reallocate resources; to be fair the contractor should have compensation for this as well as for the cost of the work change itself. The legal requirement is therefore to establish a principle for the determination of time and cost for any eventuality. It should provide for evaluation of the net effect of a number of concurrent changes and delays, as well as the net result of a multiplicity of occurrences on site performance and productivity; and it should also enable a determination to be made as to whether the contractor had the ability to perform within the contract time if various changes and delays had not been encountered. In this respect, it must not be forgotten that the law places on the contractor (subject to the contract conditions) the right to determine priorities and the sequence of operations, and that a contractor is generally hired because of any ability and experience in work of the nature included in the contract.

Probably the most valuable feature of CPM is that it enables the required principle to be established. On a project planned and controlled by network analysis, the contractor's original network program depicts the basis on which price and time for the work was determined; it therefore follows that, *if this initial network is accepted as part of the contractor's bid, any change from this program constitutes a change in the contract,* irrespective of whether the change was instituted by the requirements of the contractor or the owner or whether it arose from a cause beyond the control of either. If another network is drawn to show the effects of this change on the rest of the project, then its analysis and direct comparison with the initial network will enable all the resulting time delays to be determined. *Since only delays affecting critical paths can affect the project duration, these critical delays may thus be readily examined and their basic causes identified.* Equitable evaluations for extensions of time, in conformity with the provisions of the contract conditions, would then be available.

If, in addition, the contract schedule provided direct and indirect cost items separately for the various parts of the work, then by analysis of the network it would also be possible to derive equitable monetary evaluations for all the validly reimbursable work changes and delays that had occurred, whether these were critical or not, and to assess these within the terms of the contract. Hence fair and reasonable contract values for both time and cost may be determined for any proposed change or delay, and any advantageous replanning to suit the new conditions can be made amply clear to the parties concerned, thus enabling agreement to be reached before inauguration of the change.

To summarize, therefore, it is clear that the principle for determining compensation in time and cost, for any eventuality affecting the contract work, is the derivation by network analysis of the critical delays arising or likely to arise from such eventuality (bearing in mind the legal necessity for all parties to minimize the effects resulting therefrom), together with the calculation of the net cost of such critical delays, and the resolving of liability therefore between the parties in strict accordance with the provisions of the contract (whether these are fair and reasonable or not).

11.2 EFFECT OF A SIMPLE VARIATION

As a simple illustration of the principle stated above, consider the pipeline construction example for which the optimum solution was computed in Sections 5.7 and 5.8. The most economical duration was there found to be 102 days, with a corresponding network model shown in Figure 5.7*f*. Suppose now that this project is under construction and had reached event 5 (completion of trench excavation) when the contractor was ordered to provide an additional 50% of concrete anchor blocks because of bad ground encountered. From the utility data (Table 4.1), it is obvious that the durations and costs of activity 7–8 (concrete anchors) will be increased by 50%, becoming:

$$t_{7-8}^N = 18 \quad \text{and} \quad c_{7-8}^N = 600$$
$$t_{7-8}^C \quad \text{and} \quad c_{7-8}^C = 780$$

Examination of the network model shows that activities 5–7 and 7–8 are both critical and already fully crashed, so that any increase in the duration t_{7-8} must delay event 8 and consequently the entire remainder of the project. Since the critical activity 7–8 is to be extended by the extra-work order, it is obviously best to continue it at crash speed, thus delaying critical event 8 by 4 days only. The position is shown in Figure 11.1a, which is a reproduction of the relevant part of Figure 5.7f with the new data added; the project duration is increased by 4 days and the project direct cost by 260 units ($T_P = 106$ and $C_P = 13,050$), whereas the project indirect costs will incrase by 485 units (computed for 4 days from the indirect cost curve in Figure 5.10). At first thought, therefore, *the contractor might claim 4 days extension of time, and an extra payment of 745 cost units plus profit.*

This would not, however, be justified in this case. Notice the looped critical path 5–6–8: activity 6–8 is not compressed and hence cannot be lengthened and cheapened; but activity 6–5 can be decompressed if required by as much as 8 days at a cost slope of 50 units. It follows then that, since event 8 is only delayed by 4 days because of the extra work, the correct solution is to decompress activity 5–6 by 4 days at a cost saving of 200 units. The result is illustrated in Figure 11.1b, where $T_P = 106$ and $C_P = 12,850$ only; the additional direct cost of the extra-work order thus amounts to only 60 units, and hence the contractor's expenditure need rise by only 60 (direct) plus 485 (indirect, as above)—that is, 545 cost units. *The additional time required is confirmed at 4 days, but the correct extra payment is 545 cost units plus profit.*

An identical approach may be made for deductions ordered from contract work.

It is therefore clear that an arbitrary increase or decrease in price, based on the quantity, extended at the billed rate, may have no justification in fact. Theoretically, the only correct way is to submit the proposed addition or deduction to network analysis and to compute the true value of the alteration by optimal solution, using compression and/or decompression as appropriate. In practice this may readily be done, as discussed in Section 11.8.

11.3 EFFECT OF A SIMPLE DELAY

Another related problem is the evaluation of delays to the project, either by late performance on the subcontractor's or supplier's part, or enforced retar-

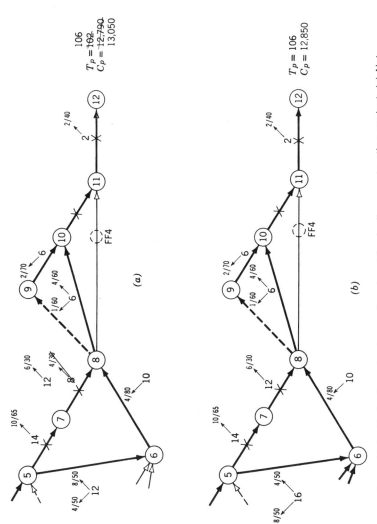

Figure 11.1 Evaluation of extra work ordered in pipeline construction project. (a) Network amended with additional work ordered in activity 7–8 (see Figure 5.7f). (b) Network adjusted to optimal solution for additional work ordered in activity 7–8.

265

dation of work to suit the owner or the contractor. When the contractor is not legally liable for such delays, it has been extremely difficult in the past to calculate and substantiate an actual monetary claim for recompense. Countless arguments have been waged between contractors and owners, or between contractors and subcontractors, as to the true evaluation on these situations regarding both the costs and the project duration arising therefrom.

With a critical path network the position is very clear, enabling the true time and cost of the delay to be logically and quickly calculated. Indeed this may well be done with advantage when such a delay appears probable, so that the consequences may be appreciated beforehand and the parties thereby forewarned.

Consider the case in Figure 11.2 showing (a) part of a fully resource-levelled network in which a skilled crew of 14 men is scheduled to carry out activities 30–31–33–34–35–38 concurrently with 30–32–37–39 as indicated. Event 30 is scheduled for achievement in day T. Suppose now that at day $T + 4$ the critical activity 29–33 is delayed one day by some cause. To maintain project duration, the work must be rescheduled from day $T + 5$ to give the network shown in (b). As a result there will be 2 mandays lost at the end of activity 32–37, event 37 is achieved 1 day early, critical event 35 one day late (35A) and critical event 38 is on time; by working overtime on the last part of activity 37–39, event 39 remains on time. The project is not delayed, no time extension is required, but the direct cost of the work has increased.

This remedial solution is only valid, however, if efficient rearrangement of subcrews is feasible. If not, the position will be as shown in Figure 11.2c. Here 10 mandays of labor are lost before event 33, and thereafter critical path I runs 1 day late; this immediately affects the parallel critical path II, where its entire crew is idle for a day at event 36. In addition, the noncritical activity 37–39 loses 4 worker-days of labor, and event 39 is also 1 day late. The ramifications of the day's delay to activity 29–33 are, in this case, more complex. Additional time as well as additional direct and indirect costs are required.

It is therefore clear that any arbitrary decision regarding time or cost can have no real justification. Even with such a simple single delay, the only correct way is to submit the problem to network analysis and determine the true effects of the occurrence. In fact, it is the responsibility of all parties to minimize the effects of all delays, and the contractor should not overlook this.

11.4   CHANGE AND DELAY REPORTING

The usual methods of reviewing a project and revising its networks during construction were discussed in Chapter 9. If, in addition to or instead of the regular updating of the original network, another network is plotted progressively as the work proceeds, this latter network provides at any given time an

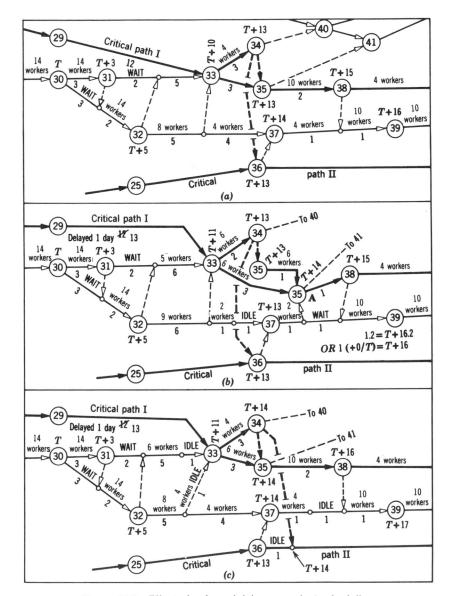

Figure 11.2 Effect of enforced delay on project scheduling.

authentic record of the work executed to date. Such a network may show, as they happen, all work changes and delays encountered, and the actual starting and finishing dates of every activity. Since it is a record of facts, it is known as a *factual network*. On finishing a project it provides a complete record of construction performances, the same as work-as-executed drawings provide a complete record of the technical details of the finished structure. When compared with the original program (initial network), it furnishes the data for the evaluation of all time delays encountered during the contract.

The initial network may be as detailed or as condensed as desired to suit the project; it may show only direct-cost activities and constraints, or it may include a series of arrows for indirect-cost services and operations. The limit of detail is that of useful practicability. Each activity is identified by its node numbers, as usual, but lengthy operations should be broken down into a series of short, consecutive activities for better status reporting, and the network should be drawn to an appropriate time scale. Practical planning of construction projects is discussed in Chapter 8, and attention is drawn to the importance of critical resource paths for realistic networks in Section 8.9.

The factual network consists of the same activities as the initial network, plus any others required because of work changes and delays encountered during construction and is drawn to the same time scale and layout. For reporting regularly on activities in progress, just completed, or about to begin—or for introducing new activities—the conventional forms shown in Figures 9.1 and 9.4 are satisfactory; but further documentation is required for recording work changes and delays in detail.

A suitable report form is seen in Figure 11.3 in which a progressive account is kept of every occurrence that may affect the time and/or the cost of any activity in the project. Delays from *all* causes, as well as work change orders, should be included, for in most cases their significance will not be apparent until later.

In Figure 11.3 the first column shows the date that the change or delay became known. The second and third columns are used to identify the occurrence (by some numerical system), and record its basic cause or category. A brief description is reported in the fourth column, followed by reference to site instructions, minutes of meetings, change order numbers, and so on.

			PROJECT _ _ _ _ _ _ _ _ _ _ _ _ _ _ _ _ _ _ Critical Path Program—Change and Delay Record Sheet						
Date	Delay No.	Category	Description	Reference	Effective Dates	Delay Time	Activities Affected	Remarks

Figure 11.3 Typical change and delay record sheet.

As soon as the change or delay becomes known, the record may be filled in to this point. The last four columns are completed subsequently when the facts are available. This record, like the activity status reports, should be kept strictly up to date; not only does it provide an authentic summary of all eventualities, but also a valuable basis for estimating future performances. It is emphasized at this point that it is quite fallacious to form any opinion about the importance of any single change or delay; this will not be known until the factual network is analyzed and its critical paths established. This analysis may be done at any time and is an essential feature of dynamic project control.

It is axiomatic that the recorded data must be realistic and indisputable; preferably all information should be mutually agreed on between the parties, or at least notified from one to the other in writing with adequate substantiating evidence. *The prime purpose of change and delay reporting is not to favor one party or the other, but to establish the facts of the matter.*

11.5 PLOTTING THE FACTUAL NETWORK

The regular status reports (Figures 9.1 or 9.4) provide the data from which may be plotted progressively, as work proceeds, the actual attainments on the site. Any new constraints arising during construction are shown as they occur, so that the network will be completely accurate. This factual network must be a genuinely legitimate record of facts concerning the times at which each event was achieved, whereas the initial network was merely a plan for doing the work. Direct comparison between the two shows the project status at a glance.

Project control and replanning as discussed in Chapter 9 can be carried out on copies of the progressive factual network at any date, extended by an estimated network to completion (realistically based on current performance). The ultimate effects of such replanning will be reflected in the factual network later as actual attainments are recorded and plotted.

If now the data from the change and delay report (Figure 11.3) are added (as they occur) to the factual network, this network will not only show the current time status of the project, but also the causes of all the time changes that happened. It has become a reliable record of work as executed, with circumstances affecting performance clearly designated.

The two ways in which changes and delays may be indicated on a network diagram are (1) by the use of new and artificial activities and (2) by codified notation on existing activities. Either or both ways may be used as convenient to indicate realistically the significant facts—unforeseen when the initial network plan was adopted—that have eventuated during the construction period to date. In Figure 11.2 new and artifical activities are used to indicate scheduled "wait" periods in each diagram and enforced "idle" periods in (b)

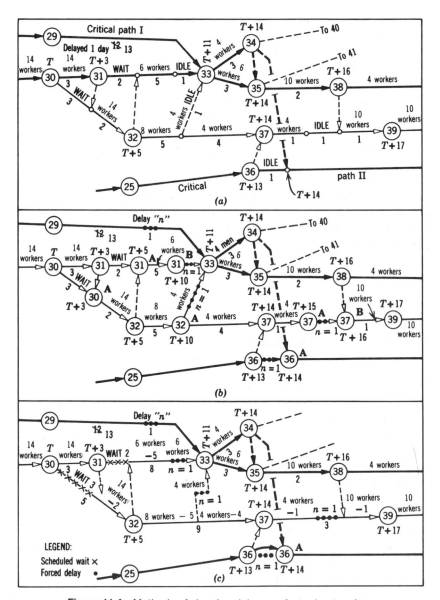

Figure 11.4 Methods of showing delays on factual networks.

and (c). Figure 11.4 illustrates the various ways that delays may be shown on factual networks.

Figure 11.4a is a reproduction of Figure 11.2c; the effect of the enforced delay of one day on activity 29–33 is shown solely by the revised timing of the events. If this enforced delay is designated delay "n," and if all its effects are shown as artificial activities, the diagram will appear as Figure 11.4b; this diagram demonstrates clearly the influence on all other affected activities of the direct delay "*n*" to activity 29–33. In Figure 11.4c the ramifications of delay "*n*" are indicated by notation on the existing activities. Different symbols are used to show different types of delay, so that the contractor's scheduled "wait" may not be confused with the enforced "idle" periods resulting from delay "*n*"; the duration of each delay is also noted beside it.

For ease of progressive plotting and for instant awareness of project status, both the factual and initial networks are drawn to the same time-scale; if the factual network is drawn as a transparency, it may at any time be superimposed on the initial network for direct visual comparison. As discussed in Section 3.4, time-scaled networks are helpful in emphasizing the relative times of achievement of the events, for on a time-scaled initial network the nodes may be numbered in chronological sequence while still preserving the conventional logical order (arrowhead number always greater than arrowtail). With the same events still shown by their initial node numbers on the factual network, examination of the actual sequence of event achievement (originally planned to be in chronological order) provides a clear, qualitative indication of work disruption. This is especially valuable with milestone events (see Section 7.3) and interfaces (Section 15.3).

Time-scaled networks also enable float calculations to be avoided, since spare time is shown by horizontal broken lines. Furthermore feasible program revisions are easily seen. Figure 11.5 shows Figure 11.1 plotted to a time-scale; the possible decompression of activity 5–6 is immediately apparent, as is also the unaffected free float in activity 8–11.

The progressive factual network is regularly updated as site work proceeds in conformity with periodical reporting of project status but at a maximum of once a week. At any stage it may be used for estimating the effects of current eventualities and for reallocation of resources as critical paths change. As soon as the contract is finished, however, the complete factual network, showing all the facts pertaining to site performance, is immediately available for a realistic determination of the total effect on the project of all the eventualities encountered during its construction.

11.6 ANALYSIS OF FACTUAL NETWORKS

Until the progressive plotting of a factual network reaches project completion, it cannot be analyzed in the usual way by calculation of earliest and latest finish times for each event. The method of determinating critical paths on a progressive factual network of this type is to proceed backward from the cutoff date, starting with that event exhibiting the greatest tardiness with respect to its original LFT, and to trace back through the network those preceding dependent activities currently having no float, until finally the start of the project is reached. Sometimes several critical paths are apparent with this procedure.

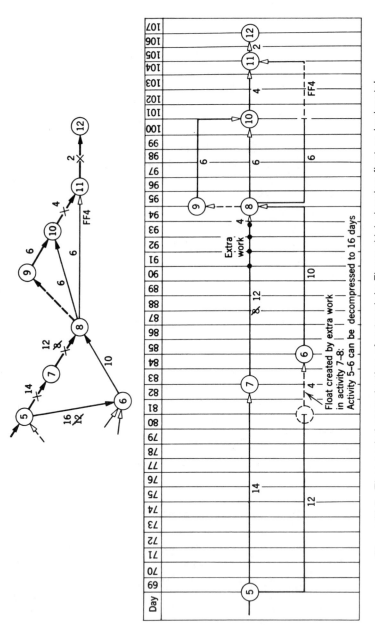

Figure 11.5 Time-scaled reproduction of network in Figure 11.1 showing float as horizontal broken lines.

A complete factual network, being fully closed from start to finish, may be analyzed in the conventional manner; or, since its final event is that showing the greatest tardiness at the date of project completion, it may with equal accuracy be analyzed in the same way as a progressive network.

If, as a result of either method of analysis, there is only one tardy critical path, it obviously controls the project duration and may be examined immediately to ascertain the delays it has experienced. If, however, there should be more than one equally tardy critical path in the network, each is examined in turn to determine its *net working duration*. The net working duration of *any* path through a factual network is found by deducting from its total duration to date the delay times of those work changes and delays (and only of those) lying on the path being analyzed. The net working duration is, in fact, the actual time in which all the activities along that path could have been completed if there had been no work changes or delays affecting that path.

Having examined all apparent critical paths between two common events in this way, the primary path is determined as that with the longest net working duration between the relevant events. The work could not have been completed in less time than this, even if the delays had not occurred. Other parallel apparent critical paths may be classified as secondary, tertiary, and so on, as their net working durations decrease; they do not control the project duration, for they have, in fact, float with respect to the primary path.

The determination of the final primary factual critical path for a project, or for a separable part of a project, cannot, of course, be made until its final event has been achieved. It is thus a reliable and accurate determination in retrospect, because it is based on facts and not on estimates, in contrast to the analysis of a progressive factual network extended by an estimated network to completion. Consideration will show that if the net working duration of a final primary factual critical path is less than the total duration allowed in the initial network, then the planned performance could have apparently been achieved if the delays had not occurred. Conversely, if the final net working duration is greater than the initial duration, then it would seem that the original activity times were underestimated. This is, however, not always so, and further discussion of this aspect appears in Section 11.7.

Nevertheless, it is possible by this technique to establish the total amount of critical delay experienced by a project to any desired date, and to say generally whether this is due entirely to identifiable causes or to possible incorrect estimation of some of the original activity durations, or partly to both factors. It is therefore practical to determine at any time whether, in accordance with the conditions of the contract, an extension of time for completion is due. If it is, it may be granted at once without awaiting application by the contractor. In fact, this may be most conveniently done in practice at the time of claims for progress payments, and the regular progress certificate may then cover both money and time simultaneously.

Furthermore, at any time after its completion, a valid analysis may be made of a project in order to identify its overall critical delays provided reliable data are available to enable a factual network to be plotted; this is called *analysis in retrospect*. It will be obvious that any path through a network may be analyzed if it is of interest to inquire why it has undergone tardiness; but it is emphasized that the overall effects of *all* eventualities to date on a project as a whole are determined solely by analyzing its primary factual critical path. The total effect of all unforeseen occurrences is the difference between the total actual duration and the net working duration of the project's final primary path. This difference comprises the total critical delays to the project. Further discussion of this aspect appears in Section 15.5.

11.7 RESPONSIBILITY FOR CRITICAL DELAYS

The responsibility for project delay up to any given date may thus be accurately determined by examining critical delays alone; no other occurrences, whatever the cause, affected the performance of the work as much as those on the primary factual critical path. By this approach, therefore, it is possible to find all the causes of protracted performance beyond the original completion date, and to apportion responsibility in accordance with the relevant terms of the contract. For instance, if the contract conditions stipulate only certain occurrences as entitling a contractor to recompense, then the proper entitlement is solely the delay times of those specific occurrences that lie on a critical path. In passing it should be observed that similar occurrences on noncritical paths have no effect on the project duration.

Clearly, theoretically the net working duration is the time taken to carry out the work with no critical delays at all. It is also clear that no contractor can expect in practice to achieve this efficiency and must allow for some lost time in the estimate of project duration in this initial network (see Section 15.5). For this reason it must be emphasized that any float included in the initial network belongs to the contractor and may not be consumed by the owner or the engineer (architect) in ordering variations without their compensating the contractor for any critical delay that may arise in consequence. And, if the contractor at any time chooses to revise the plan in order to take advantage of such available float in order to improve the resource position, the contractor is fully at liberty to do so, provided that the engineer (architect) is informed of the revised scheduling.

Consequently, if the net working duration is less than the estimated duration shown on the initial network, the contractor has demonstrated the ability to perform within the estimated time. However, if some of the critical delays are due to causes over which there was no control, or to causes for which there is no time entitlement under the contract, these critical "contractor delays" must be added to the net working duration to see whether this total exceeds the contract time. If it does not, the contractor has demonstrated the ability to perform within the terms of the contract.

Conversely, if the net working duration exceeds the contract duration shown in the initial network, then the contractor's original estimates were apparently incorrect because of arithmetic mistakes or optimistic views on site productivity. There is, however, one exception to this situation: if the delays are so numerous or complex that the general site productivity and efficiency suffered, then the apparent inability to perform may be due to this indirect effect of the delays encountered, and not to a contractor's error or underestimation. Thus there may be valid entitlement to compensation if the originating delays are reimbursable.

Inability to perform because of this reason is determined by productivity factors, and can only be assessed for the finished project. To do this, a comparison must be made between various undelayed activities in the factual network and their counterparts in the initial network to find examples where the work was, in fact, done under the anticipated conditions; hence "normal" productivity can be determined in labor-hours or plant-hours per unit. It is important to observe that this normal productivity is completely unrelated to any estimated value; it is what is actually achieved on site under the initially expected normal conditions. When similar operations have been delayed on site, their actual productivity factors may likewise be derived. This approach is not absolute, but if it can be established that productivity has decreased when there is no other valid cause, the difference represents a measure of inefficiency attributable to the overall effect of a multiplicity of delays or of enforced piecemeal working.

11.8 PRACTICAL APPLICATIONS

Since computer processing of factual networks is neither feasible nor necessary, application of these procedures must be carried out manually. This is not very laborious even on a large and complex project, but it is the only logical way to evaluate the effects of delays and work changes on the intended project construction plan.

It is not possible within the space of this book to present the details of the actual evaluation in practice of work changes and delays encountered on several major projects recently concluded. The initial and factual networks, drawn to a scale of one day to 5 mm, are too large to reproduce, and the analysis of the factual networks for the derivation of the primary and other critical paths are too voluminous.

The reader will, however, appreciate the procedures by studying Figure 11.6, which shows a series of hypothetical work changes and delays applied to the pipeline construction example of Figure 5.7*f*. In Figure 11.6*a*, the initial network (Figure 5.7*f* drawn to a time scale) shows the original plan for completing the work in 102 days. Figure 11.6*b* shows the final factual network, incorporating the occurrences designated in Table 11.1, and some expediting of certain activities where the contractor was delayed through his own fault, with the end result that the work occupied a total of 116 days. Analyses of all the factual critical paths are presented in Table 11.2, from

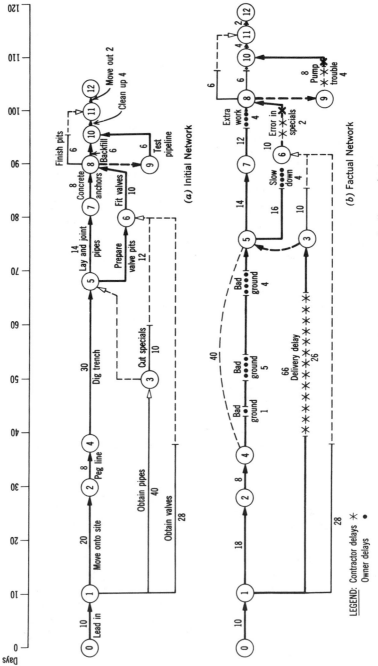

Figure 11.6 Effect of work changes and delays on pipeline project.

LEGEND: Contractor delays ✳
Owner delays ●

(a) Initial Network

(b) Factual Network

Table 11.1   Change and delay record for pipeline project

Delay No.	Category		Description	Effective Dates	Delay Time	Activities Affected	Remarks
1	Contractor		Delay in pipe delivery	Day 41–66	26 days	1–3	—
2	Owner ⎫			Day 44	1 day	4–5	—
3	Owner ⎬		Extra timbering ordered in bad	Day 51–55	5 days	4–5	—
4	Owner ⎭		ground	Day 67–70	4 days	4–5	—
5	Owner	(a)	More anchors ordered	Day 99–102	4 days	7–8	—
	Owner	(b)	Resulting decompression	—	4 days	5–6	(See Section 11.2)
6	Contractor		Error in specials delays valve fitting	Day 96–97	2 days	6–8	Fully recovered by crashing
7	Contractor		Pump trouble delays testing	Day 103–106	4 days	9–10	2 days recovered by crashing

which the primary path is shown to be 0–1–2–4–5–7–8–9–10–11–12, since it has the longest net working duration.

As a result of this analysis it is clearly established that the responsibility for delayed completion of 14 days beyond the original 102 days should be apportioned against the owner, as follows:

Delay No.	Responsibility	Activity Affected	Critical Delay
2/3/4	Owner	4–5	10 days
5	Owner	7–8	4 days
7	Contractor	9–10	2 days[1]
			16 days
(Gain)	Contractor	1–2	Less 2 days
			Total 14 days

[1] After losing 4 days, work was fully crashed to 4 days; net delay = 2 days.

If the contractor is entitled to time extensions for all causes beyond its control and to monetary recompense for all causes over which the owner has control, then a 14 day extension of the contract time should be received and an extra payment of 2772 cost units plus profit, computed as follows (direct-cost data from Table 4.1, indirect costs from Figure 5.10):

Delay No.	Time (days)	Activity Affected	Direct-Cost Change +	−
2/3/4	10	4–5	1000	
5(a)	4	7–8	260	
5(b)	(4)	(5–6)		200
			1260	200
Net Variation, Direct Costs			1060	
Indirect Costs for 14 days			1712	
Total Additional Payment			2772 plus profit	

Notice that, in this example, the contractor took steps (at some expense) to recover the time lost by management, thus avoiding liquidated damages. If this had not happened, the project would have overrun more than 14 days. In

Table 11.2 Analyses of factual critical paths, pipline project (see Figure 11.6)

Critical Activities	Total Duration (days)	Delay No.	Delay Time (days)	Delay Category	Net Working Duration (days)
0–1	10	—	—	—	10
1–3–5	66	1	26	Contractor	40
5–6	16	5(b)	4	Owner	12
6–8	10	6	2	Contractor	8
8–9–10	8	7	4	Contractor	4
10–11–12	6	—	—	—	6
	116		36		80
0–1–3–5	76	1	26	Contractor	50
5–7–8	26	5(a)	4	Owner	22
8–9–10	8	7	4	Contractor	4
10–11–12	6	—	—	—	6
	116		34		82
0–1–2–4	36	—	—	—	36
4–5	40	2/3/4	10	Owner	30
5–6	16	5(b)	4	Owner	12
6–8	10	6	2	Contractor	8
8–9–10	8	7	4	Contractor	4
10–11–12	6	—	—	—	6
	116		20		96
0–1–2–4	36	—	—	—	36
4–5	40	2/3/4	10	Owner	30
5–7–8	26	5(a)	4	Owner	22
8–9–10	8	7	4	Contractor	4
10–11–12	6	—	—	—	6
	116		18		98

such case there is entitlement to 14 days extension plus the above money less the value of liquidated damages incurred.

Although it is impracticable (because of space limitations) to present here the complete details of an actual project controlled by the factual network technique, Figure 11.7 is included to illustrate some particulars of its use on a major building construction project. This shows a portion of the factual network affected by the delays summarized in Table 11.3, and serves to demonstrate the complexity of a variety of delays of different categories, as well as the method of identifying their causes by symbols defined in the "delay legend" on the drawing; the specific symbol denoting the basic cause of each delay is plotted on or beside the appropriate activity arrow. Observe also that the original critical path through 41–47 has switched to 40–43–47A; with a multitude of delays the current critical path differs frequently from the original one shown on the initial construction network.

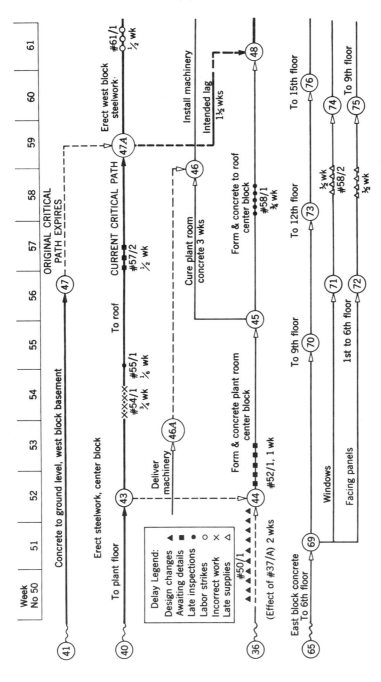

Figure 11.7 Portion of time-scaled factual network for building project.

Table 11.3   Extract from delay tabulation for building project (see Figure 11.7)

Delay No.	Category	Activity Affected	Description	References	Time Lost
#50/1	Design change	36–44	Effect of previous delay #37/A to steel causing idle time to concrete	Letter WC/1104	2 weeks
#52/1	Late details	44–45	Plant room block-out details awaited	Drg. No. 65/206	1 week
#54/1	Incorrect work	43–47A	Steelwork not plumb; remedial work delayed further erection	Site Instruction No. 103	$\frac{3}{4}$ week
#55/1	Late inspection	43–47A	Late inspection of rectification of #54/1	Letter AX/234	1 day
#57/2	Late details	43–47A	Duct support details awaited; also decision relocation of brackets	Drg. No. 65/210	$\frac{1}{2}$ week
#58/1	Late inspection	45–48	Inspector absent $4\frac{1}{2}$ days; concreting delayed	Letter AX/235	$\frac{3}{4}$ week
#58/2	Late supplies	72–75	Precast panels delivery running late	Delivery Note PC/97	$\frac{1}{2}$ week
		71–74	Window erection awaiting panel delivery	—	$\frac{1}{2}$ week
#61/1	Strike	47A–51	Steel crew on strike 3 days	Diary Week 61	$\frac{1}{2}$ week

Once accurate time evaluations are available, equitable cost recompense can readily follow, provided the contract unit rates are stated as direct costs, with additional sums for indirect costs stated separately, and profit spread throughout as the contractor may desire. A suggested subdivision for indirect costs sums would be the following:

1. Establishment and all fixed charges (move in)
2. Supervision and administration (all time-dependent charges)
3. Clean up and dispersal charges (move out)

These three items may be further subdivided if necessary, but the important feature is that the second item be expressed as a unit rate per day as well as a total amount; this daily rate is derived by dividing the total amount by the construction time shown on the initial network, thus establishing a contract rate for proper compensation to the contractor for delays occasioned by the owner or the engineer (architect).

A few overhead and indirect cost items to not lend themselves to this easy classification, but they are usually small and unimportant in comparison with the total; however, if necessary, they could be shown separately in the contract schedule.

It must not be overlooked that the detailed site management of a construction project is the prerogative of the contractor, and that (unless this is prevented by the contract conditions) the original program can be changed whenever necessary, provided the contractor completes the works, or separable parts of the works, within the specified contract time(s). Hence all such changes must be built into the factual network as a contractor's responsibility. Furthermore, many contract conditions empower the owner to order alterations in the sequence of operations on the site, as well as to vary the work to be done. This power must be preserved, but in these instances each such change must be built into the factual network as an owner's responsibility, for which the contractor is entitled to compensation if necessary.

Another problem associated with claims for extensions of time should also be clarified. Suppose that, with an owner-specified contract period of 100 weeks, a contractor submits an alternative bid with an initial network offering to complete in 70 weeks, and that this bid is accepted by the owner. In the event, suppose that with sundry changes and delays by the owner the final factual network shows that the project took 85 weeks. Is the contractor entitled to time and indirect-cost recompense for the additional 15 weeks? Under the factual network concept, where the initial network is part of the contract, the contract time accepted by the owner is, of course, 70 weeks, and the contractor is entitled to both time and money for the 15 weeks extra time occasioned by the owner. Hence bid proposals must be based on realistic construction times, if they are to be competitive, and preplanning must be tempered with practical allowances for normal delays rather than with fictitious activity durations.

The factual network concept is a powerful management tool; but it is a double-edged tool available to both owner and contractor. With the initial network as the basic contract program, all parties are aware at the outset of the time available for provision of details, samples, final approvals, material deliveries, and so on. With each updating of the progressive factual network, all parties know of the gains or losses of time in the various activities, and of the remedial measures and/or replanning necessary to maintain the desired progress. Contract rates are more realistic, yet none of a contractor's private cost information is disclosed. Disputes are minimized and hence bid prices are lower. There are better relations between contracting parties, and more efficient construction administration. Finally, at every stage of the project, an up-to-date factual network can be used as a basis for future planning in order to find the best way to finish the works from the currently existing status.

Evaluation

a. What detailed information would you consider prudent to keep in a construction diary?

b. Why is it difficult to computerise factual network calculations?

c. Taking any existing contract with which you are familiar, identify the clauses and conditions relevant to the material in this chapter.

PROBLEMS

11.1   Draw the network shown as the answer to Problem 7.3 to a time-scale as an *initial network,* with resources requirements tabulated beneath.

Suppose now that, because of work changes required by the owner, the following additional activity times and resources are imposed on the contractor

Activity	1–5	2–3	2–4	3–5	3–6
Extra time	3 days	2 days	2 days	—	—
Additional resources	A for 3 days	2 men and C full time	C for last 4 days	C full time	1 man full time

Suppose also that only one equipment A and one equipment C are available.

Determine the correct extension of time and the additional direct costs (expressed in labor-days and equipment-days) due to the contractor and draw the revised network to the same time-scale.

11.2   Draw the original network shown as the answer to Problem 9.2*a* to a time-scale as the *initial network* for this project. Beneath it, to the same scale, draw the *factual network* incorporating the following work changes and delays.

Delay No.	Category	Description	Effective Dates	Delay Time	Activities Affected	Remarks
1	Owner	Design change in culvert Ch. 96	Days 38–44	6 days crash	6–7 12–13	Remedial measures as in Problem 9.2(*b*)
2	Weather	Heavy rain; stopped work	Days 60 and 61	2 days	12A–13	
3	Contractor	Rescheduling	Day 60–78	18 days	9–10	Start on day 78
4	Contractor	Paver not on site until day 81	Days 80 and 81	2 days	10–11	Start on day 82

If the contract provides for time extensions only for weather delays and for time and cost recompense for extra work and delays caused by the owner, what is the contractors' entitlement on the completion of the project?

12

ATTITUDES, RESPONSIBILITIES, AND DUTIES

12.1 USE OF CRITICAL PATH METHODS

It is usual at this time that any formal planning and management training includes some consideration of critical path methods, so that lack of contact with CPM networks, or with someone knowledgeable with such techniques in the construction industry, would probably be limited to the novice, or to a local self-taught tradesperson venturing into small job activity for the first time.

The active and continuous use of CPM in day-to-day activities is another matter. The conditions generally thought relevant to the environment in which CPM is, or is not, used has been the subject of many industry surveys and studies. Some obvious conditions are the complexity of proposed construction activities, the criticality of interactive scheduling of documents, material deliveries, and equipment availabilities. Additionally, the technical and management expertise and sophistication of the relevant planning and management staff and the volume of information flows necessary for large projects will often force the use of systematic methods for all phases of project management from initial conception to project completion and commissioning. Some obvious examples where CPM has been effectively used are given in Chapters 8 and 9.

Sufficient for our purposes here is the observation that if CPM techniques have been found to help management, and the problems and issues now being addressed are pressing, then CPM will probably be used. Hence leaving aside the question as to whether CPM are relevant to planning and/or management situations facing a decisionmaker, this chapter addresses the professional use and requirements of CPM. There are two fundamental reasons for using CPM: (1) because its use is seen as an aid in the planning and/or management of a project and (2) because its use is requested and there is an interest in meeting this request. In many cases, of course, both of these reasons apply. Also it may happen that only the first reason applies but CPM is not used because at the time resources are not available, or the apparent advantages are not compelling. Clearly, the regular use of CPM as a management aid, or the consistent request that CPM be used, indicate positive attitudes to this technology. On the other hand, the use of CPM purely because it is being requested may only reflect a convenient passive response, but its use under these conditions and in certain circumstances may be the only way or opportunity to express a positive attitude, or it may be due to contact with a dedicated user.

If there were a concern as to the experience and competence of an apparent CPM user, and the issue warranted the effort, it would not be difficult to assess the true position by requesting evidence of previous use, and the close examination of both the proffered CPM networks and of those responsible for their preparation and use during the management of previous projects. This, of course, is the general procedure for evaluating levels of experience and competency, and could well be a focus of any prequalification investigation.

In the determination of professional competence one would expect that any professional person, contractor, or organization using CPM have a duty of care to fulfill; that is to say, any such person would be required to have professional competence at least equivalent to that of the ordinary experience of a professional in the field, and be required to have an adequate knowledge of current practice and theory. Hence a client seeking information on a potential bidder's competence in CPM technology could request samples of its use on previous projects as a means of controlling project progress and its use in the detection and resolution of project and project management problems.

Major issues that may develop in this endeavor include the establishment of an appropriate level of current professional practice in the use of CPM at a reasonable and practical level commensurate with the nature of the project in question, so that its required use in that project, at the prescribed level of sophistication, does not become an unduly onerous requirement in a contract. Care must be taken to ensure that the demand in a particular contract does not, except in exceptional cases, exceed the normal usage expected in contracts of a similar type and size.

Additionally, given that the special requirements of a particular project call for a higher than normal level of expertise and usage of CPM, there remains the problems of ensuring that specific individuals and organisations are capable of this upgraded level of management. In the event that it is considered necessary to import, or impose, expertise by nominating specific third parties, or requiring the temporary acquisition of suitably skilled staff, there remain the problems associated with ensuring the effectiveness of such arrangements, or (probably more troublesome) the resolution of issues that arise when such arrangements fail, or at best are inadequately handled.

12.2 SPECIFYING THE CONSTRUCTION PLAN

Critical path methods are well suited to portraying a construction plan and have gained considerable recognition in the industry as a suitable format for their portrayal. It is common practice to prepare a CPM network or a time-scaled bar chart form of plan for virtually any construction activity, whether under contract or not. Thus we often have the case that the contractor furnishes a CPM representation of a proposed construction plan as an informal illustration of the construction approach to the project, even when critical path methods are not formally requested in the contract. As an example of this approach, the Australian Standard 2124-1986 General Conditions of Contract indicates in Clause 33.2, "Construction Program," that "The Contractor may voluntarily furnish to the Superintendent a construction program" and also that "the Superintendent may direct the Contractor to furnish to the Superintendent a construction program within the time and in the form directed by the Superintendent." And as the introduction to this section indicates, "For the purposes of Clause 33 a 'construction program' is a statement in writing showing the dates by which or the times within which the various stages or parts of the work under Contract are to be executed or completed." Common practice, however, is to insert clauses into the contract requiring in some way the development and presentation of a network representation of the proposed construction plan. Thus a particular owner company in its Design and Construct Contract has standard CPM clauses as follows: "Unless agreed otherwise, the Project Program shall be prepared by the contractor in network form, using the critical path method, showing the planned sequence by which the Contractor intends to complete the project." The specification of the detail required in the Construction Program may range from a vague to an overly detailed shopping list of inclusions. Thus, starting with the simple schedule of dates as indicated above in AS2124-1986, "the aforesaid Design and Construct Contract continues with an extensive list of requirements on the detail to be incorporated in the network, including such items as:

- "The dates information is required"
- "EST, LST, EFT, LFT dates for main activities comprising the design and the construction of the project, including Consultants' activities"
- "The logic interdependencies of the main activities and key constraints or milestone dates"
- "The Company's intentions with respect to sequence of work and resource levels for labor and key items of construction equipment and plant"

- "Durations for activities in working days not exceeding 15 (fifteen) days for any activity"

Some contracts with government agencies require even more extensive and detailed shopping list requirements that reflect their previous experience with many and varied contractors, and the nature of the work to be carried out. Thus, as an example, one such contract states: "For the purpose of this special condition, 'construction program' means . . . ."

- "A detailed program using an event-oriented critical path technique drawn to a weekly time scale showing to the satisfaction of the Superintendent the following": (specified items) together with "The logic on which the sequence of activities defining the program relies, including whatever clarification or extra information may be required to allow the Superintendent to model the program using a (nominated) Project Management Software package." "Activities should be supplied in subnetworks of not greater than 'a specified number of' activities with details of internetwork relationships. Earliest and latest start dates and earliest and latest finish dates for each activity should be included in the program and activities with no float should be highlighted."

It is evident that a considerable range of CPM related requirements exist in current construction contracting, some much more demanding than others, and some verging on the overly demanding or onerous state. In many cases, of course, this involvement in CPM methodology is required in two successive phases; this calls for a gross formulation of the proposed construction plan at bid submission with the development of a more detailed presentation by the successful bidder within a specified time. This approach generally results in less expense to each bidder and lower ultimate project cost for the client.

When initiating a project and establishing the requirements of the contract on the contractor in respect to CPM usage, the client or consultant usually adopts one of three approaches:

1. A broad CPM network is produced as an indication of how the client wishes the project to develop and makes this requirement known to all prospective tenderers. This network could include such details as milestone dates, proposed access schedules, the sequence of desired construction of major components, interfacing requirements with other contractors, and the required commissioning dates. In this way the responsible party (whether the consultant, or project manager acting for the client) takes responsibility for the feasibility of this broad plan and all the consequences that flow from this decision; costs and cash flows determined by these time and conditional sequencing requirements are obvious examples, as are other impacts that flow from delays imposed on the contractor while awaiting others not under

the control of the contractor. For this approach a rough construction plan is necessary, with salient target dates indicated; Figure 12.1 shows a client's general requirement for a contract for harbor works (to be discussed in more detail in Chapter 15) Only the general overall requirement is shown at this stage and will be expressed in greater depth when the time comes to define the project more fully.

2. The tenderers are invited to submit their own plan on how the project is to be constructed during the nominated contract period. Here the prospective contractor has the duty of providing a plan and must take responsibility for its feasibility, and (if accepted) for the completion of the project according to this plan, subject to some form of penalty if this plan must be changed to make it feasible. However, the selection of a particular contractor and of the proposed plan in preference to others, based on the advice of a consultant acting for the client, introduces other aspects of responsibility on the part of that consultant.

3. As a variation of (1), in projects such as process plant additions, or on-line maintenance works, in order to avoid interruptions to the client's ongoing processes, a fully detailed plan for the tenderers to adopt may be specified, on which the tenderers are to base their bids, or the client may adopt a cost-plus approach with a selected tenderer. In this initial interactive situation the parties adopt positions that indicate project and contract strategies and in general expose attitudinal positions; the formal contractural agreement then establishes appropriate responsibilities and duties, depending on the approach adopted. It must be remembered that the provision of a plan for the construction of a project in the form of a CPM network as a formal part of a contract, whether client- or tenderer-initiated, carries with it (a) the assertion that the plan is feasible and (b) the responsibility to make it work or to make it good. Further, if the plan thus specified is intimately linked to calendar periods, here is the implication that adequate resources are, or can be made, available to a schedule and rate of use to achieve the planned timetable. The major location of responsibility is established by how these initial matters are handled.

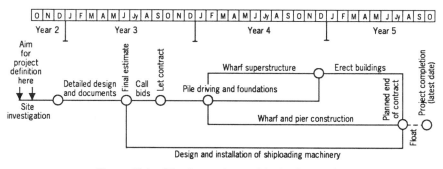

Figure 12.1   Client's requirement for harbor works.

In some contracts there are additional clauses to ensure that there are obligations tied to the provision of the original construction program that cannot be simply discarded by the provision of an overriding, updated or revised plan. For example, the contractor may submit a plan scheduled to take up all the contract period either ensure personal control in the program for the maximun possible float (whether indicated as float or submerged in excessively safe activity durations), or if the contractor really intends to finish the project in much less time than the contract period but does not disclose this intention in the tender plan.

Clauses that relate to the control of these issues are as follows:

"The Contractor shall proceed with the work under the Contract with due expedition and without delay."

"The Contractor shall not suspend the progress of the whole or any part of the work under the Contract except . . ." (under other clauses or as approved by the superintendent).

"The Principal and the Superintendent shall not be obliged to furnish information, documents or instructions earlier than the Principal or Superintendent, as the case may be, should reasonably have anticipated at the Date of Acceptance of Tender."

Such clauses (especially the last one) begin to force the contractor to provide a plan that portrays the intended schedule by effectively ensuring continuous activity on the contractor's part and restraining untimely early demands on the other parties to the contract.

Figure 12.2 presents a suggested specification summary for the proper

PLANNING AND SCHEDULING WITH BID

Each bid shall be accompanied by a time chart clearly indicating the Contractor's proposed construction program and the sequence of work activities comprising the contract. Such time chart shall be in the form of a summary network and shall be activity-oriented, showing interdependencies, and each activity shall be a clearly defined work item or items in the Contract Schedule. Only major work activities need be shown, but, in addition, the network shall include activities for design work, detailing, establishment, and procurement of materials and equipment by the Contractor, as well as the latest dates for provision of services and materials (if any) by the Owner or the Engineer. The time chart shall be drawn to a horizontal scale of time with calendar dates and shall incorporate starting and finishing dates and any other target milestone dates herein specified, together with the significant resources required for the execution of each activity in the duration shown.

Activities shall be arranged to show the "critical path" as a heavy double line, with all other activities and their float times as single full and broken lines, respectively.

This time chart shall be the construction program upon which the Contractor's bid was based and as such shall be deemed a condition of such bid.

Figure 12.2 Typical specification for "Planning and Scheduling with Bid" for Schedule of Rates Contract requiring the use of critical path methods.

requirements for the planning and scheduling clauses required to accompany each bid, and Figure 12.3 shows a similar summary for requirements to be provided by the successful tenderer after the bid is accepted.

12.3 SPECIFYING PROJECT MANAGEMENT

All parties to a contract, as well as those involved in project planning and management, and also those monitoring progress and involved in the authorization of payments, and similar, are interested in project status and progress information. Also, when large amounts of money are involved and cash flows need to be organized and monitored, there is an understandable interest in receiving frequent project progress and status updates. Furthermore, there is the need on the client's part to be satisfied that the contractor is acting in a competent and efficient manner. Consequently, considerable interest exists to ensure that adequate information be made available and that all parties become involved in the major decisions influencing project progress and status. Thus almost all contracts have clauses specifying the nature and extent of the information and documentation required to establish project progress and the verification and authorization of progress payments. In

PLANNING AND SCHEDULING ON ACCEPTANCE OF BID

Within . . . weeks from the date of acceptance of his bid, the Contractor shall present to the Engineer three copies of a detailed critical path network expanded from and in conformity with the above time chart, with subnetworks as necessary, all drawn to a horizontal time scale. Each activity shall be an item or subdivision of an item in the Contract Schedule or an activity or subdivision of an activity shown on the above time chart. All activities to be performed by Subcontractors shall be shown separately from activities to be performed by the Contractor; the Contractor shall be responsible for the performance of his Subcontractors according to such subcontract activities.

The duration of each activity shall have one time estimate only and this shall not exceed one month, except that design and procurement activities, and activities that progress uniformly from start to finish requiring no change in manpower or equipment may be shown as one activity for the required duration even if it extends beyond one month. Any such extended activity shall be justified by a narrative note on the network, including the estimate of monthly productivity.

This detailed network shall clearly show the major resources of manpower and equipment required for each activity, as appropriate, and any essential time required by the Contractor for maintenance and repair of such equipment. The head and tail node of each activity shall be identified by unique numbers in proper numerical and chronological sequence.

This detailed network shall clearly indicate the critical path and the second and third next critical paths, and shall be called the "initial network."

The initial network shall be deemed a condition of the Contract, and, whenever changed or revised, three copies thereof shall be presented to the Engineer.

Figure 12.3 Typical specification for "Planning and Scheduling on Acceptance of Bid" for Schedule of Rates Contract requiring the use of critical path methods that is compatible with the specification outlined in Figure 12.2.

fairly small projects operating under traditional management and contracts, all that is normally specified is the procedure for progress payments. This specification spells out the procedure for submitting a progress payment claim, the criteria for considering materials delivered to the site, the issuing of progress payment authorizations, and the time-related clauses for payments. On larger and more complex projects this simple procedure requires extensive documentation and is usually the subject of more demanding criteria and the focus of a more extensive specification. How this is done depends on the type of contract involved, as well as the CPM-related professionalism of the client-consultant, contractor, and of the project managers involved. Also important are the goodwill or adverse relationship attitudes between the contracting parties, and the time pressures associated with rapid construction activities and related fast-tracking processes.

A typical specification for a CPM based approach to project managent on a Schedule of Rates Contract (which is compatible with segments indicated earlier in this chapter) is indicated in Figures 12.4 and 12.5. Figure 12.4 relates directly to Project Status, the portrayal of "work as executed," "Factual Networks," and network revisions, Figure 12.5 introduces the relevent specification segments to explicitly link project start, general project management administration, and project close activities to the portrayal of the construction plan, together with special cost items for project management and the costs associated with the preparation of project networks and of the management of the project.

Normally the contractor wishes to manage the project in a certain personally preferred way using whatever information, documentation, and reporting systems the company has developed. In this environment the contractor operates with minimum interference, supplying sufficient information just to satisfy the client regarding progress for authorizing progress payments. When the needs of the project require more extensive documentation, however, as spelled out in more specific contract clauses, the contractor must consider the manner in which existing in-house procedures can be modified, upgraded, or replaced by more suitable systems. These considerations become more demanding in complex projects in which the client has extensive business and management experience and when the timing of project implementation is critical. In these cases, unless special arrangements are made, attempts by the client to ensure receipt of "full" information in adequate time to query contractor decisions affecting project progress and priorities, and/or to exert pressure for more controllable progress, meet with stiffening resistance from contractors who see their freedom to manage their project being eroded. Contractors are reluctant to adopt a level of management sophistication that they consider to be either inappropriate, beyond their capabilities with present staff, not adequately reimbursed, beyond the level of ordinary practice, or verging on becoming unduly onerous. In some contracts the client stipulates that the contractor must notify the owner of various requirements or lose advantage and payment. This seems particularly onerous. In

PROGRESS CONTROL
(1) Monthly Status Report
Once each calendar month the Contractor shall present to the Engineer copies of the Initial Network with the current status line clearly indicated and with all completed activities and the completed portion of current activities marked in color. This shall be accompanied by a narrative report summarizing matters of significant progress since the last reporting date and shall contain the following summary tabulations:

(a) Activities rescheduled or reestimated in the current report, with reasons therefor.

(b) Activities added or deleted in the current report, with reasons therefor.

(c) Activities completed since the previous report.

(d) Activities due for completion since the previous report but not yet completed, with reasons therefor; and details of any anticipated delays or problems that may affect completion dates, together with remedial action intended by the Contractor.

(2) Factual Network
The Contractor shall also submit each month with the above information a work-as-executed-to-date network to the same scale and of the same format as the Initial Network, showing all work changes and delays to date by suitable notation, accompanied by a statement of all critical delays, and the causes thereof, and the currently due total extension (if any) of the time of completion of the Contract, together with the present extended completion date. This shall be called the "factual network." It shall be updated weekly at the site to the mutual satisfaction of the Engineer's representative and the Contractor and shall be deemed to be a true record of the progress of the Contract to date.

(3) Network to Completion
Where

(i) new activities that change the critical path are inserted.

(ii) noncritical activities are delayed beyond their allowable float time and change the critical path.

(iii) the critical path becomes more than five per cent behind the time remaining for completion of the Contract,

(iv) any milestone date specified in the Contract will be affected at all by slippage within the Contract,

(v) a previously specified milestone date has been changed by order of the Engineer

a copy of the Factual Network to date shall be extended by an estimated network to completion of the Contract and shall be submitted, together with a report of action proposed by the Contractor to improve progress if necessary. Such Factual Network extended to completion of the Contract shall replace the Initial Network for the purposes of this clause.

(4) Time of Submission
All the above networks and reports shall be submitted to the Engineer in triplicate and shall be not more than 5 days out of date when received by the Engineer.

Figure 12.4 Typical specification for project progress contol.

many cases the needs of the project require an intensive cooperation between the contracting parties and the elimination of the possibility of adversary-type roles emerging during the life of the project. Often this is accomplished through the use of cost-plus contracts. Thus it may be the case that the competition generated for these more attractive contracts ensures that the contractor's capabilities are actively marketed in delivering whatever information and documantation the client wishes to receive. This collaboration is then the focus for the specification of the information and management sys-

(a) PRELIMINARY AND GENERAL EXPENSES

The Contract Schedule of Prices and Rates shall include two separate lump sums items to cover

(a) all the work and costs incurred by the Contractor in the establishment of facilities and equipment for carrying out the works at the site and

(b) the removal of the same at the end of the contract. The Contractor may further subdivide either or both of these items.

In addition, the Contract Schedule of Prices and Rates shall include another separate item to cover all the costs of supervision and administration of the contract by the Contractor (including all overhead, site expense, and head-office charges) during the performance of work on the site. The Contractor's price for this item shall be shown as a total amount and also as a rate per working day, which, when multipled by the construction time on site shown in the Contractor's construction program submitted with the bid, shall equal the total amount of this item.

Payment for Preliminary and General Expenses shall be made in accordance with the General Conditions of Contract.

(b) COST OF PREPARATION OF NETWORKS AND REPORTS

The planning and project control herein specified shall be part of the Works to be executed under this Contract, and all costs incurred by the Contractor in the preparation and submission of the networks and reports specified above in Clauses 2, 3, and 4 shall be paid by the Contractor and shall be included in the Contract Schedule as a lump sum in a separate item entitled "Network Planning and Project Control."

(c) CORRECTION OF LATENESS

If a progress report indicates the critical path to be more than 5% behind the time remaining for completion, the Engineer may order the Contractor to take such action as is necessary to improve progress. If the Contractor's action or proposed action is not satisfactory to the Engineer, he may direct the Contractor to increase work force, construction equipment, number of shifts, or take other action. The responsibility for any additional cost of such direction by the Engineer shall be determined in accordance with the General Conditions of Contract.

Figure 12.5 Typical specification for inclusion of project management activities in the construction plan.

tems to be used on a project. In some cases contractors are prepared to adapt their own management information systems to meet client requirements and even to adopt client's own or proposed project control systems. Usually, however, these favorable arrangements are the result of special types of contracts in which the contractor is adequately reimbursed for such services. Thus a variety of attitudinal, professional, and contractural issues arise influencing the range and sophistication of CPM-related management information systems and management decision processes called for in specifications or offered for use in project management contracts. The construction plan must now be prepared with regard to the detail appropriate to the magnitude and nature of the project involved and the requirements of the contract specifications. A decision must be made concerning the depiction of this plan: Is it to be an ordinary, or time-scaled arrow diagram, or a precedence circle network, or a critical path bar chart? The discussions in Chapters 2 and 3 are pertinent at this stage. The next consideration is the regular reporting methodology: How often should the data be updated, and how should project status be indicated in practice? These and related matters are discussed in Chapters 8, 9, and 10. It is also essential to ensure that such matters are given adequate attention by the contractor, with appropriate action to be taken if this is not done. Regular status reports, to be prepared and issued at appropriate intervals [weekly, biweekly (twice per month), or monthly?], and to be read by and acted on promptly by the owner's and the contractor's staff, together with the periodical site meetings to discuss problems and monitor remedial action, are an essential part of proper project management.

It may be necessary to maintain pressure on the contractor, remembering that it is this person's prerogative to manage the contract without undue interference from the owner's management personnel. Prudent record-keeping and prompt evaluation of the effects of work changes and delays, as outlined in Chapter 11, are in the interests of both parties to the contract. It is necessary to remember, however, that float usually belongs to the contractor (especially if that person prepared the detailed network), who thus can use it freely to alleviate any problems that might arise, and also to assist the owner at a price; Section 11.7 is of importance in this respect.

12.4   THE SCHEDULING ENGINEER

In general, "scheduling" refers to the activities of a head-office-based engineer involved in project planning and scheduling in a staff function for all projects and/or to one concerned with project scheduling control of a particular project and who is generally field-based. The latter is more likely to be a junior engineer, often classified as a "Field Engineer," while the former may have many functions, depending on company size, and is generally titled as a "Scheduling Engineer" in large organizations where the volume of company business and size of projects demand specialized personnel.

The general job description of the field-based engineer is given in Figure 12.6, and, basically, specifies that the young engineer be able to absorb and understand the contractor's construction plan and be capable of suggesting amendments to that plan or corrections if necessary. The field-based engineer must be capable of developing the CPM plan in more detail, especially for the next time-period of 2 or more weeks. This officer must become competent in CPM so that consistency of logic and activity size would result. This would include full knowledge and use of a company standard for node and activity symbols, and the engineer must also be aware of special company concerns on subcontractor inferfacing, heavy-equipment availability and scheduling, and critical material delivery dates, together with company strategy and operational policies influencing the availability and rate of use of company resources. The competent field engineer should be able to produce detailed networks for any planned or prospective construction activity and

A. General Functions

Schedules and coordinates.

Serves as a trouble-shooter when there is a breakdown in delivery schedule.

Maintains a constant follow-up on the schedule to ensure progress as previously planned

B. Details Functions

1 Receives plans and specifications and breaks them down by trade

2 Expedites follow-up for shop drawings or detailed drawings; checks with own staff for follow-up

3 Ensures plans and drawings reach the right people at the right time

4 Keeps in touch with subcontractors as needed

5 Writes memos as needed to superintendents, subcontractors, etc.

6 Establishes delivery times for materials, equipment, or labor; determines the lead time required for acquisitions

7 Sets up delivery schedules for job superintendent

8 Follows up daily on trouble areas, where delivery of materials may be lagging

Figure 12.6   Job description for a Field Engineer.

know whom to consult in the company in order to produce a CPM that reflects their decisions, and must also be able to ferret out the necessary detail and find out what the company wants to commit to paper.

A more detailed job description for the ''Scheduling Engineer'' is given in Figures 12.7 and 12.8, where the duties and responsibilities are divided into those associated with project planning and scheduling in the head office (Figure 12.7) and those associated with project scheduling control in the field (Figure 12.8). As indicated, the scheduling engineer has both a field and a head-office location, which are administratively linked; this engineer assists and has contact with many project staff members, and is commonly one person in a medium-sized company. At the large company level, field engineering becomes highly technical in terms of interpreting specifications, drawings, and other activities necessary to support construction, and therefore becomes an area of expertise by itself. Planning and scheduling become more specialized by virtue of the techniques and management demands for comprehensive systems, which focus on work package concepts, and therefore require a specialist engineer who would be familiar with many scheduling-related computer packages, and can make recommendations on suitable systems for company activities. Additionally, this engineer should be aware of the implications of specific contract clauses related to scheduling and reporting functions relevent to a specific project contract.

12.5   THE PROFESSIONAL OPINION REPORT

The client's consultant or project manager discusses the manner in which the project is to be managed, calls tenders, advises the client on the resulting submissions, and makes recommendations on suitable tenderers. In the initial stages of this process the main CPM-related issue is whether to give

SCHEDULING

S1 CONSTRUCTION SEQUENCE

Defining work activities
  and categories, having
  regard to impact of
  resources, materials,
  and equipment
Develop construction plan
  through logical
  sequencing of work
  activities

S2 ESTABLISH CONSTRUCTION SCHEDULE

Establish start/finishes
  dates of activities,
plan progress curves with
  milestones, and worker-
  hour content of
  activities
Compare availability of
  resources, project
  priorities, subcontracts,
  work weeks, etc.

S3 DETERMINE PROJECT AND TRADE WORKER-HOUR PROFILE

Develop total project and
  trade worker-hour
  curves, adjusted to
  match planned progress
  estimates

S4 DEVELOP WORK PACKAGE DETAILED LOGIC

Develops: detailed
  construction logic to
  match overall schedule
  time frames for work
  package activities
Identifies: planned
  progress, worker-hours,
  manpower for work
  package

A3 DETERMINE PROJECT MANPOWER CURVES

Develops: total project
  and trade manpower
  profiles
Adjusts: manpower to
  match availability,
  planned progress

P5 DEFINE WORK PACKAGE

Specifies: work package
  required in relation to
  schedule

Figure 12.7 Job description for a scheduling engineer in head office—Project planning and scheduling, and site manpower profile.

MONITORING

M1 SUMMARY PROJECT REPORTS

Generates: reports depicting progress comparisons, productivity comparisons, manpower, and progress curves to complete
Summarize: problems or potential problems

M2 CURRENT PROJECT STATUS

Establish: critical activities, progress achievements, productivity indicators, necessary schedule revisions, manpower to complete, and progress to complete
Identify: existing or potential problems

M6 WORK PROGRESS

Assists: evaluation of physical progress for work tasks
Inspects: progress achieved in work tasks

RECORDING

REVISE WORK PACKAGE SCOPE

Reassesses construction logic, completion requirements of work tasks
Reevaluate progress and manpower targets

UPDATING

DEFINE CURRENT PROJECT STATUS

Establishes: activity progress, activity forecasted completion dates, forecasted start dates, critical activities
Impacts: trends and changes in scope

M3 PROJECT FORECASTING

Forecast: manpower curves to complete, progress to complete, impact of scope changes on schedule

LABOR PRODUCTIVITY

labor productivity value for work areas

REVISE CONSTRUCTION SCHEDULE

Revises: start and completion dates, change construction logic
Forecasts: new progress and manpower curves to complete

M4 ACTIVITY ANALYSIS

Suggest: activity analysis to improve progress or performance for selected work tasks

M5 LABOR ANALYSIS

Compare actual productivity with target productivity for work areas

QUANTITIES IN PLACE

Review quantity value achieved for work tasks

Figure 12.8   Job description for a scheduling engineer in the field

tenderers an outline CPM that embodies the main sequence and scheduling concerns of the client, or to let tenderers supply their own CPM plan. If the latter is the case, then it may be necessary for the consultant to ask that each tenderer include considerations on a variety of details considered pertinent to the efficient planning and performance of the contract. Thus the consultant may require the CPM plan to explicitly incorporate major equipment, and/or material delivery logic. There also may be a need to ensure that sufficient detail is presented in the CPM plan to indicate adequately the contractor's competence in proposed project work sequences and understanding of unique issues relating to the project site and anticipated work conditions.

The consideration of these issues results in the drafting and inclusion of specific CPM-related clauses in the contract documents of the nature indicated in the previous sections of this chapter and reflect the competence and professional opinions of the consultants involved. Since these special clauses and requirements impose considerable obligations on all tenderers, and especially on the successfull contractor, it behoves the consultant to give very careful consideration to personal involvement in this process.

The contractor may have called in a consultant to prepare a CPM plan. In such case the contractor may say either: (1) "Here is a job, make me a CPM plan of how to do it," or (2) "Here is a job, and this is how I am going to build it, so represent my plan in CPM form." Additionally, in this latter case the consultant may be requested to ensure that the CPM representation of the construction plan also meet the anticipated, or known, requirements of the client's specifications. Thus the consultant takes on different responsibilities depending on which role is assumed. However, whatever role the consultant assumes, the contractor, on accepting and including the resulting CPM network on submission to the client, must accept responsibility for the feasibility of the plan. In the event that the CPM network is inadequate, or incorrect, the contractor may consider dismissing the consultant (and refuse to pay the consultant's fees), and may attempt to recoup costs by suing the consultant for incompetence or negligence; this is a drastic step and it behoves a consultant to avoid it at all costs.

As a separate case, a consultant who is required to give a professional opinion on a specific CPM network may proceed to examine the plan and determine whether it meets the contract requirements, will work, and exhibits an adequate level of competence and professionalism in its formulation and presentation. Is the network, for example, deficient in some important detail, does it fail to meet some criterion that was requested, and is it consistent in its presentation, logic, activity size, and so on? In other words, is there adequate detail, and is it clearly and professionally presented? A useful guide to this examination is the CPM-related requirements specified in the contract, which should not be excessive but should be related to the complexity of the project and its special needs.

Various issues which the compiler of a plan and schedule must consider are:

- Are clauses onerous?
- How can a specific CPM plan be shown to be inadequate?
- Is a CPM network a guide only to management or a plan to be meticulously followed?
- How gross can a CPM network be and still indicate a feasible construction plan?

If a contract clause is apparently not being met (e.g., updating is lacking), if there is failure to notify others regarding slippage, or if the contractor is not prepared to go beyond a low level of CPM usage or fails to cooperate to a reasonable extent, then steps should be taken to enforce the terms of the contract. In every case the remedy must be properly stipulated and, if necessary, enforced, or the contractor will treat the specification with disdain.

Evaluation

a. How can the use of CPM be enforced on a project?
b. How would you professionally evaluate a project CPM plan?

13

COMPUTER-AIDED CPM

13.1 COMPUTER USAGE

There is a widespread use of computers in all sectors of industry, commerce, and society. The price tag on small personal computers (PCs) is often so attractive that many individuals find it possible to have their own private computers whether for business reasons, as an intellectual challenge, or even in lieu of the common typewriter, or as a plaything. Furthermore, most PCs on the market have add-on capabilities for increasing memory size [640K (kilobytes) random access memory (RAM) and hard disks of 20- or 40-megabyte capacity are common], for improving processing speed (using special microprocessor chips), and have built-in ports for input/output devices capable of accommodating a variety of relatively inexpensive accessories such as screens, printers, modems, and cursor systems (pointers, mouse, light pens, toggles) that eliminate tedious keyboard work. These accessories considerably enhance the capabilities of the basic PC computing unit beyond simple logical, data, and arithmetic processing functions to the level of total data display and manipulation systems to suit virtually any management requirement. As an example, consider the potential of video display units (VDUs). Typical VDU screens have 640×480 addressable light- (and color-) sensitive cells of pixels, each of which can be programmed to contain and portray a selected character or shape, resulting in a possible screen display of 25×80 characters. An entire screen display can be stored in about 2K of computer memory. If the screen is set up as a preformatted template in which predetermined spaces are left for the user to enter data in response to specific template questions, a powerful input system can be generated that is very user-friendly and pertinent to the needs of the program. Such templates can be stored in condensed form in memory. Similarly, VDU screens can be preformatted to display a menu of program options that the user can choose by a simple pointer. In effect the menu becomes a multilevel program switch. Most software systems now include a context-sensitive HELP option in the menu and template displays that, when selected, display preprogrammed information and examples so that the VDU system becomes an outlet for a self-learning program of instruction. These features explain the popularity of display systems now on the market. Hence it is possible for an individual to set up a powerful computing system with capabilities considered virtually impossible only a decade ago.

The usefulness of these common and readily available tools is further enhanced by the sophistication and user-friendly nature of an ever-growing number of general purpose and specialty software systems for virtually every business and management function and decision process. In this way previous barriers to the use of digital computers, associated with the need for extensive programming and hardware knowledge, have been eliminated. The result is the emergence of computer software packages that are menu- and template-driven and that, coupled with enhanced screen displays and mouse-type pointer attachments together with built-in instructional features, enable even the novice to rapidly become a productive and efficient user of the computer.

Finally, in many social environments, generations are growing up familiar with, and using, computers by the time they reach secondary schooling. Thus it is common to find PCs in construction project site offices, or on construction managers' desks, used for many project-management-related tasks and procedures.

In this management environment it is usual to find that many existing general-purpose computer packages can be, and have been, adapted to CPM usage, as will be indicated in the next section. Additionally, because of the nature of project management, many specially dedicated CPM packages are in practical use. It is convenient in this chapter to consider these in the following sections from the point of view of the juxtaposition between their use as a result of personal initiative versus company policy decisions together with their use on special problems, through project usage, to being incorporated in a companywide data and management information system. No attempt will be made to list or describe the range of existing commercially available packages, as this is both daunting in scope and futile in accomplishment, because of the highly volatile nature of the supply of and demand for creative and useful packages.

13.2 ADAPTIVE USE OF PACKAGES

Many existing, and commonly available packages have logical, procedural, or display features that may be appropriate to the formulation, documentation, or presentation of CPM techniques and methodologies. This requires an extensive understanding of both the potential of an existing package and CPM methodologies, and the creative adaption of the package to CPM usage. The purpose is to take advantage of an existing computer package, or program, that is readily available to the user, either on the company's computer system or as a simple soft-disk program that can be loaded onto the user's PC, and which has been developed for other purposes. This requires the adaptive use of a structural feature of the package for CPM purposes through manipulation and reinterpretation.

Many general-purpose packages have built-in features that can be exploited to enable them to be used effectively as aids in the representation of CPM project plans and/or calculations. These aids take advantage of either a graphical display feature that can readily be associated with the representation of a project network or a data processing capability that can be used for a CPM-related computational process.

Hence the "outlining" feature of many hierarchically structured packages (such as those for a database dBASE III), and text file and graphical insert capabilities of word processing packages (such as WRITE and WORD) together with general disk operating systems (DOS) (such as WINDOWS and FRAMEWORK), enable the user to portray Work Breakdown Structures, and project activity dissection in a way that captures subnetworks in a multilevel form. These features facilitate the development of structured project data files, and the interactive zoom lens use of macro-, and micro-, files and networks in project formulation, presentation, and reporting. In this way the framework of a personalized project information system (MIS) can be developed, while many of these efforts may not be efficient compared with the acquisition of special-purpose systems, they often are their precursors.

A simple example for the formulation, and portrayal, of a time-scaled form of a project and its resource requirement summaries is the use of LOTUS 1-2-3. LOTUS 1-2-3 is a generally available spreadsheet package, which features individual cell addressing, and a variety of formats for cell entries (numerical, textural, combined numberical and textural, and formula) that enable general-purpose functions to be performed on rows and columns. A schematic illustration of these features of LOTUS 1-2-3 is given in Figure 13.1a. The adaptive use of LOTUS 1-2-3 to CPM presentations is illustrated in Figure 13.1b. Each individual activity can be labeled and located in the worksheet matrix, now given a time scale for the columns, according to the project schedule. A "range" of cells is assigned to each activity according to its duration and the activity attributes considered relevent. As shown in Figure 13.1b, the activity "Form & Concrete Footings N" has a duration of 5 days and requires a crew of 6 workers. Additional rows can be used to portray daily formwork requirements, concrete quantities, and Back Hoe/Bobcat hours, as well as direct costs on a daily or activity basis, if desired. In this way, by the proper location in the spreadsheet array of all project activities, a time-scaled bar chart results. Finally the contents of certain cells can be formulated in terms of functions, which can perform selective row sums that effectively generate project resource summaries and requirements. A more

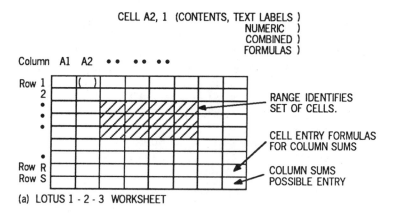

CELL A2, 1 (CONTENTS, TEXT LABELS)
NUMERIC)
COMBINED)
FORMULAS)

Column A1 A2 •• •• ••

Row 1
2
•
•
•

Row R
Row S

RANGE IDENTIFIES
SET OF CELLS.

CELL ENTRY FORMULAS
FOR COLUMN SUMS

COLUMN SUMS
POSSIBLE ENTRY

(a) LOTUS 1 - 2 - 3 WORKSHEET

SET COLUMNS TO REFLECT PROJECT DAYS/WEEEKS/MONTHS
SET ROWS TO LABEL PROJECT ACTIVITIES
SET ROW RANGE FOR PROJECT ACTIVITY DURATION
SET ROW AND COLUMN RANGE FOR ACTIVITY DETAILS
SET SELECTIVE COLUMN SUMS FOR PROJECT RESOURCE REQUIREMENT/USAGE
SUMMARY

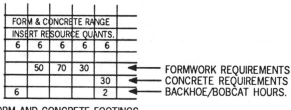

FORM & CONCRETE RANGE
INSERT RESOURCE QUANTS.

6	6	6	6	6
	50	70	30	
				30
6				2

FORMWORK REQUIREMENTS
CONCRETE REQUIREMENTS
BACKHOE/BOBCAT HOURS.

(b) FORM AND CONCRETE FOOTINGS
(5 DAYS, 6 MEN)

Figure 13.1 Adaptive use of LOTUS 1-2-3: (a) LOTUS 1-2-3 worksheet; (b) CPM representation.

useful example is with general spreadsheet-type packages that have available
built-in templates that are user-friendly in their labeling. In this way CPM-
focused generic labels can be assigned to the columns, and much more
specific activity data structure templates, designed for activity row labeling,
including appropriate data fields for pertinent activity parameters and progress status information, for quick visual absorption by project managers.
Additionally, individual activity entries can have resource parameters distributed along the activity duration cells, by selecting a relevent menu-driven
histogram. This can be done, for example, for such items as anticipated
resource usage and cash flows. These features can therefore build in realistic
representations of expected project resource usage, and produce summaries

of interest to construction managers. These examples of the manipulation of computer packages are predicated on individual initiative and the creative adaption of the package to CPM usage. Also implied is the individual's personal use of the package with usually no company interest, support, or policy, on the supply or use of CPM-related packages.

13.3 AUTOMATING CPM

The CPM procedures considered in this book can be conveniently grouped as follows:

1. Those related to project definition, including activity description and the determination of their attributes, such as durations and resource requirements and the prescribing of project resource availabilities. These may be referred to as defining project parameters.
2. The determination of the construction sequences, and the related ordering of the activities so as to indicate how construction is to proceed, and the representation of this plan in network from by either arrow diagrams or precedence diagrams. These procedures define the project network.
3. Those computational processes relating to the determination of the critical path, and activity and event floats.
4. The special procedures involved in the handling of resource limitations, constraints, and conflicts, so that a logically feasible network results.
5. The determination and portrayal of a construction schedule in project time units (or calendar form) as a data file, graphically as an annotated network, a time-scaled network, or as a Gantt bar chart.
6. The determination of special project management features, such as S-curves, cash flow curves, and overdraft requirements.
7. Those relating to the portrayal of current project status and changes.
8. Those relating to the preparation and portrayal of management reports.

All of these procedures can be automated and computerised, once decisions are made on the computer hardware configurations and range of input/output devices to be used, the manner in which information is to be displayed or formatted in hard copy, and the characteristics of the user-friendly package formulation has been decided. A full treatment of these issues is beyond the scope of this book; it is sufficient for our purposes here to consider concepts relating to automating the CPM conputational procedures, and relating the use of menus and templates to the input/output of data information and of decisions.

In order to automate CPM procedures, it is necessary to formulate each in a logical manner. Also the nature and order of presentation of the unique data required to describe any project network must be prescribed in a way that is compatible with the requirements of the CPM procedures that will operate on the data. These formulations must then be expressed in a manner that the computer can accept and process.

Consider, for example, the forward pass calculations on an activity-oriented (ADM) network. These calculations are very simply performed graphically, but require considerable logical support when the network structure is to be captured numerically. In order to understand how this can be accomplished, consider the nature of elementary calculations.

Any calculation can be considered as a collection of elementary operations, arranged in some sequence, functioning on specific data. An elementary operation depends on the processing system used and on the stimulation required to initiate the operation. Thus, addition, subtraction, multiplication, and division may be universally considered as elementary operations, whereas the determination of earthwork quantities would be considered elementary for an engineer but not perhaps for someone else. Obviously, the description or specification of a sequence of operations for a particular calculation depends on the level of detail required by the processor and the acceptable elementary operations. In some cases, a simple statement or request is sufficient; in others a comprehensive flow chart is necessary. Digital computers require a completely detailed and rigid program, listing every computational and logical step in the calculation, all expressed in elementary operations acceptable to the computer. However, once these steps have been correctly formulated and presented to the machine, they are automatically followed upon the receipt of a simple stimulus or request. In effect, the entire program becomes an elementary operation to the digital computer.

The determination of the earliest start time (EST, or T^E) for events in a network diagram (see Chapter 4) provides an excellent illustration of the way the calculations are automatically carried out by a digital computer. Figure 13.2 shows one simple flow chart for the computation of the EST for events in any network; it assumes (or prescribes) that all events are numbered consecutively from 0 at project start to N at project completion. A close study shows that the flow chart is logically correct and computationally possible, once ways are devised for finding the number of activities terminating at any event and their individual initial events. The prescribing of these ways and the development of the flow chart into a detailed computer program are beyond the scope of this book; but it will be agreed that it is possible. However, it must be understood that the correct ESTs would only be produced by the computer if the user of the program conforms meticulously, in the input data format, to the rigorous order and specification prescribed by the designer of the program. If an arbitrary presentation of the data were made, the program would either completely fail or produce meaningless nonsense. For this reason, computer companies provide manuals explaining

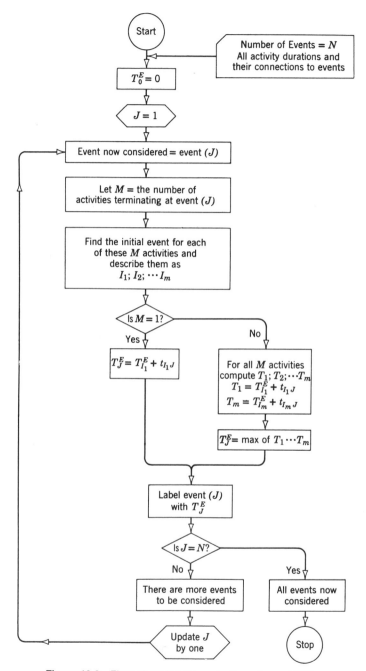

Figure 13.2 Flow chart for determination of event EST.

in detail the nature and purpose of their available programs and the method in which data are to be presented.

The reader should now be in a position to understand that, with considerable ingenuity, computer programs can be prepared for all the calculations considered in this book. To use these programs necessitates, first, determining the essential input data to suit them, and, second, entering of this data using a keyboard-type computer terminal.

The manner in which these data are to be entered, and also on how each CPM procedure is to be selected, raises the whole question as to how the user is to treat the package. That is to say, what user needs are to be addressed, and what computer size and configuration is to be assumed, and how many accessories are to be allowed for in the computer package. These issues must be resolved and explicitly stated so that the necessary formatting and input/output procedures can be prepared. This apparent rigidity must also be considered with even the most user-friendly computer package. Thus a software presentation must be devised to eliminate the need, on the part of the user, for anything but the absolute minimum knowledge of the computer itself. If the package is to be self contained, it must have a self-learning built-in manual that takes the user through the capabilities of the package. Thus considerable ingenuity and skill is required to formulate the CPM package.

The next steps must consider the manner in which the network is to be captured: input through a keyboard, using a template approach driven by a menu selection, or purely graphically using a menu of symbols and pointer devices. As mentioned previously, VDU screens can be formatted, programmed, condensed, and stored in the computer for these purposes. Once this fairly simple programming approach is implemented, considerable flexibility and enhanced graphical features can be incorporated into the package. It should be readily seen that these capabilities are feasible and only require ingenuity, considerable computing skills, and knowledge of computer hardware systems.

These and many other features need to be addressed but are clearly beyond the scope of this book.

13.4 SMALL PERSONAL COMPUTER DISK PROGRAMS

Almost every PC owner has a collection of small floppy-disk packages for a wide range of specialty functions, purposes, and computer games. Many are custom-designed for individual computers, but the majority gain their appeal by being compatible with a whole series of PC brands. While initially most have been commercially prepared and copyrighted, they soon become copied or developed as pirated versions of established packages. Usually the packages are obtained by mail order, from share or bulletin board sources, swapped for other packages, or copies are made from another set after

Package title; version; date of issue.

Ownership; copyright; right to use

General information on package purpose and scope

PC requirements; minimum RAM size; minimun DOS version; disk drive(s); printers

Booting instructions; disk copy instructions

Package capabilities; maximum number of work tasks; maximum number of resources

CPM background and concepts

Tutorial examples

Menu options: set up project parameters; set up Gantt scale; set up calendar and holidays; add, change, delete work tasks; add, change, delete resources

Critical path analysis: Gantt chart; reports; "HELP (context-sensitive); what to expect; useful tips; advanced options; disclaimer; permission to copy; registration form

Figure 13.3 Typical layout of floppy-disk CPM-type packages.

personal contact with a user. The traffic in these packages seems to have become uncontrollable, so much so that efforts have been made to encourage users to register ownership after satisfactory use, purely on the basis of being later notified of revisions, supplied at nominal prices.

Usually these packages consist of one, two, or three floppy disks that are sometimes accompanied with a small manual and ownership authorization. The typical layout of these packages is illustrated in Figure 13.3. As indicated, the packages are self-contained, and the quality of the program ranges from the purely utilitarian to the high end of sophistication in programming skill, visual displays, and user assistance. Most are menu- and template-driven, so that any PC user can become proficient in their use in a matter of several hours. Because of their size and ready availability they are very useful, can be machine-loaded in minutes without taking up permanent memory space, and are used whenever the need arises.

As an example of the capabilities of such programs, consider the following small project network executed using the HORNET package on a PC. Figure 13.4 is a typical plot of the network logic for a construction project with multiple completion dates. A listing of the activities, with durations and logic constraints, is illustrated in Figure 13.5 and may be useful for checking and updating the network logic. The printed output from network calculations may be either a listing containing early start/finish, late start/finish dates and float, if any, on each activity as shown in Figure 13.6, or a bar chart similar to

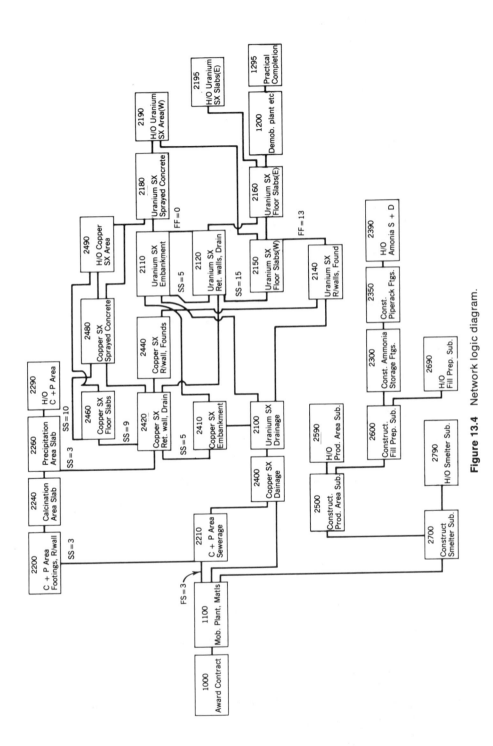

Figure 13.4 Network logic diagram.

Figure 13.5 Activity data listing.

Activity Number	Description (short) Management Code	Type	Calendar (shutdown)	Priority	Estimated min/max	Target Start/Fin.	Actual Start/Fin.	Preceding Activities Activity	Link	Delay
1000	Award Contract	A N	0	0:1					
1100	Mobilise Plant, Matls	A N	0	3:0			1000	FSG	
1200	Demobilise Plant etc	A N	0	2:0			2160	FSG	
1290	Practical Completion	A N	0	0:0	17-09-87		1200	FSG	
2100	Uranium SX Drainage	A N	0	0:3			2400	FSG	
2110	U SX Embankment	A N	0	3:3			2410	FSG	
								2100	FSG	
2120	U SX Ret. walls, Drain	A N	0	5:3			2110	SSG	1:0
								2110	FFG	
2140	U SX R/walls, found.	A N	0	1:0			2100	FSG	
								2440	FSG	
2150	U SX Floor slabs(W)	A N	0	3:3			2140	FFG	2:3
								2120	SSG	3:0
2160	U SX Floor slabs(E)	A N	0	3:3			2150	FSG	
								2120	FSG	
2180	U SX Sprayed Conc.	A N	0	0:2			2480	FSG	
								2110	FSG	
2190	H/over U SX Area(W)	A N	0	0:0	24-07-87		2150	FSG	
								2180	FSG	

Figure 13.5 (Continued)

Activity Number	Description (short) Management Code	Type	Calendar (shutdown)	Priority	Estimated min/max	Target Start/Fin.	Actual Start/Fin.	Preceding Activities Activity	Link	Delay
2195	H/over U SX Slabs(E)		A N	0	0:0			2160	FSG	
2200	C+P Area Ftgs, R/wall		A N	0	2:0	04-09-87		2210	SSG	0:3
2210	C+P Area Sewerage		A N	0	1:0			1100	FSG	0:3
2240	Calcin. Area Slab		A N	0	2:3			2200	FSG	
2260	Precip. Area Slab		A N	0	2:3			2240	FSG	
2290	H/over C+P Area		A N	0	0:0	29-05-87		2260	FSG	
2300	Ammonia Storage Ftgs		A N	0	2:3			2600	FSG	
2350	Const. Piperack ftgs		A N	0	4:3			2300	FSG	
2390	H/over Ammonia S+D.		A N	0	0:0	26-06-87		2350	FSG	
2400	Copper SX Drainage		A N	0	0:3			1100 2210	FSG FSG	
2410	Cu SX Embankment		A N	0	2:0			2100	FSG	
2420	Cu SX Ret. wall, drain		A N	0	3:2			2410 2260	SSG SSG	1:0 0:3

Code	Description	A/N	Val	Ratio	Date	Num	Type	Ratio
2440	Cu SX R/walls, Founds	A	0	1:0		2420	FSG	
		N						
2460	Cu SX Floor slabs	A	0	3:3		2420	SSG	1:4
		N						
2480	Cu SX Sprayed Conc.	A	0	0:2		2420	FSG	
		N				2460	SSG	2:0
2490	H/over Cu SX Area	A	0	0:0		2460	FSG	
		N				2480	FSG	
2500	Const. Prod. Area Sub	A	0	1:2	26-06-87	2700	SSG	2:4
		N						
2590	H/over Prod. Area Sub	A	0	0:0	01-05-87	2500	FSG	
		N						
2600	Const. FillPrep. Sub.	A	0	1:2		2500	FSG	
		N						
2690	H/over FillPrep. Sub.	A	0	0:0	01-05-87	2600	FSG	
		N						
2700	Const. Smelter Sub.	A	0	3:3		1100	FSG	
		N						
2790	H/over Smelter Sub.	A	0	0:0	17-04-87	2700	FSG	
		N						

Figure 13.6 Time schedule.

Activity Number	Description and Management Code	Durations Estimated min/max	Early Start /Fin.	Late Start /Fin.	Float	Target Start /Fin.
1000	Award Contract	0:1	23-02-87 / 23-02-87	26-02-87 / 26-02-87	0:3	
1100	Mobilise Plant, Matls	3:0	24-02-87	27-02-87	0:3	
1200	Demobilise Plant etc	2:0	16-03-87 / 12-08-87	19-03-87 / 12-08-87	0:0	
1290	Practical Completion	0:0	25-08-87 / 26-08-87	25-08-87 / 26-08-87	CRITICAL 0:0	
2100	Uranium SX Drainage	0:3	25-08-87 / 02-04-87	25-08-87 / 22-04-87	CRITICAL 2:1	17-09-87
2110	U SX Embankment	3:3	04-04-87 / 24-04-87	24-04-87 / 22-05-87	6:4	
2120	U SX Ret. walls, Drain	5:3	22-05-87 / 29-05-87	15-07-87 / 29-05-87	0:2	
2140	U SX R/walls, found.	1:0	10-07-87 / 05-06-87	15-07-87 / 22-06-87	1:4	
2150	U SX Floor slabs(W)	3:3	16-06-87 / 19-06-87	27-06-87 / 19-06-87	0:0	
			15-07-87	15-07-87	CRITICAL	

No.	Description		Date	Date		Date
2160	U SX Floor slabs(E)	3 : 3	16-07-87	16-07-87	0 : 0	
2180	U SX Sprayed Conc.	0 : 2	11-08-87 01-06-87	11-08-87 23-07-87	CRITICAL 6 : 3	
2190	H/over U SX Area(W)	0 : 0	02-06-87 16-07-87	24-07-87 25-07-87	1 : 2	24-07-87
2195	H/over U SX Slabs(E)	0 : 0	15-07-87 12-08-87	24-07-87 26-08-87	2 : 0	
2200	C+P Area Ftgs, R/wall	2 : 0	11-08-87 24-03-87	25-08-87 24-03-87	0 : 0	04-09-87
2210	C+P Area Sewerage	1 : 0	06-04-87 20-03-87	06-04-87 24-03-87	CRITICAL 2 : 1	
2240	Calcin. Area Slab	2 : 3	28-03-87 07-04-87 30-04-87	13-04-87 07-04-87 30-04-87	0 : 0	
2260	Precip. Area Slab	2 : 3	01-05-87 20-05-87	01-05-87 29-05-87	CRITICAL 1 : 2	
2290	H/over C+P Area	0 : 0	21-05-87 20-05-87	30-05-87 29-05-87	1 : 2	29-05-87
2300	Ammonia Storage Ftgs	2 : 3	30-04-87	02-05-87	0 : 2	
2350	Const. Piperack ftgs	4 : 3	19-05-87 20-05-87 24-06-87	21-05-87 22-05-87 26-06-87	0 : 2	

Figure 13.6 (*Continued*)

Activity Number	Description and Management Code	Durations Estimated min/max	Early Start /Fin.	Late Start /Fin.	Float	Target Start /Fin.
2390	H/over Ammonia S+D.	0 : 0	25-06-87	27-06-87	0 : 2	
2400	Copper SX Drainage	0 : 3	24-06-87 30-03-87	26-06-87 14-04-87	2 : 1	26-06-87
2410	Cu SX Embankment	2 : 0	01-04-87 06-04-87	16-04-87 28-04-87	3 : 2	
2420	Cu SX Ret. wall, drain	3 : 2	23-04-87 04-05-87	21-05-87 04-05-87	0 : 0	
2440	Cu SX R/walls, Founds	1 : 0	28-05-87 29-05-87	28-05-87 12-06-87	CRITICAL 1 : 4	
2460	Cu SX Floor slabs	3 : 3	04-06-87 15-05-87 12-06-87	20-06-87 29-05-87 26-06-87	1 : 4	
2480	Cu SX Sprayed Conc.	0 : 2	29-05-87 30-05-87	25-06-87 26-06-87	3 : 1	

2490	H/over Cu SX Area	0:0	16-06-87	27-06-87	1:4	
2500	Const. Prod. Area Sub	1:2	12-06-87	26-06-87	0:2	26-06-87
			04-04-87	07-04-87		
2590	H/over Prod. Area Sub	0:0	13-04-87	15-04-87	1:4	
			14-04-87	02-05-87		
2600	Const. FillPrep. Sub.	1:2	13-04-87	01-05-87	0:2	01-05-87
			14-04-87	16-04-87		
2690	H/over FillPrep. Sub.	0:0	29-04-87	01-05-87	0:2	
			30-04-87	02-05-87		
2700	Const. Smelter Sub.	3:3	29-04-87	01-05-87	0:4	01-05-87
			17-03-87	20-03-87		
2790	H/over Smelter Sub.	0:0	11-04-87	16-04-87	0:4	
			13-04-87	17-04-87		
			11-04-87	16-04-87		17-04-87

that shown in Figure 13.7. Various options exist for the bar chart, which may depict early start, late start, or resource-leveled start dates for activities with float. The precedence diagram with enlarged activity-on-node portrayal of networks has become the preferred format in such computer applications.

Many CPM-oriented floppy-disk packages exist for small PC systems. Their use is generally limited to personal initiative and choice with no company involvement, other than the supply of PCs to personnel or to the workplace, and no policy on the use of CPM methodologies.

13.5   DEDICATED CPM PACKAGES

The dedicated CPM package is usually professionally prepared, and commercially offered as a "complete" system capable of meeting the needs of construction contractors and government and related project initiating, or monitoring and controlling, agencies and consultants. They commonly can handle a number of projects simultaneously, and offer project managers many display and report preparation options. In the past these packages would have been mainframe computer-loaded systems with their associated special functions, languages, and instructions requiring considerable knowledge of computers and special training programs. Some still do, but many have been adapted to PC systems with enlarged memories and ancilliary accessories, so that most of the advantages associated with large computer installations are available for a modest investment. Some of these packages still rely on the overall presentational formats, logic, and language of the parent system, although these are now very vulnerable to exploitation, and becoming obsolete with the growth of specially prepared PC packages that feature many new innovative approaches to project management.

The selection, adoption, and use of a dedicated CPM package is generally a policy committment by an organization, although the price range of these packages is still attractive for the individual professional. Most packages in these categories have supportive training programs, service agreements with frequent package updating, and on-line professional service assistance to customers.

Usually specialist staff are trained, or hired, to run the packages in the company, and staff training and familiarization courses planned for all company personnel down to project managers, field engineers, and superintendents. Head-office managers either have desktop terminals for on-line monitoring of every personal project that the company is currently handling, or would be regularly supplied with frequent project summary reports from their own computer room. The size of the computer system that is installed is now a function of management's committment to the monitoring of their projects. Often field site offices have terminals linking directly into the system, but usually these offices run their own small PCs and send frequent disk-loaded data to head offices. Thus these companies are committed to computerization of data handling and reporting systems as part of the environment in which they operate.

The growth market in dedicated project management packages is such that considerable ingenuity and effort are devoted to the supply, and frequent updating, of packages that meet almost every use need that can be imagined. Additionally, many packages exhibit features that are very useful innovations to project managers, and that soon become pace-makers for all other packages. Typical of these features are "outlining," local-area network (LAN) support for linking the various PCs and workstations on a construction site, through the use of patchboards, or between projects and head offices through tied line systems, and menu- and template-driven resource management and report generation features. Outlining, for example, is a way of organizing, and portraying, progressively larger sets and series of activities by establishing narrowing levels of detail for lower hierarchical subdivisions, or components, for each activity in the higher-level set, if appropriate. Thus parent-child relationships similar to Work Breakdown Structures, or project area activity dissection, can be readily developed and compactly displayed on VDU screens in progressively indented formatting of the entries.

While outlining cannot display complex work task dependencies, its hierarchy of indented lines shows parent-child task relationships and the basic sequential flow of the project. If each level entry has its own network logic linkages defined, it is a simple matter to create different summary level views of a project by the proper development of subnetwork groupings and summation of resources and durations for each parent-child grouping of work tasks. In this way outlining acts as a zoom lens. Wide-angle views of the project are obtained by closing up the project levels, while close-up views are obtained by opening project levels. Higher-level tasks sum the durations and costs of those beneath them, so if all are closed up but the top level, a macrosummary of the major project phases results. Similarly, opening the levels to the lowest in any breakdown presents the information relevent to the lowest subnetworks. Outlining is integrated with the different project views, so that the Gantt charts and network diagrams change as the outline levels are opened and closed.

Typically, project networks handled by these CPM packages run into hundreds of activities, with several levels of summary reporting. No attempt will be made to portray these or the range of user-selected and formatted management reports that can be generated at will. Instead, a simple example of a "retaining wall" construction (see sketch in Figure 13.8) will be used to illustrate the network formulation stages using the PRIMAVERA[1] computer package. Figure 13.9 shows a hard copy of the precedence diagram for the retaining wall construction; and Figure 13.10, that for its time-scaled logic diagram.

---

[1] PRIMAVERA Project Planner is a registered trademark of Primavera Systems.

Figure 13.7 Bar chart.

Activity Number	Description	:23	March 87 :2	:9	:16	:23	:30	April 87 :6	:13	:20	:27	May 87 :4	:11	:18
1000	Award Contract	S---	:	:	:	///	:	:	:	/////	///	:	:	///
1100	Mobilise Plant, Matls	:SSSSS	SSSSSS	SSSSS	S---	///	:	:	:	/////	///	:	:	///
1200	Demobilise Plant etc	:	:	:	:	///	:	:	:	/////	///	:	:	///
1290	Practical Completion	:	:	:	:	///	:	:	:	/////	///	:	:	///
2100	Uranium SX Drainage	:	:	:	:	///	:	SSS------	----	/////---	///	:	:	///
2110	U SX Embankment	:	:	:	:	///	:	:	:	/////	S///SSSSS	SSSSSS	SSSSS	///SSSS-
2120	U SX Ret. walls, Drain	:	:	:	:	///	:	:	:	/////	///	:	:	///
2140	U SX R/walls, found.	:	:	:	:	///	:	:	:	/////	///	:	:	///
2150	U SX Floor slabs(W)	:	:	:	:	///	:	:	:	/////	///	:	:	///
2160	U SX Floor slabs(E)	:	:	:	:	///	:	:	:	/////	///	:	:	///
2180	U SX Sprayed Conc.	:	:	:	:	///	:	:	:	/////	///	:	:	///
2190	H/over U SX Area (W)	:	:	:	:	///	:	:	:	/////	///	:	:	///
2195	H/over U SX Slabs(E)	:	:	:	:	///	:	:	:	/////	///	:	:	///
2200	C+P Area Ftgs, R/wall	:	:	:	:	///CCCCC	CCCCCC	C	:	/////	///	:	:	///
2210	C+P Area Sewerage	:	:	:	:	S///SSSSS	------	------	-	/////	///	:	:	///
2240	Calcin. Area Slab	:	:	:	:	///	:	:CCCCC	CCCC	/////CCC	///CCC	:	:	///
2260	Precip. Area Slab	:	:	:	:	///	:	:	:	/////	///	CC CCCCCC	CCCCC	///CC---
2290	H/over C+P Area	:	:	:	:	///	:	:	:	/////	///	:	:	///S---
2300	Ammonia Storage Ftgs	:	:	:	:	///	:	:	:	/////	///	SSS SSSSSS	SSSSS	///S--
2350	Const. Piperack ftgs	:	:	:	:	///	:	:	:	/////	///	:	:	///SSSS
2390	H/over Ammonia S+D.	:	:	:	:	///	:	:	:	/////	///	:	:	///
2400	Copper SX Drainage	:	:	:	:	///	SSS---	------	----	/////	///	:	:	///
2410	Cu SX Embankment	:	:	:	:	///	:	SSSSSS	SSSS	/////	SS-///-----	------	----	///---
2420	Cu SX Ret. wall, drain	:	:	:	:	///	:	:	:	/////	///	CCCCCC	CCCCC	///C- CCCC
2440	Cu SX R/walls, Founds	:	:	:	:	///	:	:	:	/////	///	:	:	///
2460	Cu SX Floor slabs	:	:	:	:	///	:	:	:	/////	///	:	:	S///SSSSS
2480	Cu SX Sprayed Conc.	:	:	:	:	///	:	:	:	/////	///	:	:	///

318

		June 87					July 87				August 87			September 87		
:25	:1	:8	:15	:22	:29	:6	:13	:20	:27	:3	:10	:17	:24	:31	:7	:14
:	:	///	///	:	:	:	///	:	:	:	///	:	:	:	///	:
:	:	///	///	:	:	:	///	:	:	:	///	:	:	:	///	:
:	:	///	///	:	:	:	///	:	:	:	///CCCCC	CCCCCC CC	:		///	:
:	:	///	///	:	:	:	///	:	:	:	///	:	:C	:	///	: >
:	:	///	///	:	:	:	///	:	:	:	///	:	:	:	///	:
------	-----	///----	///-----	------	------	------	///--	:	:	:	///	:	:	:	///	:
: CC	CCCCC	///CCCC	///CCCCC	CCCCCC	CCCCCC	CCCCCC///--	:	:	:	:	///	:	:	:	///	:
:	:	S///SSSS	///S----	------	:	:	///	:	:	:	///	:	:	:	///	:
:	:	///	///CC	CCCCCC	CCCCCC	CCCCCC///CC	:	:	:	:	///	:	:	:	///	:
:	:	///	///	:	:	:	///CCC CCCCCC	CCCCCC CCCCC	///C	:	:	:	:	:	///	:
:	SS---	///----	///-----	------	------	------	///----- -----	:	:	:	///	:	:	:	///	:
:	:	///	///	:	:	:	///S--- ---->	:	:	:	///	:	:	:	///	:
:	:	///	///	:	:	:	///	:	:	:	///S----	------	--	:	>//	:
:	:	///	///	:	:	:	///	:	:	:	///	:	:	:	///	:
:	:	///	///	:	:	:	///	:	:	:	///	:	:	:	///	:
:	:	///	///	:	:	:	///	:	:	:	///	:	:	:	///	:
-----	:	///	///	:	:	:	///	:	:	:	///	:	:	:	///	:
---->	:	///	///	:	:	:	///	:	:	:	///	:	:	:	///	:
:	:	///	///	:	:	:	///	:	:	:	///	:	:	:	///	:
SSSSSS	SSSSS	///SSSS	///SSSSS	SSS--	:	:	///	:	:	:	///	:	:	:	///	:
:	:	///	///	:S-->	:	:	///	:	:	:	///	:	:	:	///	:
:	:	///	///	:	:	:	///	:	:	:	///	:	:	:	///	:
:	:	///	///	:	:	:	///	:	:	:	///	:	:	:	///	:
CCCC	:	///	///	:	:	:	///	:	:	:	///	:	:	:	///	:
: SS	SSSS-	///----	///-----	:	:	:	///	:	:	:	///	:	:	:	///	:
SSSSSS	SSSSS	///SSSS	///-----	-----	:	:	///	:	:	:	///	:	:	:	///	:
: SS	-----	///----	///-----	-----	:	:	///	:	:	:	///	:	:	:	///	:

319

....∍ 13.7 (Continued)

Activity Number	Description	:23	March 87 :2	:9	:16	:23	:30	April 87 :6	:13	:20	:27	:4	May 87 :11	:18
2490	H/over Cu SX Area	:	:	:	:	///	:	:	:	/////	///	:	:	///
2500	Const. Prod. Area Sub	:	:	:	:	///	:	S SSSSSS S--		/////	///	:	:	///
2590	H/over Prod. Area Sub	:	:	:	:	///	:	:	S---	/////---	///---->	:	:	///
2600	Const. FillPrep. Sub.	:	:	:	:	///	:	:	:SSS	/////SSS	///SS--	:	:	///
2690	H/over FillPrep. Sub.	:	:	:	·	///	:	:	:	/////	/// S-->	:	:	///
2700	Const. Smelter Sub.	:	:	:	:SSSS	///SSSSS	SSSSSS	SSSSSS ----		/////	///	:	:	///
2790	H/over Smelter Sub.	:	:	:	:	///	:	:	S----	///	///	:	:	///

Tozer & Associates Pty. Ltd., Chatswood, N.S.W.; Job Name: Hydromet Plant Civil; Reference: Tender Programme; Project Date: 23-02-87. Page: 1; Date: 08 Jan 89; Time: 13:33.
Legend to symbols: SSSS Activity in Progress; CCCC Critical Activity; ---- Float time; Key Date, Nonworking period.

While PRIMAVERA is essentially a package capable of handling large projects with many project activities and resources, and is essentially a large computer system package, many popular project management PC packages are available that readily handle all the CPM needs of small- and medium-sized projects. Typical of the packages in the low end range is Microsoft Version 4.0.[2] These packages offer full-service project scheduling and reporting capabilities with many appealing features and fast processing times that mainframe systems have difficulty in matching, because the latter are intrinsically tied to larger project sizes with their encumbant data handling procedures.

13.6 COMPANYWIDE SYSTEMS

CPM methodology provides procedures for formulating construction plans, schedules, and a framework on which to portray real-life project progress. The techniques themselves, being computational and graphical in nature, need to reflect, and be supported by, management and project staff action, or they become purely informative exercises that "may help" company staff in

---

[2] Microsoft Project Version 4.0 is a project scheduling and reporting program for personal computers running the MS-DOS operating system. System requirements are 256K memory, DOS 2.0 or higher, and two double-sided disk drives or one double-sided disk drive and a hard disk. Available from Microscoft Corporation, 16011 NE 36th Way, Box 97017, Redmond, Washington 98073–9717.

Figure 13.7 *(Continued)*

:25	:1	June 87 :8	:15	:22	:29	:6	July 87 :13	:20	:27	August 87 :3	:10	:17	September 87 :24 :31	:7	:14
:	:	///	S///-----	----->	:	:	///	:	:	:	///	:	: :	:	/// :
:	:	///	///	:	:	:	///	:	:	:	///	:	: :	:	/// :
:	:	///	///	:	:	:	///	:	:	:	///	:	: :	:	/// :
:	:	///	///	:	:	:	///	:	:	:	///	:	: :	:	/// :
:	:	///	///	:	:	:	///	:	:	:	///	:	: :	:	/// :
:	:	///	///	:	:	:	///	:	:	:	///	:	: :	:	/// :
:	:	///	///	:	:	:	///	:	:	:	///	:	: :	:	/// :

the course of their general duties. To be effective, CPM techniques must reflect in-house procedures in planning, scheduling, procurement, monitoring, and project control. In the fulfillment of these interacting management functions and real-life requirements lies the basic rationale for setting up a coherent, consistent, coordinated, and integrated companywide management information system (MIS).

A companywide MIS would be capable of handling all aspects of the data that are produced by, and respond to, the actions and needs of company staff at all levels and phases of each and every product that the company is handling. Such a system is very helpful because in the early phases of a project, before bid award, many decisions and assumptions are made that establish the rationale for the basis on which the company bid is built, and this information and criteria need to be available to others later on with bid award. Although this prebid activity is often regarded as being completed, and finished with on bid award, much of the information established by prebid planning, estimating, and scheduling must be made available and used by staff who now have the responsibility to manage and complete the project. This information should not disappear as the impact of different project life phases transfer project responsibilies to other company functional staff.

While many companies set up special document and worksheet filing systems, these often are incomplete, and/or very bulky, and thus become ineffective. An efficient approach is to set up some form of information system, which in time often leads to the need to establish a suitable database for company operations.

Any company wide database, and management information system,

CONSTRUCTION OF A RETAINING WALL.

A retaining wall is to be constructed in two separate sections: North (N) + South (S).
Assume that two separate shutters are to be made, one for each section of wall. The excavation is to be carried out by only one gang.

Other operations can be carried out concurrently. Use the following activity list:

Activity	Duration	Labor
Exavate N	2 days	2 men
Prefabicated wall shutters N	4 "	2 "
Prefabicated wall shutters S	4 "	2 "
Form & concrete footing N	5 "	6 "
Excavate S	2 "	2 "
Backfill footing N	3 "	4 "
Form & concrete wall N	9 "	6 "
Cure & strip N	3 "	2 "
Patch she-bolt holes N	1 "	1 "
Form & concrete footing S	6 "	6 "
Backfill footing S	2 "	4 "
Form & concrete wall S	7 "	6 "
Cure & strip S	3 "	2 "
Patch she-bolt holes S	1 "	1 "

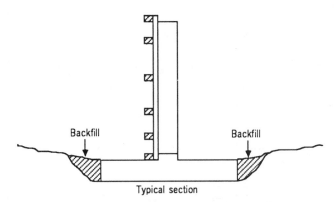

Figure 13.8   Construction of retaining wall.

should be postulated on the need to provide a holistic basis for all the company decisions and information needs regardless of whether individual project focused. To be effective, the companywide system needs to become a functional unit in that all staff who have recourse to it—whether for costing, reporting, or similar—should find their needs satisfied within the database, or by MIS procedures operating on the database. Thus, for example, project costing functions should be related to estimating, payroll, field monitoring

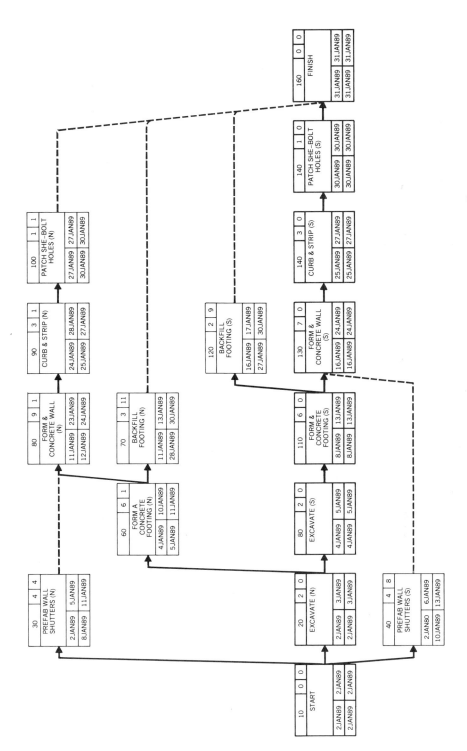

Figure 13.9 Precedence diagram for retaining wall.

323

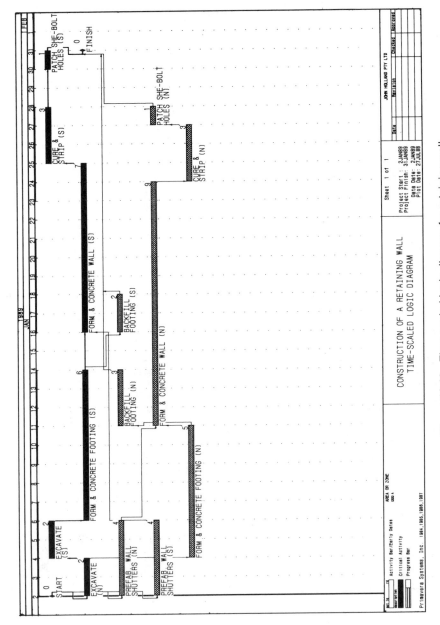

Figure 13.10 Time-scaled logic diagram for retaining wall.

and status reporting, and accounting functions through the MIS, rather than be unrelated except by tedious manual information gathering and data transfer.

Similarly, information recorded in the database during the project definition, planning, estimating, and scheduling phases should be readily available to project management for procurement, field activity scheduling, and works management. Thus the database, and MIS needs of its users, should be compatible with the different life phases of the project, and meet company requirements. A companywide database should therefore have relational capabilities, if other than a superficial use or integration is intended. Relational database systems are available that either store data in functional, or application, areas and files or have suitable linking pointer system attributes encoded into the data system. Hence the actual data becomes source material that can be accessed by many users and handled to produce management reports rather than the reports becoming part of the database. Thus CPM procedures would operate on project files rather than being an independent, or isolated, effort.

Many specially designed in-house database and integrated MIS systems exist in the construction industry, and a number of commercially available packages are in use. A typical example of the latter is ARTEMIS[1] which is used by a number of European and American contractors. The ARTEMIS system, as its promotional brochure suggests, puts the project control system in the hands of the project controller. Users are given a full set of capabilities from which to develop their own applications and tailor systems or reports. ARTEMIS systems can be linked and interwoven to serve the spectrum of management needs from day-to-day activity specification to monthly executive status reporting. Files are automatically linked for handling with a single command, or by menu options. While a special language is used, and a training program is required for the staff, it is claimed that ARTEMIS systems can be readily tailored to suit each user's unique set of requirements. Clearly such companywide systems are selected because they meet and serve company functions, rather than the specific requirements of CPM procedures. These procedures must therefore be tailored to operate on, and from, the company database. If they cannot do this, then serious doubts must be raised as to the relevence of the companywide system to the needs of project management.

13.7 THE CYCLONE-TIMELAPSE ANALYSIS SYSTEM

A new technique for the planning and field management of construction operations has been developed at the University of New South Wales. The technique, called the CYCLONE-TIMELAPSE Analysis System, has been

---

[1] ARTEMIS Project Management Information Systems. Metier Management Systems, Inc., 5884 Point West Drive, Houston, Texas 77215.

successfully used on a number of construction projects in Australia to pinpoint inefficiencies used on a number of construction projects in Australia to pinpoint inefficiencies in the field management of operations, to quantify improved operational design and changes in resource allocations, and to demonstrate the impact of different management policies on the productivity of the operations.

The CYCLONE-TIMELAPSE Analysis System has been developed for the study, analysis, simulation, and design of construction operations. It is a computer-based system[1] that processes films of construction operations and prepares a variety of work-as-executed management reports of field activity. The focus of the management reports can be selected by the user, who may be either the head-office planner or the field manager. Using the system it is possible to critically review the design and field management of in-progress or new construction operations. The impact on field productivity of different resource allocations, crew sizes, and work assignments can be documented and the effectiveness of alternative field management operating policies quickly investigated. The system is especially useful for highly repetitive, critical, and cost-sensitive operations, whether equipment-heavy or labor-intensive, and provides the user with a decision-oriented methodology for the planning and management of construction operations.

For in-progress construction operations the process requires time-lapse (movie or video) photographs of field operations. These are used in movie form as information on the use of manpower and equipment on site. The data are stripped from the film using a special computer-driven projector, and critical events are registered and stored for later use. During the running of the film, a CYCLONE model of the work in progress—in the style of a railway system network map (see Figure 13.11)—captures the technological structure of field operations. Using a colorgraphics terminal to present the model, or map, colors can be used to present a clear picture of the work situations to field managers. In this way, colored segments of the CYCLONE model trace the movement of equipment and resources through the site operations; checks can be made on availability and level of use of critical resources in interacting work sequences. Thus the detection of unproductive bottlenecks and a backup of resources can be seen at a glance and related to the effective management of different sized work crews and equipment fleets.

The CYCLONE modeling approach focuses on the portrayal of construction operations at the equipment, crew, and individual laborer work-task levels, and on the identification of work sequences and interacting processes. In this way the technological structure of the construction method is captured

[1] *The CYCLONE-TIMELAPSE Analysis System* by J. L. Knott and R. W. Woodhead, Construction Monographs No. CM 2, The School of Civil Engineering, University of New South Wales, Sydney, Australia, October 1980. See *Introductory User Manual, The CYCLONE-TIMELAPSE Analysis System* by W. F. Dorman, J. L. Knott, and R. W. Woodhead, copyright University of New South Wales, December 1980.

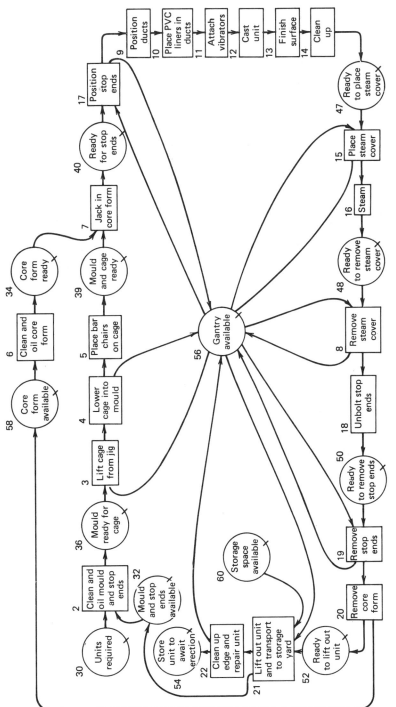

Figure 13.11 CYCLONE diagram of the gantry involvement in a bridge unit casting operation.

and, in conjunction with the time-lapse film, enables an estimating data-base of the operation to be established and used for performance prediction studies.

An example of a CYCLONE representation of the casting yard processes introduced in Chapter 9 is given in Figure 13.11. In simplified form it focuses on the involvement of a gantry crane in the field-casting and storage of bridge units. A main line sequence of unit preparation and storage management work tasks exposes the basic steps required for each bridge unit, the individual work tasks being numbered for identification. The gantry's involvement in the casting yard operation is indicated by the work assignment shown here in typical butterfly pattern. Any constraints imposed by the number of available molds or storage spaces may be indicated by simple feedback control loops on the manufacture and progress of each bridge unit through the casting and storage processes. In a given situation the ad hoc scheduling and priority use of the gantry may seriously affect field production in terms of lost work hours per unit, for example, and field management should establish guidelines for the conduct of the casting yard operation. The CYCLONE model of Figure 13.11 represents an initial attempt to explicitly portray the management problem involved.

A comprehensive software package enables the user to select and carry out multilevel planning and management studies of the construction operation and to generate a variety of user-designed management reports. These management reports can focus on simple work-as-executed field activity statements, the portrayal and summary of resource utilizations, the investigation of crew sizing details or multiple activity charts of new or existing operations, and the establishment of a data-based estimating approach to construction processes. Reports can be produced in a visual display form using colorgraphics terminals, or in hard copy form produced from a line printer.

A typical, but simplified, segment of a work-as-executed multiple activity chart is shown in Figure 13.12. It focuses on the gantry's involvement in the casting yard operations shown in Figure 13.11. Each bar is printed in duplicate for easy reading, and the legend clearly identifies the work-task number and description involved; the asterisks show idle time. The printout repeats the work-task number and description where space permits along each bar. Beneath the bars is a time scale in hours, and it can be seen that this printout covers 3 days work.

The versatility of the entire system hinges around its capability to portray and perform analyses of the impact of different planning and management policies on the way in which a construction operation is to be performed or managed in the field, preferably before these decision and policies are brought into effect. This is readily done for in-progress operations, but for new operations it requires estimates of work-task durations. Once these estimates are available, a simulation of the movement of resources through the construction operation according to management policies may be per-

```
BRIDGE UNIT CASTING OPERATION

                17       22          56 GANTRY AVAILABLE      17         22         56 GANTRY AVAILABLE      17      22      56 GANTRY AVAILABLE      17    22    56 GANTRY AVAILABLE
GANTRY          BBBC*DDEEEG***************************BBBBC*DEEEFG*H****************************BBBBC*DDEEEF**H*****************BBBBCCDDEEEF**H*************
                BBBC*DDEEEG***************************BBBBC*DEEEFG*H****************************BBBBC*DDEEEF**H*****************BBBBCCDDEEEF**H*************

                                                 58   34   40 CORE IN POSITION                           17  9  12   16 STEAM
                17       12         16 STEAM
CORE FORM 1     BBBIJKLLLMNNOOOOOOOOOOOOOOOOOOOOOOOOOFF**CQ***R***S***************BBBIKKLLLMNNHOOOOOOOOOOOOOOOOBBBIKKLLLMNNHOOOOOOOOOOOOOOOOOOOO
                BBBIJKLLLMNNOOOOOOOOOOOOOOOOOOOOOOOOOFF**CQ***R***S***************BBBIKKLLLMNNHOOOOOOOOOOOOOOOOBBBIKKLLLMNNHOOOOOOOOOOOOOOOOOOOO

                     5  40 CORE IN POSITION      17            16 STEAM                                                  22
BRIDGE UNIT 3   T**GUU****************************BBBBIJKLLLMNNHOOOOOOOOOOOOOOOOOOOOOOOOFF*CCDDEEEE
                T**GUU****************************BBBBIJKLLLMNNHOOOOOOOOOOOOOOOOOOOOOOOOFF*CCDDEEEE

                     2   40 CORE IN POSITION                          17  9  12   16 STEAM
BRIDGE UNIT 4   TT*FGUS******************************BBBIIKKLLLMNNHOOOOOOOOOOOOOOOOOOOOOOOOO
                TT*FGUS******************************BBBIIKKLLLMNNHOOOOOOOOOOOOOOOOOOOOOOOOO

                :   .  .  :  .  .  :  .  .  :  .  .  :  .  .  :  .  .  :  .  .  :
               0.00         8.00        16.00        24.00         8.00        16.00        24.00

LEGEND
------

TTT  2 CLEAN AND OIL MOULD         FFF  3 LIFT CAGE FROM JIG         GGG  4 LOWER CAGE INTO MOULD
UUU  5 PLACE CHAIRS ON CAGE        RRR  6 CLEAN AND OIL CORE FORM    SSS  7 JACK IN CORE FORM
AAA  8 REMOVE STEAM COVER          III  9 POSITION DUCTS             JJJ 10 PLACE PVC LINERS IN DUCTS
KKK 11 ATTACH VIBRATORS            LLL 12 CAST UNIT                  MMM 13 FINISH SURFACE
NNN 14 CLEAN UP                    HHH 15 PLACE STEAM COVER          OOO 16 STEAM
BBB 17 POSITION STOP ENDS          PPP 18 UNBOLT STOP ENDS           CCC 19 REMOVE STOP ENDS
QQQ 20 REMOVE CORE FORM            DDD 21 LIFT OUT UNIT              EEE 22 CLEAN UP EDGES AND REPAIR UNIT
*** 34 CORE FORM READY             *** 36 FORMS AVAILABLE            *** 40 CORE IN POSITION
*** 50 READY TO REMOVE STOP ENDS   *** 56 GANTRY AVAILABLE           *** 58 CORE FORM AVAILABLE
```

Figure 13.12 Typical work-as-executed printout for Figure 13.11.

329

formed and displayed visually on the colorgraphics terminal. During this process the software strips out operational data in a manner similar to that obtained from construction films. Hence resource utilization and management reports may be produced from the simulation and used to support the selection of specific resource mixes and the adoption of desirable management policies for the planning and management of critical new operations. In this way realistic activity durations can be established and entered on the project CPM network.

Evaluation

a. Locate some currently available CPM-based computer package, and list some of their advertised selling points.
b. Taking any existing contract with which you are familiar, write a critique of its approach to computer processing.

14

SELECTION OF TECHNIQUE

14.1 WHY CPM?

All engineering procedures depend on some fundamental basic assumptions and some intelligent approximations. The measure of the accuracy of such engineering procedures is the reliability with which the results may be used.

In construction planning and scheduling and in site control of projects, it is obviously not possible to determine, with absolute precision, the correct site procedures to give the minimum cost nor the fastest economical time—for construction itself is an art. This art is, however, founded on applied science, from which it is certainly possible to devise mathematical procedures that can evaluate optimum cost-time relationships. The techniques of CPM and PERT (see Appendix B, PERT concepts) are two such mathematical procedures.

Sound development of costing and estimating in civil engineering and building construction has removed the former uncertainty that plagued this important industry. Generally speaking, therefore, construction costs and performance times can be forecast with sufficient accuracy to satisfy the basic fundamental assumptions of the CPM technique, and hence the probability feature of PERT is usually not relevant. Furthermore, PERT technique (with its various time predictions) requires extra computational effort and hence additional cost.

PERT is, however, most valuable when very little background information is available, so that the utility data are not sufficiently accurate, or when many of the activities require considerable research and development. PERT is concerned with circumstances which are totally different from those of CPM. Thus both techniques have their place in industrial management;

however, for general construction CPM techniques are so accurate and reliable that the additional cost of using PERT is not warranted.

Indeed, where uncertainty does exist in performance times and costs, the astute planner can appraise the project with the usual CPM procedure and add a probability quality by using "expected times" and corresponding "expected costs" in his utility data.

14.2   HOW MUCH INVESTIGATION IS WARRANTED?

Assuming then that the CPM technique will be used, consideration must be given next to the amount of investigation to be undertaken, especially at the planning stage. It can be fairly claimed that practically all jobs (even the smaller ones) warrant the drafting of a network diagram, followed by a simple scheduling and the production of a CPM bar chart program (showing floats); even resource leveling may be worthwhile on the smaller projects. Very little additional effort is required thus far in comparison with the former procedure for a conventional bar chart program; in addition, the timing of works operations will be much more accurate.

If the project is reasonably large, investigation into alternative methods is warranted. This necessitates preparing several networks (all at the all-normal level) together with their comparison and evaluation as to relative total costs, taking account of the effects of the different direct and indirect costs applicable to the various project times deduced. When the final decision has been made, after completing these comparisons, the scheduling and CPM bar chart program or time-scaled network are drawn up for the selected method. Investigation to this stage adds very little to the cost of bidding for any civil engineering works.

To date, the majority of construction works submitted to critical path analysis has been investigated only this far. This is probably because such investigations can be done readily by manual methods and also because of lack of knowledge of additional computerized CPM applications to complex problems. Be that as it may, CPM is a most valuable tool to the construction planner, even if its use ceases at this point; and the site manager has a much more accurate plan from which to work.

The usefulness of such applications of CPM will be enhanced if resource leveling is added, even on the broadest basis and with approximate requirements determined by manual methods only. This provides the site manager with a reasonable appreciation of the resource needs and permits the planning for the apparent high and low spots—or to resort to detailed resource leveling when warranted.

It is far more valuable, however, if the procedure is taken further in order to investigate the effects of compression, and thus to provide the optimum solution. Nevertheless, it must be appreciated that this is a more difficult task, involving considerable additional effort. For this reason it is only warranted perhaps on major works, although some benefit must be gained for any project submitted to such a time-cost analysis. The decision to expend this extra effort can, however, be arrived at only after careful consideration by management of the many factors involved, not the least of which are availability of skilled personnel to carry it out and the time to do it. It should not be forgotten, of course, that this may possibly be the most fruitful field for a probable reduction in the overall cost for the project.

Finally the adopted construction plan should certainly include the major resource requirements and constraints, and should be presented as a time-scaled network, showing all the work items together with other critical requirements such as major procurement orders and dates for approvals.

14.3 MANUAL OR COMPUTER PROCESSING?

Throughout the construction industry, considerable attention is being paid to the use of CPM for solving many of the problems that formerly construction engineers and management approached from the basis of experience and judgment. Today this is being done by the sound mathematical and logical approach of CPM, using both manual methods and computer processing. In addition, recent developments in computer technology have produced very powerful and portable minicomputers suitable for field-office situations. This, coupled with the fact that students now take computer courses as a regular component of their education, means that a rapidly growing pool of expertise in computer processing exists on almost every construction project.

Although the use of computers has many strong advocates, it must be pointed out that manual methods still have much to recommend them, if only because they force the user to think logically through the entire project. Undoubtedly, there are disadvantages in manual methods, including a considerable amount of arithmetical effort and its consequent opportunity for error, a reluctance to pursue network changes to their final logical conclusion because of the extra effort required, and, finally, the necessity on the site to review the schedule regularly and to repeat considerable computations as remedial measures are taken. A large amount of the tiresome routine may, of course, be overcome by the intelligent development of standardized records and report sheets for job usages, the adoption of sufficiently accurate approximations to fit the case, and the evolution of simplified checking procedures to suit the particular project.

There is no doubt that manual methods of CPM can be used for quite large projects. If the number and interrelation of activities are such that the complete project appears at first to be too complex to comprehend, it is simple to begin with a series of summary fragnets and prepare the first network model on this basis. When the optimum solution has been found, each of these fragnets may then be further broken down and analyzed until the limitation of improvements is reached. The final plan may then incorporate all the individual detailed diagrams. Network models of a project are undoubtedly better when done entirely by manual methods, if only to provide the planner with a completely detailed appreciation of the problems involved in the construction. However, computer programs are useful for large networks, especially when minicomputers enable the construction staff to divide their efforts between summary fragnets for long-term assessments and detailed fragnets for immediate future planning.

Once the network is finalized, its scheduling is a purely mechanical operation involving the compilation of data by simple arithmetic. Even on a complex project this may not warrant processing by an electronic computer; for again the manual planner gains the advantage of detailed insight into the problems that can be adjusted at will. Nevertheless, computers have here a major advantage in their ability to revise the results with great rapidity whenever input data are amended; and, of course, the chance of arithmetical error is eliminated. Computer programs are also available to handle the shifting of activities within available float and to schedule labor skills and equipment requirements.

With complex time-cost relationships, computers may well be essential. Even though it has been shown that manual methods can be used for this stage of the task, it is here that the greatest opportunity lies for project cost reductions, and consequently there is much computation work even in the simpler projects. On the other hand, it must not be forgotten that the straightforward computer solution requires that the input data for each cycle remain as set out for it at the start; with manual solutions these data may be revised at any time during each cycle, if necessary.

Such revisions are indeed often required. If, for example, the solution requires that a particular activity be crashed from an 8- to a 9-hour day, practical performance on the site would surely dictate that some of the related and simultaneously performed activities must also be similarly crashed for overtime (whether necessary or not) unless industrial trouble is to ensue. Hence this single change in the first activity will necessitate changes in concurrent activities; this the computer is unable to take into account, unless the original input data are varied accordingly and the whole process repeated. Another similar situation arises when an additional piece of equipment is brought onto the job to assist in crashing one activity but is also available for others; the cost slopes of all activities using this additional equipment will change, some increasing and some decreasing. With manual methods such changes in utility data are readily made, even if the task is somewhat laborious. With computers it is not possible to revise input data part way through a cycle.

To sum up, it is well to remember, when considering whether to adopt manual or computer processing, that computers certainly provide speed and accuracy, but can never supply judgment or ingenuity. The electronic machine provides rapid computation and comprehensive scheduling very efficiently; but the construction planner still has to do the real brainwork.

14.4 HOW LONG DOES IT TAKE?

Although every project presents its particular problems, whose solutions require variable amounts of effort, some indications of the cost of applying CPM techniques may be gained from surveying a number of appropriate examples. The cost of application of CPM, to be realistic, must be reckoned as the *extra cost* over and above that normally considered by an estimator in preparing a conventional project estimate. However, it must be understood that the preparation of a CPM construction plan often saves many costly mistakes arising from lack of coordination once the project is under way, and, to be fair, the extra cost involved in a CPM application must be credited with the increased confidence gained from the production of such a CPM schedule.

Manual Methods

With a reasonably experienced planner, the preparation of the first network diagram for a project of about 50 activities, up to the stage of critical path determination, will require about 6 hours; subsequent networks showing alternatives and improvements will require 3 to 4 hours each. With 100 activities, the first network may take 8 to 12 hours, and subsequent ones 4 to 6 hours each. Scheduling of floats may take 1 or 2 hours. As specific examples, the three-span bridge analysis (presented in Chapter 8) required 10 hours effort for the four networks plus another 6 hours for scheduling and the sketching of the bar chart. The rock-fill dam (also presented in Chapter 8) required a total of 18 hours work, from beginning the first network until conclusion of the compression calculations. The annual budget plan (presented at the end of Chapter 8) represents approximately a day's work.

Simple compression calculations, as discussed in Chapter 5, depend on the nature of the network. After preparing the utility data, the seven stages of compression calculations for the network model of Figure 5.7*a* to *g* required less than 4 worker-hours. Compression of complex networks will take much longer. Decompression (see Chapter 6) requires about the same time as compression; calculations of the type discussed in Sections 6.8 and 6.9 require about half an hour per stage. The optimization calculations set out in Section 6.10 required between 2 and 3 worker-hours.

Computer Methods

The major cost of computer processing is the initial preparation of data for the machine. For example, with a network of 500 activities, the drawing of the network diagram and the codification of activities may consume 80 to 100 worker-hours by the planner. For a project with 1800 activities, the initial planning by an inexperienced team took nearly 600 worker-hours, whereas an experienced planner should not require more than 200 to 250 worker-hours. Perhaps a reasonable estimate for the preparation and codification (ready for the computer) of large and complex projects would be one day per 50 activities by the planner.

Once the computer programme is set up, the cost of computer operations is quite small.

Compression calculations are very variable but are completed in a matter of seconds of computer time for a network containing 500 to 600 activities, after which the complete printout is similarly produced in a matter of minutes.

In conclusion, it has been estimated that the total cost of providing complete CPM coverage to major projects, including detailed preplanning and resource leveling, with regular monthly updating for project control, should not exceed 0.50%, and with project cost control 0.75%, of the contract price. This is indeed a small sum to pay for the advantages to be gained.

14.5 IS IT WORTHWHILE?

There has been criticism that the planning and scheduling of construction projects and the site control of the works in progress take considerably more time and paperwork with CPM than with the former familiar bar chart program; and furthermore, that factors totally outside the construction manager's control (and which cannot be foreseen at the planning stage) upset the schedule and necessitate continual revision to the critical path network during the construction phase. The accuracy of such criticisms is not denied. The essential point to be considered, however, is whether CPM provides a better answer than the former coventional controls to the multitude of problems encountered on every project. There can be no doubt that it does. The very nature of CPM procedure insists on an evaluation of every known constraint, of every project activity, and of every time-cost question, at the planning stage. Complex and unpredicted site problems certainly necessitate numerous revisions to the program during construction; under such circumstances, the network must be constantly revised—but so must the conventional bar chart. At least, with CPM the revisions are made logically and in mathematical relationship to all relevant activities, with reappraisal and reevaluation of all the constraints and time-cost problems then current. Some rescheduling problems, in such cases, create considerable additional work for the CPM user; but at least a logical answer can be provided. The conventional bar chart user has no basis from which to begin; each revision depends only on the user's skill and judgment.

To regard the precise CPM diagrams, neat schedules, and accurate computer printouts as being infallible answers to all the construction problems likely to be encountered is patently fallacious; it is a similar fallacy to make the same claim for conventional bar charts. To regard the CPM optimum solution to a time-cost problem as absolutely precise is most certainly unsound; but it is assuredly more reliable than a works schedule based on a conventional works estimate. The principal point to remember is that a program is only a prophesy and, consequently, should incorporate some tolerance for error; it is the provision for practical tolerance (such as allowances for lost time) that makes the program feasible and flexible.

Two things a conventional bar chart cannot show with any certainty are the correct date to start each site operation and which of those operations are critical in timing so that the whole job may flow smoothly (at least theoretically) to its logical conclusion, in the projected time and at the least overall cost. Planning with CPM and replanning when unknown factors appear on the site have been introduced to the construction industry for these very purposes.

When used with common sense and flexibility, critical path methods enable the construction manager to be forewarned of dangers ahead and to be better equipped to meet them as they arise, at very little extra cost in comparison with the losses that may otherwise be risked. It must be added, however, that the planning and scheduling on which the decisions are to be made, as well as any revisions, should be undertaken only by a competent officer, skilled in construction engineering and well versed in the specific problems of the project. The original network, and all necessary revisions, must be the prerogative of a construction planner with adequate experience in the type of work involved.

The use of CPM in construction practice is sometimes described as "management by exception"; that is to say, there is no management problem while everything proceeds as planned. The great value of this technique is that, once something goes wrong, it becomes an exceptional occurrence, which immediately makes itself apparent and enables management to concentrate on its correction. Other advantages concurrently obtained may be summarized as follows.

1. The necessity for analytical and logical thinking about all phases of the work.
2. The appraisal and determination of those construction operations which will control the time of completion of the works.
3. The production of the most economical timetable for all operations involved.
4. The quantitative evaluation of leeway (float) in the timing of all site activities.
5. The visual representation of the project by a diagram, and hence the prevention of omissions of any elements necessary for its compilation.
6. The determination of the optimum project duration, and hence the best completion date.
7. The most economical selection of crew sizes and equipment to meet this completion date.
8. The rapid evaluation of alternative schemes.

9. The quantitative assessment of the effect on the whole project of variations in any single operation.

10. The logical determination of the remedial measures necessary to correct any adverse situation which may arise during the execution of the works.

11. The rationalization of construction costing and financing.

12. The quantitative evaluation of the time and cost of work changes and enforced delays.

Under these circumstances the specification and use of CPM in the construction industry are certainly worthwhile.

14.6   HOW IS IT SPECIFIED?

Critical path methods have been in use on civil engineering and building works since 1960, largely on an experimental basis by contractors for the first 4 or 5 years, but more recently as a mandatory requirement by owners, especially on large and complex enterprises. As a result, it has become necessary to specify in contract documents that the preliminary planning and scheduling, and the control of the project throughout its construction, shall be carried out by critical path techniques.

In preparing such a specification care must be taken not to overload the contractor with unnecessary work. Only the minimum of detailed procedures to provide essential data should be demanded. The choice of manual or computer processing should be left to the contractor to decide; the specification should merely stipulate the frequency of reporting and the information required.

Evaluation

a. By discussion with your colleagues ascertain their reasons for (i) adopting a CPM approach to project planning and/or management; (ii) use of manual, or computer, processing of networks.

b. Taking an existing contract with which you are familiar, set out the estimated time required to prepare an initial network, in approximate meaningful detail, stipulating the various necessary steps required.

15

INTEGRATED PROJECT DEVELOPMENT AND MANAGEMENT

15.1 PROJECT CONCEPTION AND INITIATION

The procedures for the initiation and development of construction projects—whether by government, statutory authority, or private enterprise—are well established. The owner of the proposed works engages or employs and explains to a professional engineer or architect the project conception and requirements; the engineer or architect then arranges for a thorough investigation into the technical and economic ramifications of the proposal, so that the owner's requirements may be properly examined and developed. This investigation includes a comprehensive assessment of the merits and problems of the project and of the various feasible alternative designs applicable to it. It requires considerable study of previously assembled data. For this reason, it is essential that sufficient time be allowed at the outset, for time properly spent at this point will later save both time and money during the design and construction stages.

The development of a project, from the original idea to its completion ready for operation, may be divided into five consecutive stages. These comprise the detailed investigations, the final definition of the works, the detailed design and estimates of cost, the planning of its construction, and finally, the construction work itself. The detailed investigations include the collection of essential data pertinent to the owner's conception of the project, the preparation of preliminary designs for each practical solution to its problems, the selection of the best method to adopt, and the provision of all necessary physical information for the detailed design and construction of the works. It is during this stage that two determinations must be made: first, whether the project is truly worthwhile and second, which of the various alternatives for carrying it out will be the best. The work involved in making these determinations is known as a *feasibility study*.

Obviously, the collection and interpretation of all the data in a feasibility study for a project must be objective and impartial, so that the conclusion reached will be completely honest with regard to both the favorable and unfavorable aspects of the proposal. Obviously, such a study also consumes time and money. Hence, in order to attain maximum efficiency and economy, the detailed investigations for any project must be properly and adequately planned. For this reason alone, the application of CPM to construction works should begin at their initiation. Only in this way can there be fully integrated planning and control of the entire development of such projects. This initial planning will take the form shown in Figure 15.1, as discussed in the next section.

15.2 PROJECT DEFINITION AND DESIGN

Project definition is the decision to adopt a particular scheme for a project from the various feasible alternatives available. It establishes the general arrangement of the works and adopts a specific conceptual design from which the detailed design will follow. This definition of the project includes not only the clear delineation of exactly what the owner wants, but also the timetable by which it is to be carried out. This necessitates the preparation of a detailed design brief and a practical program for the development of the final details of the works and their subsequent construction.

At this point there is a definite site with its specific characteristics and problems, and there is a firm decision concerning the sizes and types of structures required. It is therefore practicable to prepare a program for the execution of the works so that the owner may know how long it will take (and how much it will cost) to reach the objective or, alternatively, may demand completion by a specific date; this presents the problem in reverse. All stages of the development must be considered, and ample time allowed for each, notwithstanding the fact that the owner is probably pressing for accelerated action. The time and cost of each stage of the work require careful consideration so that the planned program may be as realistic as possible. Even when the estimated durations of various operations are only approximate, a rough program at this stage is better than no plan at all; the tendency to postpone early planning must be strongly resisted.

The time required for each operation in the various stages of the development of a project will depend on the intensity of personel and finance that can be provided and on the complexity of the works themselves. From two to three years is by no means too long a period to allow from the initial conception to the completion of the design of a major construction project. There must then be time also for considering which parts of the works shall comprise separate contracts and which parts (if any) shall be executed by day labor. With contract works there must be adequate periods for the provision of bid documents, the preparation of bid proposals, the receipt of bids, and their analysis. Finally, the time allowed for the construction stage requires considerable deliberation, for an unduly short period may seriously increase the cost. Various factors will influence the durations of these operations, and adequate consideration must be given to each, so that the planned program may be a feasible and realistic one.

Figure 15.1 shows the general preliminary planning for the development of a $300,000,000 mining venture in a remote part of Australia. It allowed 18 months for the detailed investigations and feasibility studies, which comprised one for the mine and treatment plant and another for the harbor and town. Following project definition it was estimated that $4\frac{1}{4}$ years would be required for the complete design, construction, and commissioning of the works. With this preliminary plan, the development of the project began.

Control of progress in the investigation stage of any project is always difficult, and due allowance must be made in the planning for contingent delays and unexpected problems. Once project definition is reached, however, planned progress is considerably easier; the times required for detailed design, specification writing and quantities, preparation of the final estimate, procurement of bids, bid analysis, and site construction are then more readily forecast.

This feature is demonstrated in Figure 15.2, which shows the planning fragnet for the harbor installation forming part of the project shown in Figure 15.1. Following the feasibility study, this marine facility was defined as a bulk-loading pier for export of mine products and a general cargo wharf with adjacent transit buildings for imported cargo. It is now seen that design and preparations for letting a contract are expected to occupy 12 months, whereas site construction has an estimated duration of 24 months, including 2 months float.

Planning will obviously be expanded as project development proceeds. The construction stage will be subdivided into more detailed operations, so that when bids are invited, a realistic contract time may be specified. In this regard, it is wise to adopt a construction contract period perhaps 10% less than that required to meet the milestone completion date, so that there will be enough float available to permit valid extensions of the contract time in the event of unforeseeable delays or unexpected extra work. If this is not done, the owner's date for starting operations may be overrun. Thus, in Figure 15.2, the contract period ends on July 31 of year 5, whereas the milestone date for testing the shiploading machinery is September 30. Such inbuilt float must, of course, be provided from the inception of the project when the general development network is being planned, as discussed in Section 15.4.

With a fragnet diagram similar to Figure 15.2 for each part of the entire project, the detailed design will proceed. From each of these diagrams, the engineering design office can prepare its own detailed network so that all its

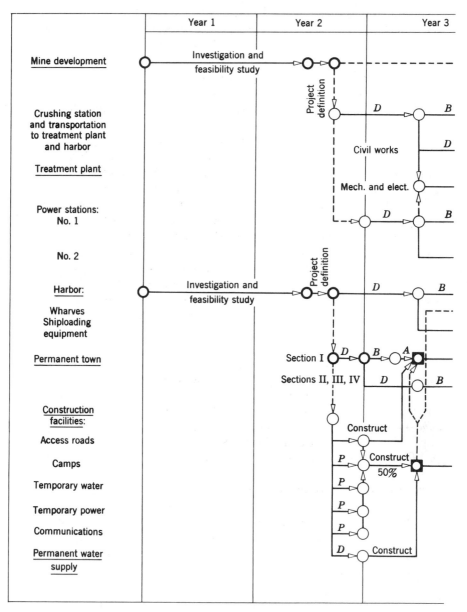

Figure 15.1 Preliminary planning network diagram for development of mining venture.

tasks are logically programmed and its resources allocated accordingly. Figure 15.3 shows the summarized engineering design network for the project illustrated in Figure 15.1. The sequences of design work and preparation of specifications, and quantities and final estimates, are arranged so that the invitations to bid will be issued in conformity with the dates required in

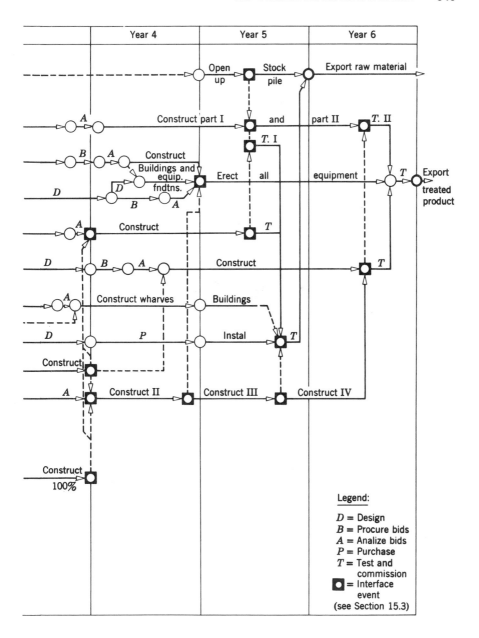

Figure 15.1. Needless to say, the same procedure is followed whether the various parts of the project are to be constructed by day labor or contract; but in day labor, no time is required for bid preparation and analysis.

Similarly, detailed fragnets may be developed for many other aspects of the project, particularly the procurement of materials, fabrication and testing

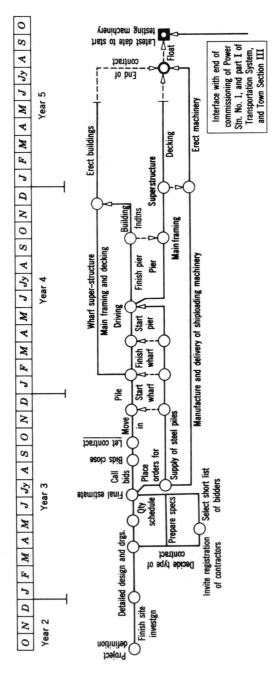

Figure 15.2 Planning fragnet for harbor works after project definition.

344

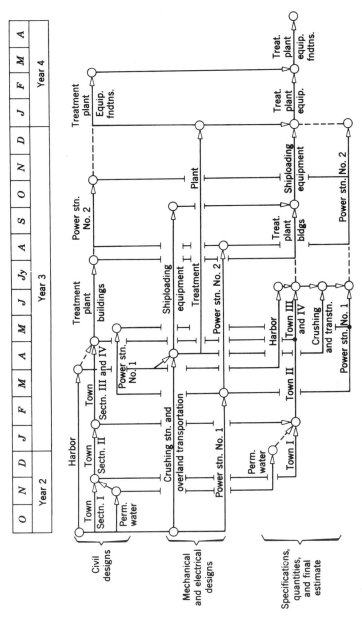

Figure 15.3 Engineering design network for project in Figure 15.1.

of custom-build items, delivery of various materials and equipment, and so on. Details depend on the complexity of the project.

15.3 CONSTRUCTION POLICY

There are two systems for the construction of any project: the day labor system and the contract system. In the day labor system, the owner carries out the work with in-house forces, under the control of in-house construction engineers, and pays all the costs thereof as a direct commitment. In the contract system, the owner selects a construction contracting organization to do the work at a mutually agreed price, and has a construction engineer supervise the work of this contractor. In both systems working drawings and specifications will be required, but the choice between contract and day labor work must be made before these are completed so that they may be appropriate to the selected system of construction. In addition, the subdivision of the overall project into separate parcels of work, and the size and nature of each parcel, must also be determined before the drawings and documents can be finalized.

In planning this subdivision of work, attention must be paid to the necessity of defining *interface events;* these are critical events common to two or more networks. For example, in Figure 15.1, the first pair of interface events occurs in the middle of year 3, where 50% of camp construction must be achieved before construction of the permanent town can begin (in order to provide sufficient accommodation for workers); then at the end of year 3 there is a group of four interface events, so that the completed camp and Section I of the town will provide enough accommodation for the increase in personel required to start work on the construction of No. 1 power station and Section II of the town; the next interfaces occur at the end of year 4 when all the civil works (buildings and equipment foundations) of the treatment plant, together with Section II of the town, must be finished before a contract is let for erecting treatment plant equipment. Similarly, the completion of the various parts of the project must be coordinated in order to provide for orderly testing and commissioning; thus, in the middle of year 5, mine development (to provide ore), No. 1 power station (to provide power), and part I of the crushing and transportation system must all finish together so that the last two may be tested and commissioned prior to testing the shiploading equipment; and the latter requires completion of Section III of the town to provide permanent accommodation for operating personnel. Finally, the testing and commissioning of the finished crushing and transportation system require completion of No. 2 power station in the middle of year 6 so that there will be ample time to test and commission the treatment plant before export of treated product begins at the end of that year.

The number and juxtaposition of interface events will influence the size and number of parcels into which the project is divided for construction purposes, and it will also affect the duration of inbuilt float required. A study of all the networks for the various parts of the project, compiled after project definition (that is, all the networks similar to Figure 15.2), will enable a logical determination to be made—not only about the separate parcels of work, but whether they should be done by day labor or contract. A diagram of the type shown in Figure 15.4 can then be compiled to act as a broad control diagram for the construction of the works. This diagram shows the inbuilt float to each interface event and also the actual contract time allowed for site construction (SC) in each parcel of work.

The decision to carry out work by day labor or by contract will be determined by the owner's construction competency, ability to move speedily onto the site (thus saving time), and desire for self-protection against high contract prices for unknown risks. Consequently, early preliminary works are often best carried out by the owner, as are also hazardous tasks for which complete data are not available, provided always that the owner has the ability to carry out such work economically and expeditiously in accordance with the program.

Once the construction policy has been finally determined, the construction stages of the networks of the type seen in Figure 15.2 may be expanded into more detailed diagrams, similar to those illustrated in Chapter 8.

15.4 CONSTRUCTION PLANNING

The preparation of network diagrams for the practical planning of construction works and the compilation of utility data for this purpose were discussed in Sections 8.1 and 8.2; these techniques have been applied to the examples quoted in Sections 8.3, 8.4, and 8.5, and the method of allowing for normally anticipated lost time in determining practical durations for activities was demonstrated in Section 8.6. There is therefore no need to consider these aspects of construction planning here; this section will be more profitably concerned with management planning and its effect on the adopted network diagram.

The degree of success that attends any construction project depends not only on the accuracy pertaining to the estimate of its cost, but also on whether the plan adopted for its execution is realistic and practical. A successful plan is the direct result of considerable ability in predicting the various conditions and contingencies likely to be found during the work on the site, for these have a significant influence on the duration of any activity that they affect, and also on the method and resources necessary for its completion in the most efficient way. Practical construction planning and the successful integration of project development with management therefore include not only planning and scheduling as defined in Section 2.1, but also establishing realistic judgment of all the circumstances pertinent to the project.

Judgment and CPM planning go hand in hand, for both demand logical

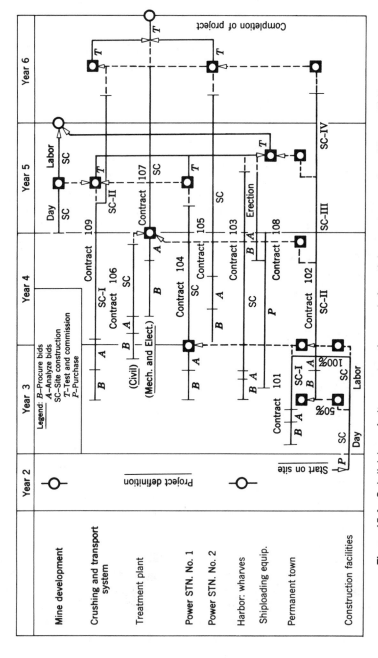

Figure 15.4 Subdivision of site construction work for mining venture project (Figure 15.1) showing inbuilt float to interface events.

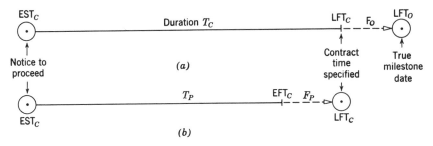

Figure 15.5 Planned contract times and floats: (a) by the owner; (b) by the contractor.

thinking through a series of steps, first, with a view to defining the basic objective of the exercise, second, with a view to listing all the pertinent factors involved in achieving this objective, and finally, by a study of these factors, to determining the optimum course of action to be adopted.[1]

Consider, for example, Figure 15.5. Diagram (a) shows the duration of a contract from the owner's viewpoint. As already discussed in Section 15.2, the contract time T_C is specified to provide some undisclosed built-in float, F_O, as a precaution against unforeseen valid extensions of time; these may arise because of delays due to causes beyond the control of either party, or delays caused by the owner (see Section 11.1). Diagram (b) shows the position as the contractor plans it; in order to ensure that liquidated damages are avoided for late completion from self-induced acts, additional float F_P is provided such that $T_C = T_P + F_P$. In the general case, T_P will be the optimum solution determined, as illustrated in Figures 2.6, 5.10, and 8.8; in special circumstances it may be derived by the methods discussed in Section 6.10 (Figure 6.17) or Section 7.7 (Figure 7.15).

In both instances the owner and the contractor have shown judgment in their respective approaches to construction planning of the necessity for precautions against unforeseen circumstances for which they may be legally responsible under the contract. Hence the effect of these judgments is that the contract may actually be finished as early as EFT_C or as late as LFT_O, depending on circumstances. This feature of construction planning must be constantly borne in mind by the owner when considering the relationship between the various interface events in the overall project. Considerable experience is necessary in determining the ratios of F_O to T_C and F_P to T_P if inconvenient gaps are not to occur in practice; needless to say, these ratios depend largely on the size and nature of the project, and on the types of hazards pertaining to the contract concerned.

Another aspect of practical construction planning is the use by the owner of prebid networks similar to Figures 15.2 and 15.3 as aids to project bud-

[1] See *Engineering Management*, by James M. Antill & Brian E. Farmer, McGraw-Hill, Sydney, 1990.

geting, and also in the determination of appropriate items to be shown separately in the contract bill so that they may have individual bid rates or prices. By means of these prebid networks and project cost estimates, the owner's engineer can prepare works expenditure charts similar to Figure 10.2, but showing instead the owner's expenditure commitments for investigations, designs, construction planning, and the cost of the construction itself (including its supervision and administration). As the development of the project proceeds and as the networks become consequentially more refined, these budget charts will become progressively more accurate and reliable. The owner's management will thus have an increasingly valuable measure of cash flow requirements as the project is being developed. When bids are invited, each contractor may be required to submit his own estimate of progress payment requirements (again similar to Figure 10.2, but showing contractor's income only), so that the owner's budget chart may be again updated at the time of letting the contract.

In addition, each bidder should be required to submit the prepared construction program network (as a summary), thus providing the owner's engineer with a fair and reasonable means of assessing the contractor's performance during the execution of the contract work. In this connection it must not be forgotten that, provided the work is performed within the contract time and in conformity with the drawings and specifications, it is the contractor's prerogative to determine the working plan and resource requirements. When simultaneous contracts contain common interface events, the construction planning by the owner must, of course, take cognizance of the working plans of the respective contractors involved.

From the contractor's viewpoint, good management will insist that, when there are several subcontractors involved, the main contractor will have a series of network diagrams depicting different plans to be adopted in the event of one or more of the subcontractors being delayed in starting and/or early or late in finishing. From these networks the effects of nonperformance by subcontractors may be predicted, and liability for liquidated damages assessed accordingly. As a corollary, the main contractor could also prepare networks for these subcontractors, showing the effects on each subcontract of early or late access to its work. By these means, alternative plans can be available to cope at short notice with contingent delays in the most economic and reasonable manner for all parties. This approach is, of course, not unrelated to the concept of factual networks extended by estimation to project completion (Chapter 11).

In conclusion, it must be emphasized that CPM construction planning does not provide a magic method of precise forecasting. It does, however, place the planned schedule of work to be done in a clear perspective, and it focuses attention on the various critical, near-critical, and noncritical paths, thereby enabling management to take advantage of all areas of flexibility and to concentrate resources and effort on all areas of criticality. Furthermore, CPM planning enables cumulative budget costs to be assessed for the two extremes of earliest possible finish (EFT_C in Figure 15.5b) and latest permissible finish (LFT_O in Figure 15.5a), so that the limitations of the owner's cash requirements may be investigated. Provided CPM planning also takes account of progress payments during the execution of the work and recognizes the difference between cash costs and accrual costs, it is an extremely valuable aid to the contractor in his cash management problems.

15.5 CONSTRUCTION MANAGEMENT

The effective employment of all construction resources is the most important function of construction management. The general works organization and the system of site administration and supervision must be devised with this primary objective in mind, so that the entire construction staff is aware of the planned targets and how closely or not these are being achieved. The most vital aspect of resource superintendence is the control of site labor, which will be successful only if the work force is made constantly aware that management is particularly interested in productivity and costs. Regular inspections and discussions with immediate supervisors and their crews are important so that people not only know what they are expected to do but why, and the consequences of its not being done. Control of construction materials is a basic problem of logistics; the correct quantity of the required goods must be in the right place at the desired time, and it must be of the specified quality. A similar approach applies to the control of construction plant, with the added necessity that it must be skillfully operated and properly maintained in first-class order. Control of construction finance will readily follow if the other resources are properly administered.

Control of construction resources is related closely to control of construction progress, and includes not only the proper supervision of the work and adherence to the planned or replanned program, but also constant vigilance in comparing actual and estimated performance and the provision of remedial measures whenever delays and unforeseen contingencies occur. Such dynamic control is exercised in accordance with the techniques explained in Chapters 9 and 10. The owner's and the contractor's respective administrative and supervisory staffs must be organized so that the overall project management will be efficient, economical, and successful—and capable of adapting itself to variations in circumstances as construction progresses.

Construction programs and CPM network diagrams are, of course, only estimates for the execution of the work; hence they cannot be expected to apply throughout the construction stage without changes. A well-conceived plan, however, is generally capable of performance without radical alterations, provided always that the project was efficiently engineered so that work changes and delays are minimized; the greatest cause of loss of project control is the advent of a variety of change orders and related hindrances.

Obviously, work changes can never be eliminated, because the exact nature and conditions of any site are not known until after the ground is broken and construction begins. Likewise, unforeseeable delays will sometimes occur. All these hindrances impede progress, and their administration occupies a considerable portion of a construction manager's time. Disputation concerning work changes and delays can, however, be minimized (if not entirely eliminated) by the procedures outlined in Chapter 11. Nevertheless, it must be remembered that all such occurrences create problems for both the owner and the contractor and that successful solutions will be obtained only if both parties maintain flexibility in their approach to the necessary replanning of the work.

The general case is illustrated in Figure 15.6a and b, which show the owner's and the contractor's planned construction network diagrams; Figure 15.6c shows the factual network with delays D_a and D_a' (due to neither party, for example, floods), D_b (due to the owner, for example, work changes), D_c and D_c' (due to the contractor, for example, insufficient resources). Liquidated damages apply for completion later than LFT_C or valid extensions thereof, and time extensions apply for all causes beyond the contractor's control (that is, for D_a, D_a', and D_b).

Now, in Figure 15.6c, the factual critical path has changed from the initial one, and the project duration is the following:

$$T_C^F = T + D_c + D_a + D_b \tag{15.1}$$

$$= T_P + D_c' + D_a' + F_P' \tag{15.2}$$

where T_C^F may be greater or less than T_C and greater or less than $T_C + F_O$ (= LFT_O); but liquidated damages can only apply if $T + D_c > T_C$ (notwithstanding that now on the initial critical path $T_P + D_C' > T_C$), because it is the factual critical path (not the initial one) that determined the actual contract duration. On the factual critical path, T is, of course, the networking duration defined in Section 11.6.

With $T + D_c > T_C$, the period to which liquidated damages apply is obviously $T + D_c - T_C$, irrespective of whether $T_C + F_O$ is exceeded or not. Furthermore, it will be clear from Figure 15.6 that under no circumstances can D_c and D_c' be cumulative, since they affect different paths through the network. Similarly, D_a and D_a' are not cumulative for the purpose of computing extensions of time nor would be any number of delays of type D_b when they affect different paths. In all cases the delays must be plotted and each path analyzed, as discussed in Section 11.6.

When a delay of any type occurs, it is legally incumbent on the parties to take reasonable steps to mitigate the resulting loss (see Section 11.1). Frequently, it will be feasible to replan the work by rearranging sequences or constraints, by suitable compression of critical activities, and/or by decompression of crashed activities now no longer critical. The last was demonstrated in Figures 11.1 and 11.5. However, an alternative plan for construc-

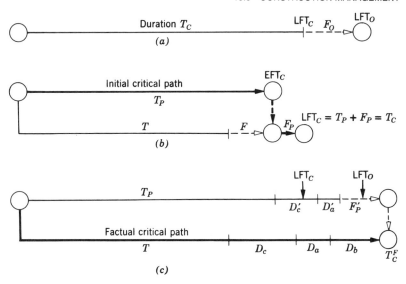

Figure 15.6 Planned and factual critical paths. (a) Owner's plan. (b) Contractor's initial network. (c) Factual network.

tion of this pipeline, which could have been implemented had the extra work in activity 7–8 been ordered *10 days earlier* (instead of at event 5), is shown in Figure 15.7; with this replanning *the initial project duration is not exceeded* ($T_P = 102$; see Figures 5.7f and 11.1a), and *the direct cost C_P is reduced to 12,660*. The prime importance of the timing of the advent of change orders is thus amply demonstrated; they should always be issued as early as possible to give ample flexibility for replanning. Finally, the maxim that there is always more than one plan for the execution of construction works is clearly exemplified by comparing Figure 15.7 with Figure 5.7f, since both of them require the same project duration, but with different resource requirements and hence different direct costs. All such feasible alternative plans are therefore worthy of development and careful consideration before one is finally selected.

One other point must be made in conclusion. Since management is concerned with efficiency, the actual layout adopted for CPM network diagrams should itself be efficient. From the use of CPM in construction practice over the last several years, it is apparent that large sheet drawings showing the entire project network are not so easy to use and to understand on the site as a series of subnetworks on smaller sheets. It is recommended that the size of drawing sheet be limited to metric A1 size. Interface events between different subnetworks on different sheets should have the same event number, with suitable notation on a dummy entering or leaving to minimize searching for the connecting subnetwork (thus: "From Sheet . . . ," or "To Sheet . . ."). Preferably all site networks should be drawn to scale (usually 5 mm to

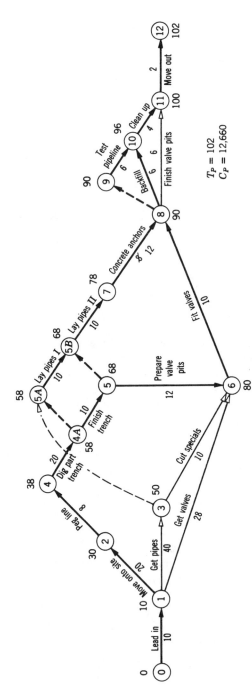

Figure 15.7 Replanning of pipeline construction (Figure 5.7*f*) to provide for 4 days extra work in activity 7-8 without increasing project duration (compare with solution in Figure 11.1*b*).

354

a shift is suitable), showing all dummies and constraints, with events numbered chronologically, yet with the number at the arrowhead greater than that at the tail ($i - j$ system); this assists in rapid location of an activity on a network diagram. Provided, however, that network diagrams are efficiently compiled and processed, the use of CPM is a valuable aid in the integration of project development and construction management.

Evaluation

a. Discuss in some detail the presentation to management of an integrated approach to project development.
b. Suggest suitable contract fragnets for the project development you have adopted in (a) above. How does this influence the management of the contract in practice?

16

CPM—A SYSTEMS CONCEPT

16.1 THE SYSTEMS CONCEPT

The emergence of management science within the last two decades reflects to some extent the rapid development of complex problems and situations in our modern society. It has led to the awareness of the need to take comprehensive overall views of many of the problems facing construction managers. In many instances a narrow definition, isolation, or modeling of a problem leads to ignoring important interactions, influences, or feedbacks implicit in the real-life situation.

The emphasis on narrow, local views of a problem is often the result of several causes. First, it may reflect the fact that little personal professional expertise or technological knowledge may exist except within a certain area of the problem. In these cases there is understandably a reluctance to consider more than a minimum outside one's proficiency. Second, it may be the consequence of the lack of suitable synthesizing concepts, models, and mathematics. It is unfortunately true that in the past mathematical modeling has tended to concentrate on the component aspects of problems. The resulting emphasis on component issues leads to the formation of a component mentality with a marked reluctance to consider overall views, and it therefore leads to adopting suboptimal policies.

In contrast, recent developments in the use of network models, combinatorial mathematics, digital computers, and management science have produced a favorable setting for systems concepts. It is now possible and feasible for certain cases to attempt the synthesis of individual components into a system or unified whole, and to adopt a system mentality by formulating overall policies and objectivies for the collection of components. This approach is often fruitful in discovering unsuspected opportunities and rewards, apart from the obvious advantages derived from taking overall views of a professional problem or area.

The idea of a system as an integrated assembly of interacting components designed to perform jointly a predetermined purpose has been obvious since the dawn of the industrial age. The system components were physical objects and their interaction mechanical and physical. More recently it has been obvious that logical, organizational, informational, or social components can be components of a system. In fact, a system can be any collection of components on which an ordering, or structuring of their interactions, can be imposed. The network model of a construction plan is a collection of activity components ordered relative to each other by a construction logic structure; hence the model is a system model.

The systems concept has led to two different developments, systems theory and the systems approach. Systems theory uses systems as a noun and thus considers the formal description and properties of a system. Consequently, it assumes that the component properties and behavior are known, and concentrates on determining the effect of component interaction on the system. System theory can therefore be described as being concerned with the analysis of behavior or the prediction of response given a specific system. Systems approach refers to the activity of determining the nature and structure of a problem and the constraints under which management must operate in accomplishing its solution. It is concerned with the identification of all the relevant components of the problem and the extent and nature of their interaction. Thus the systems approach is directed toward the discovery and definition of a suitable systems model (for which hopefully a systems theory can be formulated), and of a feasible and practical approach for its solution and implementation.

System theory leads to the CPM network model concept and information flow of a construction project. The systems approach leads to the development of complete organization models and raises issues relating to the effective implementation of CPM concepts within a construction enterprise. In the following sections these considerations will be explored first with respect to the network model itself, second with management's involvement with the network model, and finally with the description and location within a construction organization of CPM-based information systems.

16.2 THE NETWORK MODEL AS A SYSTEM MODEL

A construction contract usually calls for the construction of a collection of objects such as buildings or roads, or calls for the carrying out of a certain number of tasks such as site clearance or soil disposal. Very often the individual objects fit together in a unified way to produce a physical system such as a building and its access road. It may happen, of course, that the contract calls for the construction of a collection of isolated, unrelated objects so that in no way can they be considered to produce a physical system.

In both cases the adoption of a construction plan, which specifies the order in which the contract items will be handled relative to each other, produces an ordering, that is, defines interrelationships, between the items at an organizational level. Thus a construction plan can be considered as a management system model because it implies that an internal structure exists that ties the system components (that is, contract items) together. Consequently, in this view the network model is a representation of an organizational system with the system structure defined by the construction logic.

The consideration of an isolated project activity may, in some cases, be a systems view and in others a component view. If the activity itself is viewed as a collection of smaller tasks and operations, with the interweaving of manpower, materials, and equipment, then it can truly be considered as a system[1] and, for example, its productivity analysis becomes an activity design or a systems analysis problem; and its field management becomes a systems behavior or management system problem. On the other hand, if we consider the activity embedded within the project, yet plan its duration, resource requirements, scheduled start, and crashing in isolation from the project, then in the perspective of project scheduling a component mentality only has been adopted. Clearly, for this situation, one that is commonly met in construction practice, a local view results that should be recognized as being possibly suboptimal.

In the same manner at a higher level a complete project may be viewed as a component in some larger system such as, for example, companywide operations. Whether a particular approach corresponds to a systems or component view depends on the policy and objective of the viewer. Thus the superintendent's desire to finish an activity as soon as possible by smoothly integrating and expediting the planning and management functions, the handling of crews and crafts, materials supply, and so on may be a fully comprehensive and professional system view but, for this particular activity, just a component view for the company supervisor, who sees this activity relative to the needs of other activities and general project behavior and assigns resources and schedules for this activity subject to some overriding policy. Clearly, a hierarchy of goals and policies may exist which will influence the approach to be used in each case.

Frequently, obvious system problems emerge. The determination of the project duration, for example, is a system problem because it is a characteristic of the network model and depends on the interaction of the project activities for its evaluation.

Referring to the small network model shown in Figure 16.1, the system structure and component attributes may be described as follows:

[1]See, for example, the CYCLONE modeling approach described in Section 13.5.

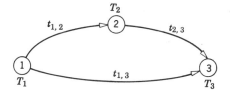

Figure 16.1

Project (Figure 16.1) consists of three activities:
    Activity 1,2 has duration $t_{1,2}$, and
                is followed by activity 2, 3;
Activity 2, 3 has duration $t_{2,3}$;
Activity 1, 3 has duration $t_{1,3}$.

                System
                Description

A system problem can then be posed as

    Given $T_1 = 0$ (or calendar day);   System
    Find T_3 Input

    We can consider the solution for the project duration as a system theory problem, and its value as an output from the system, or produced by the presentation to the system of some form of stimulation or input. We can then sketch the network model as shown in Figure 16.2 in the typical "blackbox" model form of a system. The actual network model is, of course, known precisely as well as the specific activity durations, so that in this sense we do not have a blackbox system with unknown internal structure. Figure 16.2 is presented purely to emphasize the system concept of the project duration calculation.

    The determination of critical path events and activities and free floats can be similarly viewed as systems problems because their determination requires an analysis of the systems model as a whole. However, although the determination of total and independent floats may be the result of a systems analysis, their conception and possible use in scheduling corresponds to a component mentality.

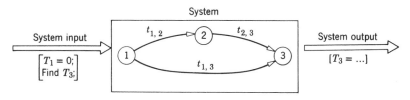

Figure 16.2

The decision to make across-the-board crashing of activities to recover lost project days can be viewed as expressing the component mentality, because it is nearly always possible to crash selectively subsequent critical path activities at lower cost by adopting the systems mentality approach.

Another common situation that may be construed as implying a component mentality is the adoption of earliest start schedules. If such schedules are adopted without thought, then it is implied that the system's concept of activity shifting within float to optimize a system resource problem is rejected. In some situations, of course, an earliest start schedule may be required because of the general urgency of the situation, and the "free-float system concept of a future safety factor" is adopted as a system policy.

It should be clear from the preceding examples that the network model of a construction plan is the representation of an organizational system, and many of the calculations shown in this book require a system theory for their evaluation and solution and a management system philosophy for their initial consideration and final implementation. These latter two aspects are considered in the following sections.

16.3 A MULTISTAGE DECISION APPROACH TO CPM

The wide spectrum of construction endeavor, the varying demands made for technical and managerial intervention in projects, together with the usual variations in interest and expertise of construction personnel, have resulted in a wide range of involvement in CPM by the construction industry. A further complication results from the interaction of the planning and scheduling functions in construction and project management, and by the variety of CPM network-oriented heuristic approaches to resolving this conflict.

It is possible to identify many different activities and decision levels involved in the normal basic concept and use of CPM. Table 16.1 lists 30 activities that can be considered as being relevant to the development of a project network model and schedule. The list is not intended to be exhaustive. These activities have been modeled in Figure 16.3 using circle notation, with a collection of connecting branches. The branches are intended to represent many of the possible logical interdependencies that exist in professional practice. They have been intentionally shown as undirected arrows so that they are able to represent many procedural and heuristic solution possibilities, as well as allowing for the interative recycling processes that exist in the planning and scheduling stages. Finally, various groups of activities have been collected together to indicate common areas of interest, as well as identifying the various departments within the construction organization responsible for undertaking the activities.

Naturally, not all of these activities or branches are relevant for any one CPM package or specific management situation; if so, they should be removed from the model. Those that do exist can be used to indicate both the

Table 16.1   A multistage decision approach to project management with CPM.

1. Examine the contract documents and define the project.
2. Examine the contract documents and determine progress payment conditions.
3. Define the overall policy or project strategy.
4. List the major contract items.
5. Define the construction technology.
6. Define the safety and other managerial logic.
7. List the specific activities.
8. List the resources required by types.
9. Determine the resource availabilities by quantity and duration.
10. Define the resource logic.
11. Define the construction logic.
12. Prepare the network diagram (linear graph).
13. Define the activity methodology.
14. Estimate the activity durations in working days.
15. Estimate the direct costs of major activities.
16. Prepare the network model (linear graph plus attributes).
17. Carry out the CPM calculations for the critical path, project duration, and floats.
18. Determine the project resource requirements by listing manpower, crew, trade, equipment totals.
19. Resource leveling within floats.
20. Heuristic resource leveling with extended project duration.
21. Establish activity utility data.
22. Define practical compression logic.
23. Carry out time-cost tradeoff calculations.
24. Establish indirect costs and determine total project cost curve.
25. Define scheduling policy.
26. Establish cash flow and overdraft demands.
27. Establish the project finance requirements.
28. Define project start date.
29. Adjust activity durations to suit climatic or special conditions.
30. Schedule project in calendar days.

level of involvement in CPM and the general way in which the planning and scheduling functions are handled. Using this approach it is possible to consider Figure 16.3 as representing a multistage decision approach to CPM. Three examples are shown in Figure 16.4, which illustrates the effect on CPM calculations of management's attitude to resource levels and availabilities. In each case the activities are identified from Table 16.1.

Figure 16.4a corresponds to the assumption of infinite resources, to the extent that resource considerations do not even enter into the listing of activities for the project. Similarly, Figure 16.4b assumes infinite resources, but the addition of arrow 5–7 implies that the list of activities for the project includes both resource procurement activities as well as more specifically

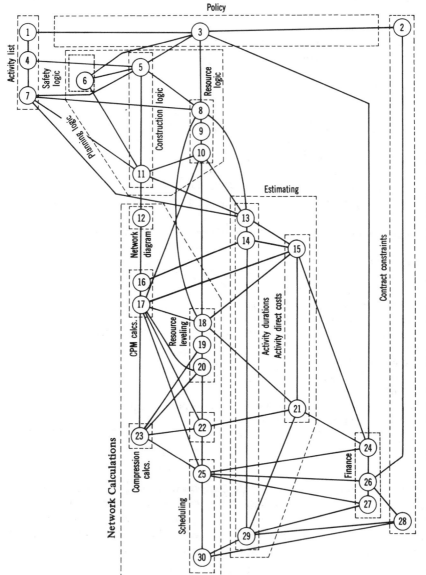

Figure 16.3 A multistage decision approach to CPM scheduling.

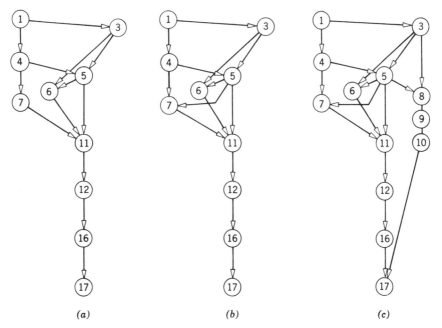

Figure 16.4 Effect on CPM calculations of Management's attitude to resource levels and availabilities.

resource-oriented activities. Finally, in Figure 16.4c both resource logic chains of activities 8, 9, and 10 are added, and the arrow 10–17 which may imply that a simulation CPM calculation is intended to ensure compatability of CPM calculations with resource availabilities.

Many other professional attitudes to CPM can be modeled from Figure 16.3 and Table 16.1. The reader may wish to draw contours on Figure 16.3 corresponding to successive stages of involvement in CPM. Naturally, to some extent these will be subjective and bear witness to the fact that decision making is necessarily not objective.

16.4 IMPLEMENTATION OF CPM IN A CONSTRUCTION ORGANIZATION

The implementation of CPM in a construction enterprise is a major undertaking which should not be begun lightly. Careful consideration should be given to whether any commitment should be made to CPM and if so to what extent; these issues should be resolved before any implementation is planned. Once a commitment to CPM is made, the implementation should be thoroughly planned in advance, with frank and open discussion and involvement if success is desired. Any failure will have long-term repercussions, severely handicapping later attempts at implementation.

Experience has indicated that a staged introduction is more successful than a sudden heavy overcommitment that cannot be supported in the field nor economically sustained from benefits achieved. A staged involvement is a more stable plan, which readily permits a heavier involvement as confidence and enthusiasm in CPM grows among the construction personnel.

The basic motives and objectives for an involvement with CPM should be stated clearly by management, who has the responsibility of ensuring that a feasible working system is evolved within the organization, and which is liberally supplied with the resources to produce results. Light, spontaneous enthusiasm will rarely suffice.

Items requiring careful consideration in a construction company are its goals, the number and competency of its various competitors, the nature of the work performed by the company, and the level of involvement in CPM considered. It is wise to consider future needs as well as current needs, but the implementation should be soundly directed at the current level of work management and aim at proving itself within a very short exposure time.

Vital factors include experience and capabilities of company personnel in general, notwithstanding the fact that a CPM expert may be incorporated into the planning section of the company. Experience has shown that the failure of many construction companies to accept CPM is due in part to the reactionary attitudes of field personnel, an issue that can be fairly easily avoided with a proper plan of implementation. Unless the field personnel respond to CPM, the implememtation may dwindle to a paper game played in the planning section, divorced from reality by the lack of application and feedback from the field. It must be clearly understood that CPM must prove itself in the field before any company success can be claimed.

Radical changes may be necessary in the company procedures if success is desired. It may be essential for job superintendents to take part in the network model formulation and development.

A minimal condition is that planners achieve an understanding of the perspective and problems of the site superintendent. It must be understood that no amount of good office planning will cover up lack of production and efficiency in the field.

The too-ready adoption of automatic CPM computer methods may be fatal unless careful consideration is given the format, quantity, and transfer of information from the computer to the understanding of the site superintendent. Nothing will alienate more quickly the busy superintendent from CPM than the too frequent delivery of hundreds of meters of computer output from a project model produced in a remote head office. In many instances superintendents assume that this material is an intrusion on their professional responsibilities. Obviously, superintendents must be supplied with information in a familiar form that they can readily see will help them in his decisionmaking, and which is based on a network model that applies to their situation as they see it in the field.

The basic data relating to work rates, crew sizes, level of representation of activities, and so on, must come from the realities of the field works. Production rates, hence activity durations, must be consistent with the capabilities of the site work force if meaningful network models and calculations are to result. No amount of planning and estimating will supply "in-field" accuracy. The use of field data, of course, is good practice and ensures that feedback to the planning section updates the historical files so necessary for the correct estimating and costing of projects.

Once company personnel can be made aware that the network model is to be used to help gain an understanding of reality, rather than as a constraint on construction, and that the network is not an end in itself, they will gain confidence in its use. Although emphasis should be placed on the necessity to build in logic correctly for the understanding of the planners, there is no need to always insert the most abstruse logic. Very frequently, this implies too fine a model description and is often correctly interpreted as restricting the managerial capabilities of the site superintendent. Good management will ensure that site personnel have room for individual action and initiative.

It should be clear that CPM can be used in a company at strategical and tactical levels; thus many different levels of models may exist of the same project at the same time within a construction organization.

The critical path method is capable of serving as the basis for an integrated companywide management system, ranging from prebid feasibility studies through estimating, planning, cost accounting, progress control, and financial control. In fact, it has such great possibilities in so many vital areas that failure to implement CPM may place the company in an inferior competitive position.

A network model for the implementation of company CPM is shown in Figure 16.5 with the various activities defined as in Table 16.2.

16.5 THE SYSTEMS APPROACH TO PROJECT MANAGEMENT

An interesting development of the systems concept is the systems approach. The systems approach attempts to determine the structure of the environment associated with any specific problem in such a way that all relevant

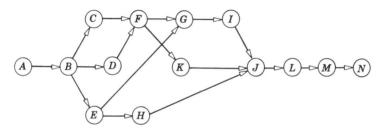

Figure 16.5 Implementation of company CPM: A network model.

Table 16.2   Implementation of company CPM—a network model

A. Define goals:	What does the company need, and how can CPM possibly help?
B. Feasibility study:	What sort of system is needed, and what level of involvement in CPM is suitable?
C. Define company organization:	Define the organizational structure, identify decision makers, lines of communication, information sources and supply, define policies and arrange support.
D. Performance standards:	Determine the minimum level of competence required by staff and job personnel.
E. Staff education:	Train staff in CPM methods covering the whole spectrum of the company's involvement, with emphasis on the significant contributions staff will be expected to make to work effectively with CPM.
F. Planning section:	Set up planning section, prepare for network building, prepare standard subgraphs.
G. Methodology:	Decide who will be responsible for calculations and how they are to be done and checked.
H. Employee education:	Train employees in CPM methods with emphasis on the vital part they play in making the CPM implementation a success.
I. Equip departments:	Provide departmental checking methods and data acquisition processes, emphasize the part each department plays and its relationship to other departments within the company organization.
J. Implementation:	Implement CP on a new project and staff with selected superintendent and engineers.
K. Updating system:	Decide updating system, reporting formats, handling of change orders and revisions.
L. Project running:	Carry out real time revisions and conferences with all concerned in the project.
M. Review and evaluation:	At project completion hold exhaustive conference on the CPM aspects of the project and suggest improvements.
N. Confirm implementation:	Formally commit company to a specific level of involvement in CPM.

influences that affect the problem are considered. If, for example, we are concerned with project management at the job superintendent level, then the systems approach would be concerned with structuring all the processes that focus on the job supervision.

Figure 16.6 indicates some of the processes that impinge directly on the job superintendent. The various processes are indicated as chains of head-office

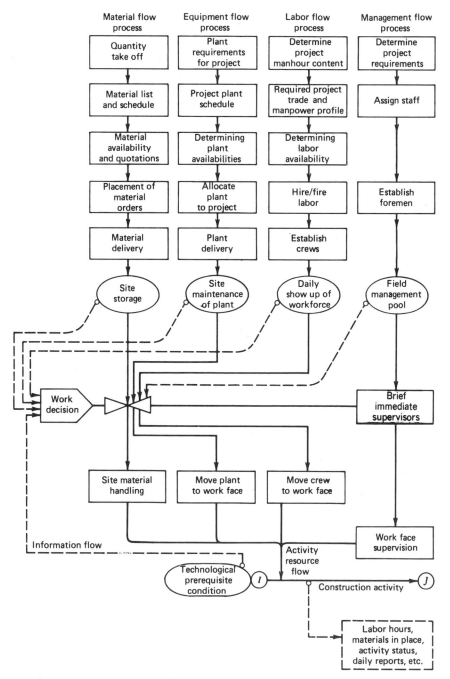

Figure 16.6   Basic job site processes.

and field functions leading up to the activity work decision of the job superintendent. Specifically, for example, we can trace the chains of management functions required to ensure that the activity can commence. We have the material chain commencing with project definition and estimating functions, the equipment chain with the plant requirements of the construction plan, and a crew chain with the project worker-hour estimate. The decision to start an activity requires that the activity resources be available; consequently, for any specific activity all the aforementioned management function chains must have been fulfilled. To make the decision to start an activity, the job superintendent must ensure that the material, equipment, and personnel required for the activity are available and, consequently, must ensure that the information paths that have forwarded this information are properly established, and that the material delivery notes, plant use reports, and the work force payroll list chains indicate that the various resources are available.

The efficient job superintendent operates within such an environment either specifically through information notes received or through information personally perceived. Whether he operates on one basis or another is immaterial. The fact remains that for efficient management all the above activities must be accomplished.

It is possible to carry out a complete systems analysis for any one of these chains, for any one means of information flow, or for any one problem requiring a decision. Thus, for example, if we are concerned with the information flow relating to project management or, more specifically, relating to CPM updating, we would be interested in establishing models similar to those indicated in Figure 16.7, which shows a specific construction company organization; thus interest would focus on the specific form the information flow takes between the various agents or decisionmakers in the construction company.

Attention to detail, to the manner, quantity, and type of information transferred among company agents, and the manner in which it is transferred, and the format and the layout of the various documents to be used will amply repay the effort expended in designing information systems.

16.6   CONCLUSION

The selection of the best organizational structure, information flow system, or technique to use for the solution of a specific problem will always rest with the management and will be based on the skill of the planning personnel and the intricacies of the problem itself. The choice of manual or computer methods will require individual consideration. Furthermore, the factors affecting a choice will vary not only with the problem but with the passing of time.

There is no doubt that computer methods will become more universal as better programming is devised to suit the construction industry's problems

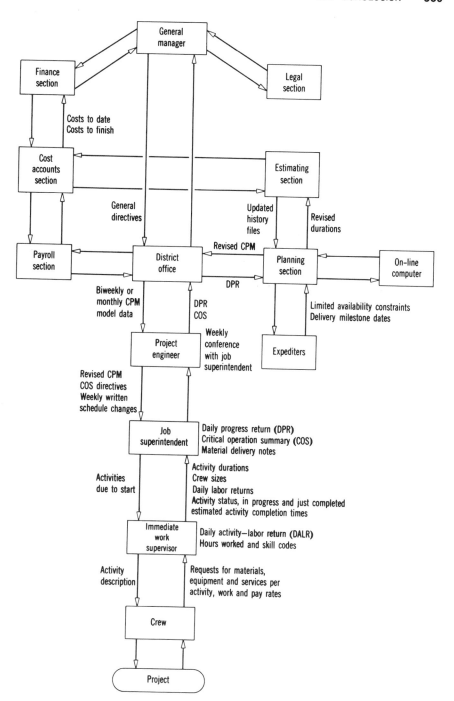

Figure 16.7 Information flow system.

and as construction managers become more familiar with computer methods. Indeed it is foreseeable that all time-cost problems may eventually be handled cheaply and effectively by computers, both in the planning and in the construction stages of a project, once the shortcomings referred to earlier in this chapter are overcome. Both CPM and PERT have already gained widespread support and have been the subject of considerable research and development in the computer field.

One thing is certain, critical path methods, in general, are here to stay. The wise planner will certainly take full advantage of them and of their future development.

Evaluation

a. How would you introduce CPM concepts into your organization?

b. How would you introduce a "total information system" to your organization?

APPENDIX A

CPM AND LINEAR GRAPH THEORY

A.1 INTRODUCTION

Construction managers need a comprehensive information system to assist them in the decisionmaking processes they use for the planning, scheduling, controlling, and management of a construction project. The success CPM enjoys is due partly to its unique capability for representing construction projects and partly to the wide range of information that can be generated from the CPM model. In addition, the CPM concept provides a basic structure for a complete companywide project-integrated management information system. The method has such a potential in so many critical areas of project and company management that it is informative to consider its mathematical basis.

Critical path method is essentially a graphical process. The network diagram is a graph of the project activities that portrays the logic of the construction plan. The project activities are collected and synthesized into a connected whole using a graphical process to show the construction logic. The network model uses labeling techniques to assign a variety of attributes to the model in such a way that computational algorithms directed toward a wide spectrum of project management questions become graph theory problems. Thus a study of basic graph theory may lead to fundamental insights into CPM and its computational processes.

Linear graph theory is a section of the mathematical theory of topology, and is the name given to the study of the discrete arrangement and connectivity of objects. The concentration on "connectivity" enables linear graph theory to provide simple tools and concepts for model construction. It is this area of concentration, and its severe simplicity, that makes linear graph theory capable of serving a wide variety of uses. Connectivity interpretations can be developed for organization, logical, and combination concepts that enable graphical models to be developed.

Thus, for example, if A and B are related to each other in some way, linear graph theory enables this fact to be modeled in two ways (see Figure A.1). In Figure A.1a, the objects of discussion A and B are modeled by "nodes," and the relationship R between A and B is portrayed by the connecting line between the nodes. In Figure A.1b, A and B are modeled by "lines," and the relationship R is portrayed by the node connecting the two lines together. Care should be taken when interpreting Figure A.1 to recognize that the figures represent a relational structure between A and B, and the figures themselves should not be confused with the concept associated with the figures. Both representations of the relationship between A and B are valid in linear graph theory and are easily recognized in the circle and arrow network models of CPM.

Graph theory is called "linear" because it is concerned with the connectivity of lines (branches, edges, arcs) and nodes (points, end-points, vertices). Graph theory is an abstract concept; it is not concerned with the physical and geometric aspects of lines and nodes. Consequently, linear graph theory strips away from a problem obscuring material and exposes the basic logical structuring of the problem.

A.2 LINEAR GRAPHS AND CONNECTIVITY

Linear graphs are composed of nodes and branches. A specific linear graph is made up from a certain number of nodes (at least one) connected together in a unique way by a certain number of branches (possibly zero). Each branch is used to connect only two nodes, if the looping of branches from a node to itself is excluded from consideration. The specific way in which the branches are used to connect some, or all, of the nodes defines the graph structure. Consequently, to describe a linear graph it is necessary to describe the graph structure as well as describing in a unique way each node and branch.

The unique description and identification of nodes and branches is a simple labeling problem that has been illustrated many times in the text. In general, both nodes and branches are given numerical and alpha-numerical labels as descriptors.

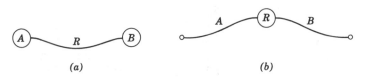

(a)   (b)

Figure A.1 Different graphical representations for "A and B are related" that is, ARB.

In circle notation graphs, the number of nodes is equal to the number of activities (plus two if start and finish activities are added). The number of branches is equal to the number of relational statements that must be made between pairs of nodes to define the construction logic. Thus, referring to Figure A.2, there are six nodes, comprising four project activities *A*, *B*, *C*, and *D*, to which has been added a start activity *S* and a finish activity *F*. In Figure A.2*a* the relational statement modeled is that of "precedence," which is a transitive relation, and twelve statements are required, whereas in (*b*) the relational statement modeled is that of "immediate precedence" and only seven statements are required to define the same logic. Notice that although the two graphs model the same project, they are different graphs, and the more restrictive relational statement demanded in CPM logic produces a minimum number of relational statements, and hence a minimum number of branches.

The graph structure can be indicated by using a mapping matrix, as shown in Figure A.3. This matrix can be readily condensed into the form of Figure A.4—a representation that is common for input data to circle notation computer programs. Notice in Figure A.3 that directional relationships are implied by the labeling of the rows as "nodes" and the columns as "immediately following nodes."

In arrow notation graphs the number of nodes and the number of branches are not specifically determined, because simplifications of the graph are usually introduced. Specifically, if we use the modeling concept corresponding to Figure A.1*b*, where each activity is modeled as a branch with its own initial and terminal node, then the number of branches in the generic graph is equal to the number of unique activities associated with the project plus the number of "immediately preceding" relational statements that can be made between any two activities. Thus referring to Figure A.5 that shows an arrow notation model corresponding to the project of Figure A.2, notice that branches corresponding to activities *A*, *B*, *C*, and *D* are included, and relational statements *R*2 (immediately preceding) appear relating the four activi-

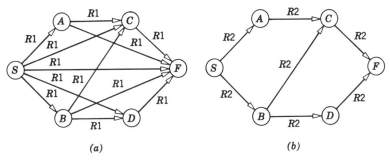

(a)   (b)

Figure A.2   Circle notation linear graphs for a network model. (*a*) Relational Statement *R*1 (precedes); (*b*) relational statement *R*2 (immediately precedes).

Nodes \ Immediately following nodes	S	A	B	C	D	F
S	F	T	T	F	F	F
A	F	F	F	T	F	F
B	F	F	F	T	T	F
C	F	F	F	F	F	T
D	F	F	F	F	F	T
F	F	F	F	F	F	F

Figure A.3 Graph structure for relation R_2 and Figure A.2*b*. *T* indicates that the R_2 relation holds (that is, true) and *F* that the R_2 relation does not hold (that is, false).

Node: Immediately following nodes;

S: *A*,*B*;

A: *C*;

B: *C*,*D*;

C: *F*;

D: *F*;

F: ;

Figure A.4 Condensed description of the graph structure of Figure A.3.

ties. In addition, relational statements *R*3 "logically concurrent" are modeled as branches at the project initial start node and final finish node. Thus the final arrow notation graph has ten nodes and eleven branches, of which four are activity branches, three are *R*2 relational branches, and four are *R*3 relational branches. In common practice the *R*2 relational branches connecting activities *A* and *C*, and *B* and *D* are implied by superimposing the terminal node of *A* and the starting node of *C* and similarly by superimposing the terminal node of *B* with the initial node of *D*. Thus the relational branch *R*2 connecting *B* and *C* becomes the familiar dummy activity *E* of Figure A.6. Finally, the relational branches *R*3 are implied in Figure A.6 by superimposing the initial nodes of activities *A* and *B* with the start event and similarly by superimposing the terminal node of activities *C* and *D* with the finish event. When this is done, the normal CPM graph of Figure A.6 results.

The graph structure can be indicated in a similar way by using a mapping matrix, as shown in Figure A.7. However, the directional properties must now be included because the mapping matrix has rows corresponding to branches and columns corresponding to nodes. A condensed form of the graph structure for Figure A.6 is shown in Figure A.8 and corresponds to the common input form of the arrow notation computer programs.

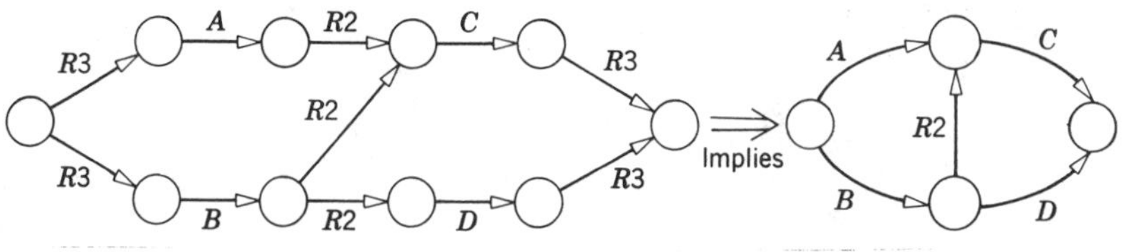

Figure A.5 Arrow notation linear graph for a network model. R2—immediately precedes. R3—logically concurrent.

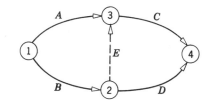

Figure A.6 Arrow notation linear graph.

A.3 LINEAR FRAPH PATHS AND CPM

The modeling of project networks in CPM produces directed linear graphs in both arrow and circle notations. In arrow graphs the directed arrows indicate the direction of time progress through the activities, whereas in circle graphs the directed arrows point to following activities. Consequently, connected and directed paths exist in CPM linear graphs, and all these paths commence at the project start node and terminate at the project finish node.

The number of different directed paths spanning the graph, from start to finish nodes, depends on the construction logic (that is, on the graph structure). The graph structure has the effect of connecting, intermingling, and dividing the various unique paths that exist in the graph.

In normal CPM calculations, interest focuses on isolating and identifying only one of these paths, the critical path. Specifically, time attributes are associated with the branches and nodes throughout the graph, and the critical path is identified as the path requiring the largest quantity of time between the project start and finish nodes. All noncritical paths have free and total floats distributed along their length, and, if necessary, all paths could be ranked according to their maximum total float or summed free floats.

In some cases it is important to identify all the paths that exist in the graph by indicating the branches that together form each specific path (see Chapter 11). If a mapping matrix is used to indicate the path graph structure, it is commonly referred to as the *path matrix*.

	Node labels			
	1	2	3	4
A	From		To	
B	From	To		
C			From	To
D		From		To
E		From	To	

(Branch labels)

Figure A.7 Matrix of graph structure for Figure A.6.

Branch labels:	Node labels	
	From	To
A :	1	3
B :	1	2
C :	3	4
D :	2	4
E :	2	3

Figure A.8 Condensed form of graph structure for Figure A.6.

Figure A.9a portrays a linear graph with branch and node identification labels, and (b) indicates the same linear graph with path labels. The path matrix for the graph is shown in Figure A.10. A path column is needed for each unique path, and branches (or alternatively nodes) lying on the specific path are identified by a label (in this case by 1) located in the row corresponding to the branch segment (or alternatively nodes) of the path and the column being considered.

Figure A.10a gives the path matrix for the graph of Figure A.9a using a branch identification for the paths, whereas Figure A.10b uses a node identification for the paths for the same graph. Obviously, the path matrix of Figure A.10a corresponds to an arrow network path concept and that of (b) to a circle network path concept. Normal CPM calculations do not require the explicit formation of a path matrix.

A.4   LINEAR GRAPH TREES AND CPM

Critical path method calculations require the determination of earliest and latest times for the linear graph nodes. In the forward pass algorithm, each node is labeled with the summed time durations along that specific path from

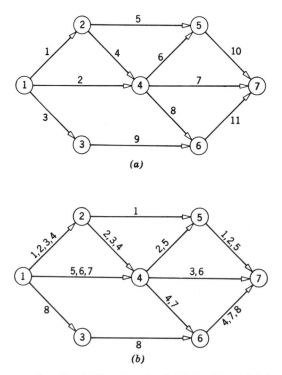

(a)

(b)

Figure A.9 Linear graph paths. (a) Branch and node labels; (b) path labels.

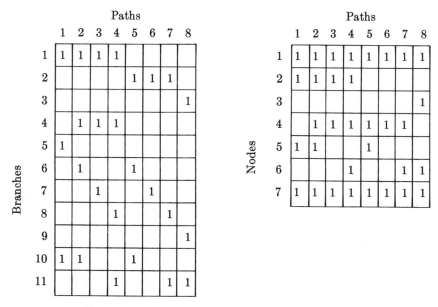

Figure A.10 Node-path matrix for graph of Figure A.9.

the start node to the node considered for which the node time label is largest. Consequently, in the forward pass algorithm, a family of paths is determined, each commencing (or rooted) at the start node and uniquely terminating at each node in the graph. The collection of paths identifies the earliest start tree graph (a subgraph of the original linear graph), so-called because it is similar to a tree in appearance. Similarly, in the backward pass algorithm, a latest start tree graph can be identified rooted at the finish node of the graph.

Figure A.11b shows the earliest start tree, and(c) the latest start tree for the linear graph of (a). These trees are generated automatically during the forward and backward pass computations and can be easily documented. Thus, for example, in Figure A.11b the timing of node 2 is determined by branch 1–2; hence node 2 can be labeled "$T_2^E = 10$, from 1." The additional label "from 1" indicates that portion of the tree graph terminating at node 2. In the same manner each node can be back referenced to the preceding node utilized in determining the earliest start times. The earliest start tree is then readily found, since any node in the graph will have pointers directed back toward the start node along the tree graph.

Branches of the linear graph not included in a tree subgraph are called *links*. In CPM arrow networks, each link branch has float time because its duration is not the determining factor in the labeling of nodes. It is important to realize that the earliest start and latest start trees graphs are not intrinsic properties of the linear graph model, for changed activity durations may alter the paths determining node times.

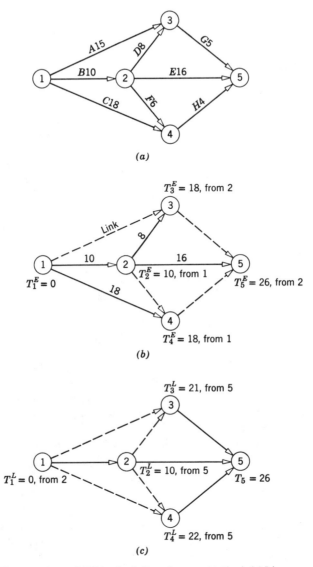

Figure A.11 Tree graphs and CPM calculations (arrow notation). (*a*) Linear graph of model. (*b*) Tree graph for earliest start times (rooted at start node). (*c*) Tree graph for latest start times (rooted at finish node).

It is obvious from the tree graph structure that no loops or circuits exist in a tree. The addition of a link branch, however, will always produce a circuit. Thus referring to Figure A.11*b* the addition of link branch 1–3 produces a closed path or circuit made up of branches 1–2, 2–3, and 1–3; and if time attributes are considered, the link branch 1–3 with duration 15 must be included as well as its free float of 3 to give the node 3 timing of 18.

Links

	A	F	G	H
A	1			
B	1	1		1
C		1		1
D	1		1	
E			1	1
F		1		
G			1	
H				1

Branches (rows A–H)

(a)

Links

	A	C	D	F
A	1			
B	1	1		
C		1		
D			1	
E	1	1	1	1
F				1
G	1		1	
H		1		1

Branches (rows A–H)

(b)

Figure A.12 Circuit matrices for Figure A.11*a*. (*a*) Earliest-start cirtuit matrix. (*b*) Latest-start circuit matrix.

Since each link branch produces a circuit when added to its tree graph, it is possible to describe the graph structure of a linear graph by using a mapping matrix for circuits and branches. Figure A.12 indicates circuit matrices for both the earliest and latest start trees of Figure A.11. The circuits are identified by the link branches. Each circuit column shows the branches included in the circuit by inserting the symbol 1 in the relevant branch rows.

A.5 CUT SETS AND CPM

A measure, although negative, of the connectivity of a graph is to determine those branches that, when removed from the original connected graph, break it into two separate graphs. The collection of branches producing such a separation is called a *cut set*. The number of possible selections (that is, cut sets) is reduced considerably if certain additional requirements are imposed.

A simple requirement is that the selection be done in such a way that, of the branches removed producing the separation, only one should be a tree branch of a given reference tree subgraph of the original connected linear graph. The earliest start and latest start trees are convenient choices for the reference tree graphs. Figure A.13 illustrates all the possible cut sets for the graph of Figure A.11 using the earliest and latest start reference trees for that network model. Note that in each case only one tree branch is affected by each cut set. Thus in Figure A.13*a* cut set *B* (identified by the tree branch *B*) requires that links *A, F,* and *H* be removed as well as the tree branch *B*. Again, cut set *D* requires that links *A* and *G* be removed as well as the tree branch *D*. In this latter cut set of the two graphs produced, one consists of node 3 alone.

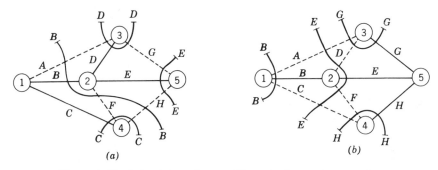

Figure A.13 Cut sets based upon different reference tree graphs.

A cut set mapping matrix can be used to describe these properties by indicating the branches required in each cut set. Figure A.14 gives two such cut set matrices; that shown in (a) is based on the earliest start tree of Figure A.13a and that in (b) on the latest start tree of Figure A.13b. The cut sets in each instance are identified by the reference tree branch used.

Another requirement for limiting the number of possible cut sets, and one of interest for CPM compression calculations, is to specify that the graph be so divided into two that one portion contains the start node and the other the finish node. Under this light requirement more possibilities exist, since no reference tree is used, as can be seen from Figure A.15.

If, in addition, the original graph is limited to those activities that are currently critical (that is, noncritical activities are considered as having already been removed), the cut set concept can assist in the understanding of

		Cut sets			
		B	C	D	E
Branches	A	1		1	
	B	1			
	C		1		
	D			1	
	E				1
	F	1	1		
	G			1	1
	H	1	1		1

(a)

		Cut sets			
		B	E	G	H
Branches	A	1	1	1	
	B	1			
	C	1	1		1
	D		1	1	
	E		1		
	F		1		1
	G			1	
	H				1

(b)

Figure A.14 Cut set matrices for Figure A.13.

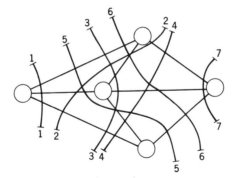

Figure A.15 Cut sets based on separation of start and finish nodes.

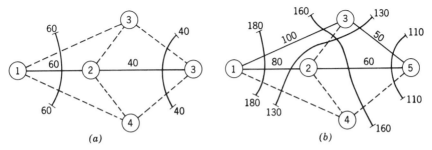

Figure A.16 Compression stages, cut sets, and effective cost slopes.

network compression logic. Clearly, at each compression stage the number of critical activities is fixed, but this number will increase steadily with successive compression stages as noncritical chains have their free float reduced to zero, and their branches then qualify for admission into the graph. Hence if the branches in each cut set have their cost slopes added, each cut set can be evaluated and the cheapest selected. Branches in this cheapest cut set are thus the branches requiring compression in this next compression stage.

Figure A.16 illustrates this concept by showing in (a) the single critical path with its two cut sets and cut set cost slopes, and in (b) a later stage in which now two parallel critical paths exist and four cut sets are now possible.

APPENDIX B

PERT CONCEPTS

B.1 INTRODUCING UNCERTAINTY INTO THE NETWORK

Planning and scheduling with CPM require a reasonably accurate knowledge of time and cost for each activity, for the CPM network model is essentially deterministic. In many situations, however, the duration of an activity cannot be accurately forecast, any time estimate being subject to doubt. If such an activity lies on a noncritical path having considerable float available, the usual CPM calculations remain valid, but there will be a local uncertainty in resource leveling and in the scheduling of men and materials, equipment, and finance. If, on the other hand, the activity lies on a critical path, the project duration and the scheduling of at least all subsequent critical activities become uncertain, unless sufficient resources are made available to crash the activity at will; if this is not possible, uncertainty becomes a factor in the construction project and must be reflected in the network calculations.

PERT (Program Evaluation and Review Technique) introduces uncertainty into the time estimates for activity and project durations. It uses an activity duration called *the expected mean time* (t_e), together with an associated measure of the uncertainty of this activity duration. This uncertainty may be expressed either as *the standard deviation* (σ_{t_e}) or *the variance* (v_{t_e}) of the duration. The expected mean time is intended to be a time estimate having approximately a 50% chance that the actual duration realized will be less, and a 50% chance that the actual duration will exceed it. The determination of such activity data necessitates using a probability distribution curve for the activity completion times and to ensure tailoring this distribution curve to the circumstances of each individual activity, three engineering time estimates are made and embedded within the theoretical curve. These three estimates of the activity's duration enable the expected mean time, as well as the standard deviation and the variance, to be derived mathematically.

The optimistic time (t_a) is an estimate of the minimum time required for an activity if exceptionally good luck is experienced; it is not a crashed time (unless the three time estimates are being made for the crashed completion of an activity).

The most likely time (t_m) is based on experience and judgment, being the time required if the activity is repeated a number of times under essentially the same conditions.

The pessimistic time (t_b) is an estimate of the maximum time required if unusually bad luck is experienced; it may take account of an initial failure or delay, but should not be influenced by major hazards (such as floods) unless these are inherent in the activity.

The general shape of this curve is shown in Figure B.1, where it is seen that there is a peak (or mode) corresponding to the most likely time (t_m). This peak may take up any position within the range of the distribution to conform to the characteristics of the activity under consideration; this range is roughly that defined by the optimistic (t_a) and the pessimistic (t_b) times, because these time estimates represent extreme cases having little chance of being realized—and therefore these two times have very small probability.

A probability distribution curve that can represent this situation is called the beta-distribution. In this curve mathematically simple (and slightly conservative) approximations can be made for the activity's expected mean time and its standard deviation.

The expected mean time is derived from the following equation:

$$t_e = \frac{t_a + 4t_m + t_b}{6} \qquad (B.1)$$

The standard deviation (the statistical measure of uncertainty being the spread of the distribution curve about its mean value) is given by

$$\sigma t_e = \frac{t_b - t_a}{6} \qquad (B.2)$$

Finally, the variance is defined as the square of the standard deviation, so that

$$v_{t_e} = (\sigma_{t_a})^2 = \left(\frac{t_b - t_a}{6}\right)^2 \qquad (B.3)$$

From equation B.1, except for a symmetrical distribution (where $t_a + t_b = 2t_m$), the expected mean time (t_e) will be different from the most likely time

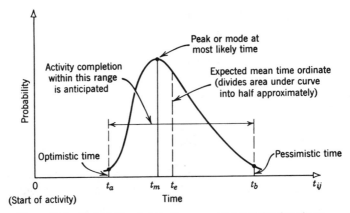

Figure B.1 Probability distribution curve for activity durations.

(t_m), as indicated in Figure B.1. Furthermore, since the optimistic (t_a) and pessimistic (t_b) times have small probabilities of attainment (1% chance is usually assigned to these two time extremes), they approximately define the range of distribution of feasible activity durations. The standard deviation, measuring the uncertainty, is one-sixth of this range.

By adopting activity expected mean times, the critical path calculations proceed as before. Associated with each duration in the PERT approach, however, is its standard deviation or its variance. The timing of events thus computed will be expected mean event times, and consequently subject to doubt; the measure of this doubt in the timing of the events requires the derivation of event standard deviations. The computation of event standard deviations and variances is amply illustrated by considering the project completion event as follows.

The project duration is determined by summing the activity expected mean times along the critical path and will thus be an expected mean duration. Since the critical path activities are independent of each other, statistical theory gives the variance of the project duration as the sum of the individual variances of these critical path activities. Therefore, for a project duration of expected mean time T_X^P,

$$V_{T_X^P} = (\sigma_{T_X^P})^2 = \sum v_{t_e} = \sum (\sigma_{t_e})^2 \tag{B.4}$$

from which the standard deviation of the project duration is easily determined. If more than one critical path exists, the project duration variance is taken as the maximum of those summed along the various independent critical paths. From this it is clear that the variance of the expected mean time of any event is the sum of the variances of those activities along the most time-consuming path (in terms of expected mean times) leading to that event.

Once the *expected mean time for an event* (T_X) and its standard deviation

(σ_{T_x}) are determined, it is possible to calculate from probability theory the chances of meeting a specific *event schedule time* (T_S). To do this the event completion time is considered to have a normal probability distribution with the mean value T_X and a standard deviation σ_{T_x}, determined as before from the series of individual activity beta-distribution curves. This hypothesis implies that the effect of adding a series of independent beta-distribution curves gives a curve of normal distribution; this is true only for an infinite series but is approximately true in practice with reasonable-sized networks.

Hence, to calculate the chances of meeting the time T_S, it is necessary to plot a normal distribution curve centered on time T_x, as illustrated in Figure B.2. With this curve, the probability of meeting the desired schedule time T_S is obtained by determining the percentage of the area cut off by this time from the total area beneath the normal distribution curve as shown.

Instead of plotting a normal distribution curve each time, the practical approach is to use standard probability tables prepared for normal distribution functions, of which a condensed version is given in Table B.1; it is emphasized that determining probability to the nearest 1% is more accurate than is generally required in construction practice and hence condensed tables are adequate. To use this approach, the difference between the scheduled and expected mean times for the event is scaled down to the standard curve by computing a factor Z, where

$$Z = \frac{T_S - T_X}{\sigma_{T_X}} \tag{B.5}$$

Using this computed value of Z, a direct entry into Table B.1 gives the probability of meeting the scheduled time T_S (interpolating in the table, if warranted) as will be demonstrated shortly.

An equivalent form of equation B.5 enables the scheduled time for an event to be determined, based on a given risk level; thus

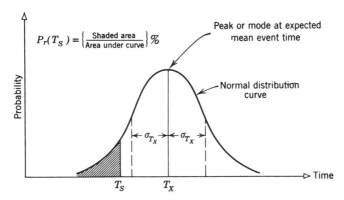

Figure B.2 Probability of meeting scheduled time T_s.

Table B.1   Approximate values of the standard normal distribution function

Z	Probability	Probability	Z
−2.0	0.02	0.98	+2.0
−1.5	.07	.93	+1.5
−1.3	.10	.90	+1.3
−1.0	.16	.84	+1.0
−0.9	.18	.82	+0.9
−0.8	.21	.79	+0.8
−0.7	.24	.76	+0.7
−0.6	.27	.73	+0.6
−0.5	.31	.69	+0.5
−0.4	.34	.66	+0.4
−0.3	.38	.62	+0.3
−0.2	.42	.58	+0.2
−0.1	.46	.54	+0.1
0	.50	.50	0

$$T_s = T_X + Z\sigma_{T_X} \tag{B.6}$$

where the value of Z is obtained from Table B.1 for a specific probability or risk level acceptable to management.

B.2   PERT Network Scheduling

Although the time analysis of an *activity* requires the evaluation of three time estimates, from which the expected mean time, standard deviation, and variance are computed, the PERT time analysis of a *network* uses only the expected mean time and the variance of each activity; this is sufficient to introduce into the network diagram the uncertainty in each individual activity. Because a single time estimate is used, PERT calculations for network analysis differ from the corresponding CPM calculations only in the introduction and handling of the activity variances. The probability of meeting a specified or selected scheduled time for any event can be found, once the event's earliest expected mean time and its variance are available (see Figure B.2). There is a theoretical difficulty for events close to the start of a project, because these event times and variances can hardly be considered as having a normal distribution; nevertheless, for want of better information, this assumption is made in practice, and (although in error) gives an adequate indication of the magnitude of the probabilities involved.

It will now be clearly understood that the scheduling of *particular* event times for a project requires an assessment of the uncertainty in the project and the acceptance by management of the risk levels in the desired schedule. As a preliminary to the actual scheduling of particular event times with specific risk levels, it is often an advantage to determine the latest expected finish time (T_{XL}) for each event, based on a neutral or 50% chance. The corresponding event variances $(\sigma_{T_{XL}})^2$ are derived similarly to those for the T_{XE} variances, starting from the project completion event. Examination of these "time boxes" will show that, in some cases, the T_{XE} and the T_{XL} values are different; this difference is called the *event slack,* and is expressed mathematically as

$$\text{Slack} = T_{XL} - T_{XE} \qquad (B.7)$$

Slack in PERT corresponds to the total float concept in CPM, and is a measure of the flexibility available in a project schedule. An event with zero slack must therefore lie on the *expected critical path.*

Two variances are now applicable to each event, $V_{T_{XE}}$ and $V_{T_{XL}}$. The first measures the uncertainty in the most time-consuming path (computed on t_e for each activity) up to the event under consideration; and the second measures the uncertainty still to be encountered along the most time-consuming path from this event to project completion. It follows then that for events on the expected critical path, the sum of the two variances must be constant; that is,

$$V_{T_{XE}} + V_{T_{XL}} = V_{T_X^P} \qquad (B.8)$$

Suppose now that a project duration of 122 days is not acceptable and that completion in 117 days is demanded. With CPM this reduction would necessitate the crashing of one or more activities by applying more resources to the project; with PERT, however, this 5-day reduction *may* be available without activity compression if favorable conditions are met in completing various activities. Hence, if a schedule is adopted on the assumption that these favorable conditions can be expected, then it contains the risk that, in fact, these conditions may not occur; and the magnitude of this risk may be computed from the statistical time data for the project.

B.3 PERT Critical Paths

The actual time consumed in completing an activity is only known when the activity is finished. If informed time estimates were available originally, the actual activity duration realized in practice should lie within the range covered by its optimistic and pessimistic times. However, its actual duration, relative to its most likely time and its expected mean time, depends on circumstances not easily foreseen and generally beyond control. Consequently, the expected critical path, based solely on activity expected times, cannot always be realized; it is, however, of great assistance in estimating the overall expected project duration and for testing the feasibility of various scheduled project durations.

The number of activity chains which may, in fact, become the *actual critical path* for the project depends on the form of the network diagram and the characteristics of the individual time estimates for the activities. Figure B.3a illustrates a very simple network from which the time when event 3 can be reached, along the three possible chains, is readily determined in Figure B.3b. It will be apparent that each chain can be completed within a large and different time range. The completion of the project requires the finishing of all three chains, and hence the feasible range for project completion is a combination of the three time ranges for the three chains. It can be easily shown that the *optimistic critical path*, the *expected critical path*, and *most likely critical path* all follow chain I activities, although the *pessimistic critical path* follows chain II activities; in fact, it is not difficult to imagine valid scheduled durations that would permit any one, or two, or all three of these chains becoming critical paths through the project. This fact emphasizes again the need for continuous reviewing techniques in PERT project control.

In some cases certain chains of activities cannot influence the project duration and therefore cannot become critical. However, in many instances considerable doubt exists as to which chain will lie on the critical path, because of the complexity of the risks involved. But it is always possible to compute (with reasonable approximation) the probability that a given chain,

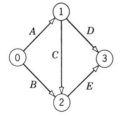

A_{ij}	t_a	t_m	t_b	t_e
A	4	4	4	4
B	6	7	14	8
C	2	4	12	5
D	9	12	33	15
E	4	7	22	9

(a)

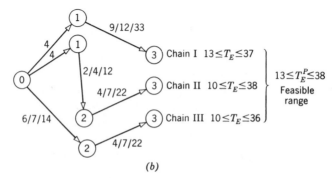

(b)

Figure B.3 Determination of critical paths. (a) Network diagram. (b) Feasible range of project duration.

or portion of a chain, will become critical—and then to assess the situation with respect to the relative chances.

The timing of event 5 in the pipeline network of Chapter 4 provides a simple example. Figure B.4 shows that event 5 is reached along the expected critical path (chain I) in an expected mean time T_{E-I}^5 of 68 days, with a variance of 46.65. Event 5 is also reached along the expected noncritical chain II in an expected mean time T_{E-II}^5 of 50 days with a variance of 177.59. The problem is to determine the probability that chain II becomes the critical path, and hence the probability that it will determine the timing of event 5.

If the probability distributions for chains I and II approximate, or may be assumed to approximate, normal distributions, the problem is simple. Statistical theory shows that the difference of two independent normal distributions is itself a normal distribution, with a variance equal to the sum of the two component variances. The probability curve for the time difference $(T_{E-I}^5 - T_{E-II}^5)$ is a normal distribution, located symmetrically about the time difference value of 18 days, with a variance of 224.24, and hence a standard deviation of approximately 15, as demonstrated in Figure B.4b. It should be

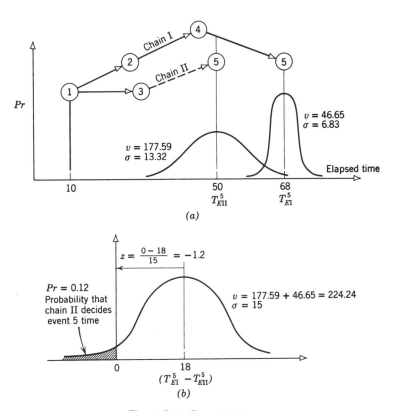

(a)

(b)

Figure B.4   Event 5 timing.

noticed that the two chains are considered only from event 1 (the branching node) and that variances are considered relative to that node.

The probability that chain II becomes critical is then the probability that the time difference $(T^5_{E-I} - T^5_{E-II})$ becomes negative. From Figure B.4b and Table B.1, the area under the curve on the negative side of the zero value is 12% of the total area under the probability curve. Therefore the expected noncritical chain II has only a 12% chance of becoming the critical path to event 5; and hence the expected critical path (chain I) has an 88% chance of determining the timing of event 5.

If three or more chains terminate in an event, the determination of the probability that any one of these will be the critical path is much more complex; the problem is simplified somewhat if the paths are not interconnected and are independent, but the statistical treatment of these situations is beyond the scope of this book. If desired, an approximate solution can be obtained by selecting the expected critical path as a control and comparing separately the other chains with it, one at a time, as discussed previously.

APPENDIX C

ANSWERS TO PROBLEMS

CHAPTER 3

3.1.

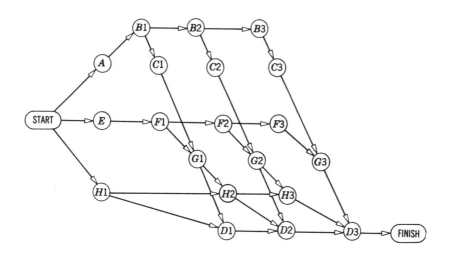

3.2.

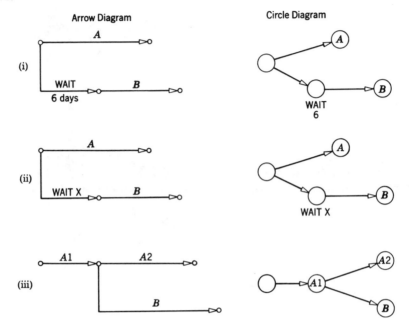

3.3. (*a*) If Figure P3.2 is a logical segment:

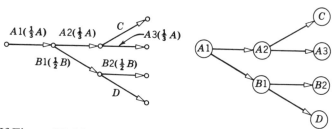

If Figure P3.2 is a complete project:

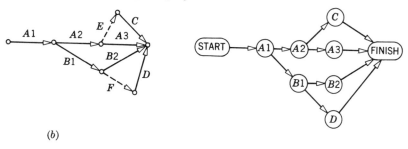

(*b*)

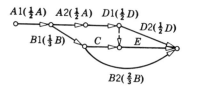

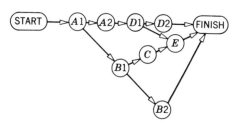

3.4.

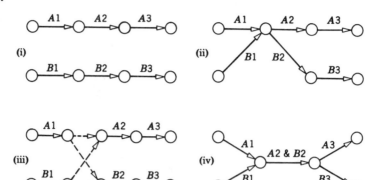

(i)

(ii)

(iii)

(iv)

3.5. (*i*) Arrow notation:

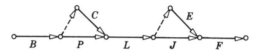

Circle notation:

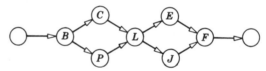

(*ii*) Circle notation

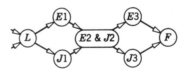

(*iii*) Circle notation:

3.6. (*i*) Assuming activity *F* is the last activity:

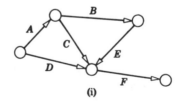

(i)

3.6. (*ii*) Assuming the diagram is a network segment:

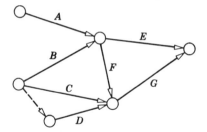

CHAPTER 4

4.1.

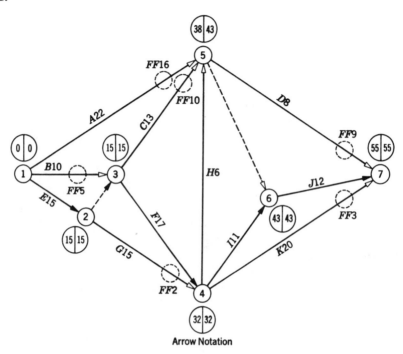

Arrow Notation

Activity	Total Float	Free Float	Interfering Float	Status
A	21	16	5	
B	5	5	0	
C	15	10	5	
D	9	9	0	
E	0	0	0	Critical
F	0	0	0	Critical
G	2	2	0	
H	5	0	5	
I	0	0	0	Critical
J	0	0	0	Critical
K	3	3	0	

4.2.

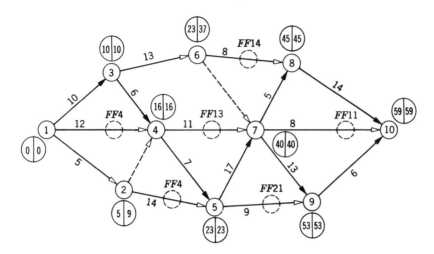

Activity		Duration	EST	EFT	LST	LFT	TF	FF	Status
1	2	5	0	5	0	5	4	0	
1	3	10	0	10	0	10	0	0	*
1	4	12	0	12	0	12	4	4	
2	4	0	5	5	9	9	11	11	
2	5	14	5	19	9	23	4	4	
3	4	6	10	16	10	16	0	0	*
3	6	13	10	23	10	23	14	0	
4	5	7	16	23	16	23	0	0	*
4	7	11	16	27	16	27	13	13	
5	7	17	23	40	23	40	0	0	*
5	9	9	23	32	23	32	21	21	
6	7	0	23	23	37	37	17	17	
6	8	8	23	31	37	45	14	14	
7	8	5	40	45	40	45	0	0	*
7	9	13	40	53	40	53	0	0	*
7	10	8	40	48	40	48	11	11	
8	10	14	45	59	45	59	0	0	*
9	10	6	53	59	53	59	0	0	*

4.3.

Activity	Duration	EST	LST	EFT	LFT	TF	FF	Critical?
1–2	4	0	0	4	4	0	0	Yes
1–8	17	0	7	17	24	7	7	
2–3	4	4	4	8	8	0	0	Yes
2–4	5	4	9	9	14	5	0	
3–4	0	8	14	8	14	6	1	
3–5	8	8	8	16	16	0	0	Yes
4–6	2	9	14	11	16	5	5	
5–6	0	16	16	16	16	0	0	Yes
5–9	3	16	26	19	29	10	5	
6–7	8	16	16	24	24	0	0	Yes
7–8	0	24	24	24	24	0	0	Yes
7–9	0	24	29	24	29	5	0	
8–10	10	24	24	34	34	0	0	Yes
9–10	5	24	29	29	34	5	5	

4.4.

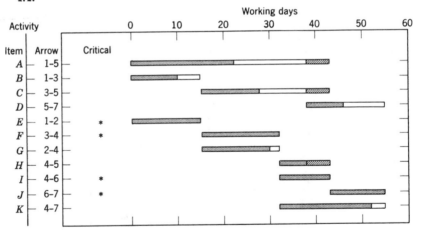

CHAPTER 5

5.1.

$$T_N = 40; \quad C_N = \$4500$$
$$T_1 = 30; \quad C_1 = \$4800$$
$$T_C = 15; \quad C_C = \$6825$$

5.2.

$$
\begin{aligned}
T_N &= 104; & C_N &= \$10{,}350 \\
T_1 &= 94; & C_1 &= \$10{,}450 \\
T_2 &= 89; & C_2 &= \$10{,}550 \\
T_3 &= 85; & C_3 &= \$10{,}670 \\
T_4 &= 83; & C_4 &= \$10{,}800 \\
T_5 &= 75; & C_5 &= \$11{,}720 \\
T_6 &= 67; & C_6 &= \$13{,}160 \\
T_C &= 57; & C_7 &= \$15{,}860
\end{aligned}
$$

5.3. (*a*)

T_P	C_P
27	1310
26	1320
25	1330
24	1350
23	1370
22	1390
21	1410
20	1430
19	1455
18	1480
17	1505
16	1545
15	1615

(*b*) $T_P = 17$; hence maximum crash $= 10$.

(*c*) Total cost is constant for $T_P = 20$ to 25; hence optimum $T_P = 20$, leaving 5 days leeway for resource leveling (if required at no extra cost).

CHAPTER 6

6.1.

T_P	C_P
38	4500
34	4700
32	4800
28	5040
24	5480

6.2.

T_P	C_P	
50	4000	Method I
38	5200	
38	4800	
35	5100	Method II
25	8100	
18	10550	

6.3.

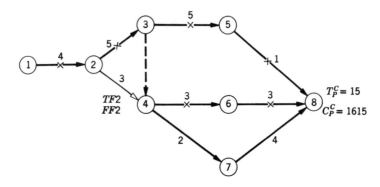

Least-cost crash solution (all-crash has cost = 1755). The coordinates of the project direct-cost/time curve are given in the solution to Problem 5.3.

6.4. (a) By decompression of all noncritical activities, deduce $T_P = 39$; $C_P = \$46,200$. Then by alternate decompression and compression:

T_P	C_P	
40	$44,650	
39	45,050	
41	43,150	
39	43,950	
40	43,400	
39	43,700	
41	42,700	
39	43,500	
40	42,950	(optimal)

(b) $T_P^N = 47$; $C_P^N = \$40,100$.

(c) $T_P{}^C = 29$; $C_P{}^C = \$54,400$.

(d)

T_P	C_P	
47	$40,100	(normal)
46	40,400	
44	41,200	
41	42,400	
40	42,950	
39	43,500	
36	45,600	
34	47,500	
32	49,600	
29	54,400	(least-time)
29	60,700	(all crash)

CHAPTER 7

7.1.

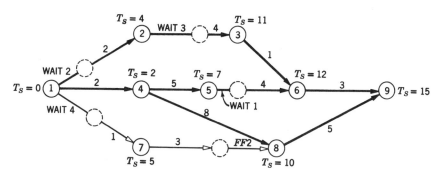

One equipment, A, is used continuously throughout. One equipment, B, is used continuously from day 3 to day 15 inclusive, and a second B is required for days 11 and 12. Personnel required = 103 worker-days, with a maximum of 9 workers on days 6, 7, and 8.

7.2.

(a) 19 days; critical activity 8–9 waits 4 days to start.
(b) 21 days; critical activity 1–4 waits 4 days to start; 4–8 does not start till day 9.

7.3.

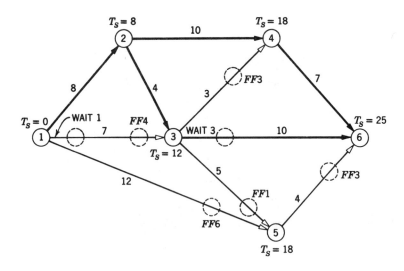

Daily	Maximum	Minimum	Average
Workers	22	9	18.8
A	1	0	0.88
B	2	1	1.6
C	1	0	0.56

CHAPTER 8

8.1. (*a*) Earliest start schedule:

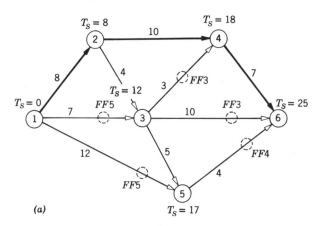

(*a*)

(b) Latest start schedule:

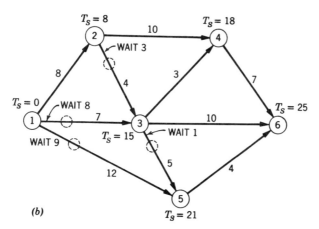

(b)

In bar chart form:

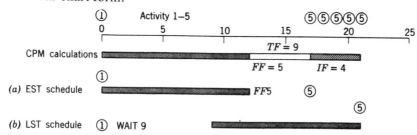

8.2.

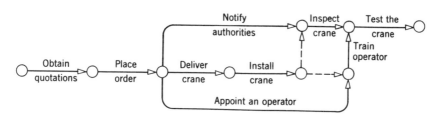

8.3. This problem demonstrates the correct use of dummies to preserve logic: (A) Clear 2nd km depends on Survey 2nd and Clear 1st; but Excavate and Deliver 1st km depends on Clear 1st only. (B) Excavate and Deliver both depend on Clear, but not on each other in any one km. (C) Lay always depends on both Excavate and Deliver. (D) Subsequent operations both depend on two preceding ones; initial succeeding operation depends on one only. (E) Backfill depends on Test; nothing depends on Backfill.

The network will be as shown on p. 404. The minimum construction duration is $11\frac{1}{2}$ weeks.

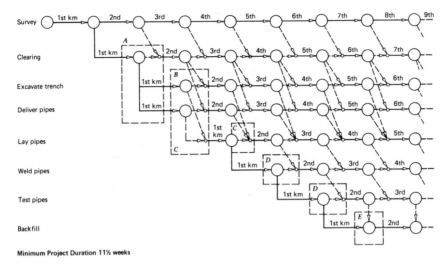

Survey | 1st km | 2nd | 3rd | 4th | 5th | 6th | 7th | 8th | 9th

Clearing | 1st km | A | 2nd | 3rd | 4th | 5th | 6th | 7th

Excavate trench | 1st km | B | 2nd | 3rd | 4th | 5th | 6th

Deliver pipes | 1st km | 2nd | 3rd | 4th | 5th | 6th

Lay pipes | C | 1st km | C | 2nd | 3rd | 4th | 5th

Weld pipes | C | 1st km | D | 2nd | 3rd | 4th

Test pipes | 1st km | D | 2nd | 3rd

Backfill | 1st km | E | 2nd

Minimum Project Duration 11½ weeks

Start project

6 km   4km/week = 1½ weeks   40 km to go = 10 weeks

8.4. (*a*) 101 days; (*b*) day 68 (event 4).

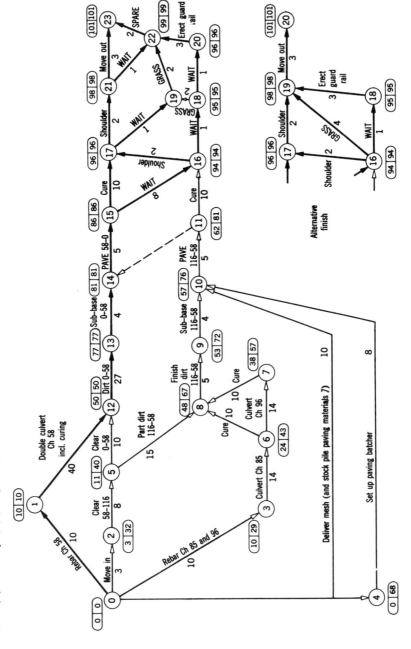

8.5. To the network of Problem 8.4, simply add a dummy 9–12, and reschedule the project. (Event 12 is now 53/53, adding 3 days to the duration). $T_P = 104$ days. LST for earthmoving plant is event 5: day 33.

8.6.

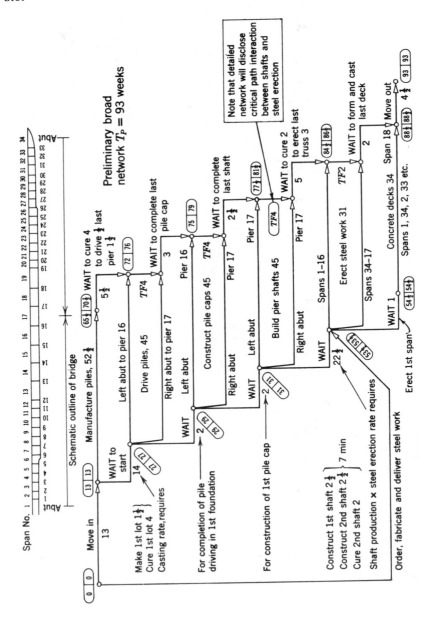

CHAPTER 9

9.1.

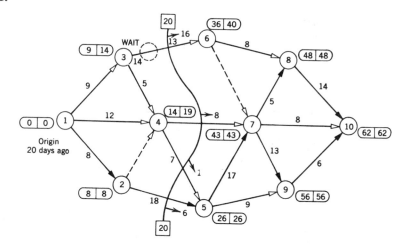

(*a*) Project delayed by 3 days. Activity 2–5 critical; activity 4–5 not now critical.

(*b*) Event 5 can be reached along 4–5 on original schedule. Entire project affected by *activity 2–5; crash if possible to gain 3 days; if impossible, crash 5–7. Commence activity 3–6 now,* since with project duration of 59, event 6 has only one day float. Remainder of network unaffected at this stage.

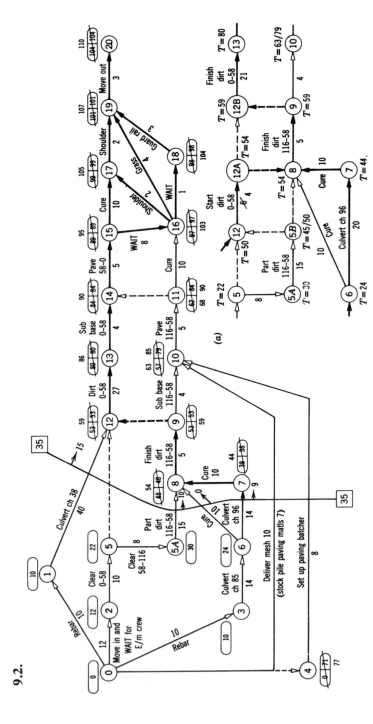

(a) Original network ($T_P = 104$) updated to show status at day 35; $T_P = 110$. Project delayed 6 days by activity 6–7.

(b) Remedial measures: complete activity 5A–8, then start activity 12–13 on day 50; crash first 6 days work to 4 days, then on day 54 move to activity 8–9; and so on. Status retrieved by event 13.

9.2.

CHAPTER 10

10.1.

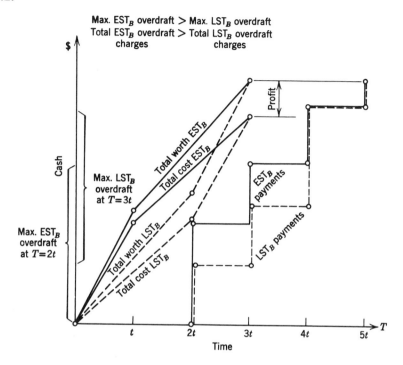

10.2.

(a) Cumulative expenditure is approximately $40,000 at day 40, $56,500 at day 60, and $73,400 at day 80. Income does not alter. Hence the maximum overdraft of approximately $57,000 now occurs just prior to the second progress payment at day 60.

(b) There is no change in the original overdraft requirement, because internal rental payments for equipment landed on site are merely advance book entries, unless the construction plant is owned by a subsidiary company, in which case the answer is the same as (a) as far as the construction company is concerned (i.e., debit construction, credit plant company) with no change in overdraft from contractor's viewpoint.

11.1.

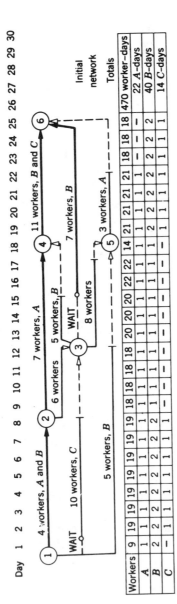

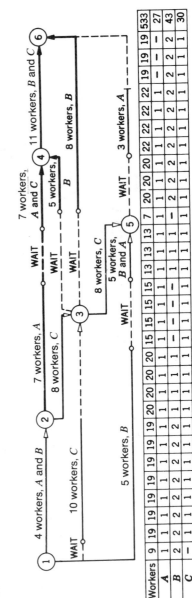

Day 1 2 3 4 5 6 7 8 9 10 11 12 13 14 15 16 17 18 19 20 21 22 23 24 25 26 27 28 29 30

Initial network

Totals: 470 worker-days, 22 A-days, 40 B-days, 14 C-days

Correct time extension = 5 days

Additional direct costs = 63 workerdays plus 5 A-days, 3 B-days, and 16 C-days

11.2. The primary critical delays are Nos. 1 and 2. Arising from these the contractor is quite entitled to reschedule activity 9–10 and so obtains continuity of work through 9–10 and 13–14; a dummy 10–13 should be introduced to show this. Similarly, there is no need for the paver to start before day 82 on activity 10–11, to preserve continuity through 10–11 and 14–15, because event 14 is running 2 days late due to delay No. 2.

The contractor is entitled to an extension of time of 2 days (for delay No. 2), and to all extra costs required to complete activity 6–7 and to crash subactivity 12–12A (for delay No. 1).

INDEX

Page numbers followed by (*n*) indicate references contained in footnote.

DATE DUE